W9-BYY-443

LANDMARKS

OF SCIENCE

FROM THE COLLECTIONS OF
THE LIBRARY OF CONGRESS

Leonard C. Bruno

Foreword by Daniel J. Boorstin

New York • Oxford

TO *Jane*

Facts On File, Inc.
460 Park Avenue South
New York NY 10016
USA

Library of Congress Cataloging-in-Publication Data

Bruno, Leonard C.
[Tradition of science]
Landmarks of science: from the collections of the Library of Congress / Leonard C. Bruno: foreword by Daniel J. Boorstin.
p. cm.
Originally published under title: The tradition of science.
"First published by the Library of Congress in 1987"—T.p. verso.
Bibliography: p.
Includes index.
ISBN 0-8160-2137-6 (alk. paper)
1. Science—History. 2. Library of Congress. I. Library of Congress. II. Title.
Q125.B87 1989
509—dc20 8933579
CIP

First published by the Library of Congress in 1987.

British CIP data available on request.

Designed by Stephan Kraft
Manufactured by Arcata Graphics, Inc.
Printed in the United States of America

10 9 8 7 6 5 4 3 2 1

This book is printed on acid-free paper.

Contents

Foreword

In this book the reader, while becoming acquainted with landmarks in the history of science in the West, can discover some of the meanings and guises of tradition. Leonard C. Bruno's lively and "friendly journey through the Library of Congress" and its collections in the history of science gives us a sense of what tradition means in the sciences. He has selected only epochal works that shaped or defined man's view of the natural world, and not those that commented on existing titles. He might have called this volume *The Makers of Scientific Tradition*. He does not purport to survey all the influential books that have been written about the sciences, but only to give an account of those that happened to be in the collections of the Library of Congress, the nation's greatest library. By making his selection in this way he has been true to the nature of tradition which—unlike a textbook or an encyclopaedia—is selective in the most unpredictable ways.

The history of the tradition of science is quite different from the history of science. The history of *science* is a story of displacements—of Ptolemy by Copernicus, of Aristotle by Newton, of Newton by Einstein, etc. A modern scientist has noted that the importance of a work of science can be measured by the number of earlier works it makes obsolete. But the history of a *tradition* is a story of accumulation. Even though the ideas in a book of science may be displaced, still if it is a momentous work, like those of Aristotle and Ptolemy and Newton, it remains a landmark in the tradiiton. Sometimes it even becomes more interesting and seems more intelligible to us laymen because it appears bizarre and outlandish by contrast to the conventional wisdom of our own time. This book introduces us to the landmarks in the Western tradition of science.

The discovery of this tradition, of how earlier works have been displaced, can make us less complacent about the conventional wisdom of our time or the finality of our own views. It can also remind us of the infinite and unpredictable ways in which people have been led to new views of their world. In an age obsessed with the grandeur of our own science, the tradition of science reminds us of the continuity of the human quest. The tradition reveals how the earlier works provided a foundation for the later works—or if not a foundation, then perhaps a target. And we see how targets, too, have been indispensable to man's search to know, and to put his knowledge in books.

All this makes the collections of the Library of Congress the best possible terrain for Leonard C. Bruno's scholarly journal in search of the tradition of science, for the Library of Congress is one of the world's greatest collections of human error. The working scientist, we are told, must cleanse his mind of the errors and dogmas and wild sallies of his predecessors. But the student of culture and of the progress of knowledge must remind us that grand misconceptions have influenced humankind as much as the powerful new, more accurate, visions that have marked human progress. These reminders are the books in our great libraries, and no library provides a richer store of them than the Library of Congress.

Scientific tradition can, of course, serve us in our time by reinforcing our belief in progress. It can cheer us along by showing how science has always somehow moved on from one way of thinking to another. Some might be tempted to Friedrich Engels' view that "tradition is a great retarding force, the *vis inertiae* of history." But the modern humanist will prefer John Donne's vision that "No man is an island, entire of itself; every man is a piece of the continent, a part of the main . . ." Tradition gives to John Donne's vision a chronological dimension. The tradition of science, extending far into the dim past, is the endless mainstream that we glimpse in this book, and in the books to which it leads us.

Daniel J. Boorstin

Preface

This book can perhaps best be described as an elaborate bibliographical essay. Invoking an author's prerogatives, I have taken liberties of style and approach and have attempted to blend bibliography with history—without, I hope, doing injury to either. The intended result was to achieve the purpose of any bibliography—in this case, to provide information about major works of science in the collections of the Library of Congress—in a manner that would both interest and inform the reader. Thus I have tried to treat each scientific work in a broad historical context, discussing its relation to other major works, to its own particular discipline, or to the history of science in general. Such treatment carries with it the danger that this book might be read either strictly as history or strictly as science, when in fact it hopes to be something different from either.

As to overall purpose, this is not a survey of the science collections of the Library of Congress. That is, it is not an attempt to assess the strengths and weaknesses of those collections as a typical survey would. Nor is it a guide to the collections that offers descriptive general information about the content of any particular collection. Finally, it is not a history of each scientific discipline in any real scholarly sense.

Rather, this book is one person's trip through the Library's collection of Western scientific treasures. The works chosen for discussion, although the result of a subjective selection process, would probably appear on most lists of the great books of science. Technology as such is given no separate or extended treatment. These selected works are organized here by traditional scientific discipline and are treated in a typical historical, usually chronological, manner. This treatment, however, is by no means a critical review, nor is it meant to be in any way inclusive of all major works. On the contrary, it is a friendly journey through the Library of Congress and thus through history. It is written for the interested general reader by a generalist, with no authority claimed. Because of this, little mention is made of either interpretive theories of history or of ongoing scientific controversies.

Little need be said of methodology. For the most part, when a book is mentioned, it is as a primary source in the history of science. No secondary works are discussed. Although a particular work may come from one of the Library's special collections—for instance, the Lessing J. Rosenwald Collection or the Joseph M. Toner Collection—no reference to that collection will be made in the text, since the emphasis here is always on the individual work itself. The bibliography provides information about particular Library collections to which a book or manuscript may belong, as well as specific bibliographical information. Title translations will also be found there.

Because of the great expanses of time and the wide sweep of ideas, both of which were necessarily compressed in this brief treatment of a mammoth subject, it sometimes became easy to impose patterns on the past or to treat a complex subject in a somewhat facile manner. Taking refuge in the house of conventional history became an unavoidable—though not always unwelcome—habit. Further, I have allowed myself to indulge in the anecdotal. This was deliberate, as I decided early on to try to write not only about the merits and significance of each scientific work but also about the person behind the work. Despite the brilliance and the heights of intellectual achievements recorded here, what often is best remembered are the small, personal details of either greatness or frailty, success or failure—human details that remind us that the intellectual tradition of science partakes of a larger, more encompassing tradition of being human.

The idea for this book came from Joseph W. Price, chief of the Science and Technology Division of the Library of Congress, who in 1981 asked that I study the Library's science collections and determine which of the major works in the history of science the Library actually possessed in first edition. It took little vision on my part to elaborate on this idea and to produce this volume. What was necessary however was Mr. Price's concurrence both that such a book as this was needed and that I would be allowed the time to write it. I thank Mr. Price for the opportunity and the continuing support, as well as for his patient reading of each draft chapter as it was produced. I hope I have met his expectations.

Very early in the project, the director of the Library's Publishing Office, Dana J. Pratt, was consulted and he joined in support of the idea. He too read each chapter and offered not only steady encouragement but concrete help in molding the entire book. His good advice and much-needed criticism and correction were cheerfully and positively given and were invariably helpful. He proved a wise counselor.

Once it was written, the book's editor, Evelyn E. Sinclair, took charge and gave the manuscript a rigorous shaping-up. Throughout the lengthy production process I was encouraged to see her editing ability and sound judgment complemented by superb organizing skills. I am very grateful to have worked with such a competent and sympathetic professional.

I am very proud of the book's bibliography—the product of Ruth S. Freitag's hard work. Although I selected all the works to be included there, the remainder of the job was left to her admirable bibliographic skills. The result is a typical Freitag job—absolutely correct and totally consistent. I can boast here since it was not my work.

There were also many others throughout the Library, too numerous to list here, whose cooperation and assistance made my job easier. Among those to whom I must have been a regular burden, however, and to whom I owe a special thanks are Kathleen T. Mang, librarian for the Rosenwald Collection, for her guidance, patience, and general indulgence, and Clark W. Evans, senior reference librarian, Rare Book and Special Collections Division, whose reference help was always forthcoming and dependable.

The final note of appreciation goes to Brenda W. Presbury, who typed nearly the entire manuscript, and to Linda W. Carpenter, who not only helped with the typing but with the proofreading as well.

At this point, most authors usually murmer mea culpa in advance—taking responsibility for all errors in fact and judgment that the attentive reader will no doubt find. This is all the more necessary in my case. Since I enjoyed full autonomy not only in deciding which books to include or omit and in what I wrote about each one but in the selection of illustrations as well, I am especially deserving of the two-edged sword of authorial responsibility. I hope, however, that whatever the shortcomings of this one-man approach, they are more than balanced by the unity derived from the continuity of personal selection and interpretation.

Fr. Bacon's Novum Organum

FRANCISCI
DE VERULAMIO,
Summi Angliæ
CANCELLARIJ,
Instauratio
magna.
Sim: Pass: sculpsit
Multi pertransibunt & augebitur scientia.
Anno
LONDINI
Apud Joannem Billium
Typographum
Regium.
1620.

1st ed.

Introduction

That we can learn from the past is no less true for science than for any other field. Any real understanding of where we stand scientifically today and where we are headed depends to a great extent on an awareness of how we reached those scientific insights. The increasing impact of science and technology on our lives makes such an understanding even more important.

The origins of science and its historical development and advancement are fraught with all the contradictions, complexities, frustrations, successes, and failures that accompany any human endeavor. As with any good story, the history of science has a very real and fascinating tale to tell—beginning with how man came to be able to identify, distinguish, and eventually predict the true effect of a particular cause. Early man probably did not see the slightest connection between the simplest and most obvious natural events. Did a succession of overcast days mean that the sun had gone away forever? Through a repeated set of what might be called experiments, or at least observations, he eventually concluded with a great degree of certitude that indeed the sun would rise every day. This very natural process of gaining experience about the physical world—simply standing and watching the same thing happen over and over—provided our early ancestors with a modest degree of usable, predictable knowledge. Such realizations marked the primitive beginnings of the scientific method.

The natural curiosity of our species seems almost to guarantee that method's success, for man is constantly asking the question, "What would happen if I do this?" and then usually acting upon his native curiosity. The major external curb to this basic inquisitiveness about the world we live in has usually occurred when a rigid, all-encompassing authoritarian teleology offers otherworldly explanations for natural events or regards the investigation of all natural phenomena as useless or even wrong. Such was the mind-set during medieval times. But even during those most unappealing of times, some science was being done.

The unique intellectual gift of mankind is our ability to learn from the experience of others, to be able to accept and to use the cumulative wisdom of our predecessors without having to actually experience what they did. Each new generation can benefit from the work and experience of its predecessor to the degree that each participates actively in a real tradition of science. No less an original thinker than Isaac Newton acknowledged this tradition when he said, "If I have seen further than other men, it is because I stood on the shoulders of giants." The tradition of science can be understood to mean the successive transmission of knowl-

Opposite page:
The frontispiece to Francis Bacon's magnum opus shows a representation of the Strait of Gibraltar flanked by the mythical Pillars of Hercules past which, before Columbus's journey, there was traditionally believed to be nothing. Here Bacon's ship embarks on a voyage past these same pillars, beyond which, Bacon seems to be saying, lies all manner of discovery. *Instauratio magna*, 1620. Francis Bacon.

edge, but it also has particular denotations, such as an established way of doing things, an inherited body of principles, standards, and practices, a developmental and historical continuity, or the force exerted by the past on the present. The central reality of the tradition of science links them all: every aspect of scientific knowledge is interrelated and contributes to an essential unity.

The division of science into categories, disciplines, or branches is a necessary but nonetheless arbitrary device. The following chapters, which separately discuss the history of each discipline, reflect this general or traditional method of scientific classification and make no mention of the subtle variations and specializations that exist in today's science. Each major work discussed in these chapters marks a high point of understanding for a particular discipline. All of these works are represented in the collections of the Library of Congress, most in their original editions. The collections have been enriched with such treasures sometimes by the donation of entire personal libraries and other times by the purchase of certain specific items or collections. Great works of science are not found only in book form, and the Library has manuscripts, papers, and letters of great significance as well as important maps and photographs. But overall, the Library's great strength in science is its unique book and scientific journal collections. This survey therefore will focus mainly on the book in all its variations.

Each of the scientific works that is discussed in the following chapters is a landmark of human achievement. Each is a moment of intellectual magic captured on paper. Whether they be works of insight and genius or works of diligence and perseverence, these are works of man at his best—in a sense at his most human. Doing science—discovering nature—is so distinctly human an activity that one is tempted to stretch Descartes's famous rationalist maxim "Je pense donc je suis" to read, "I seek to know, therefore I am human."

Behind each of these great works is a person, but more important is the essential humanity underlying almost all of them. Newton's *Principia* and a few other forbiddingly complex works aside, many of these landmarks of science become, upon translation, quite understandable to the nonscientist. The wit and natural writing ability of Galileo and the simple directness of Charles Lyell make this point. One has only to browse Darwin's *Origin of Species* to marvel at how approachable a work of real genius can be. And Francis Bacon's *Novum organum* has an enthusiasm that would inspire young scientists of today. These and all the works discussed here are at the vanguard of the intellectual heritage of the West. As such, it would be easy to treat them as secular tablets handed down by a scientific Moses—their principles our commandments, their formulators our priests. Such reverence would prove as counterproductive as the opposite extreme—to regard these works as in themselves responsible for all of society's ills. Neither extreme provides the kind of sensible respect that these works of man for man ought to engender. Each symbolizes a signpost in the never-ending maze of nature. Each offers a significant piece of the infinite puzzle of the cosmos.

Among these puzzle pieces—each a primary source in the history of science—intriguing contrasts often appear. These works vary greatly in how long it took to produce them, how soon their inherent worth was recognized, the effect they had on the author's life, and what the scientific nature of the works themselves was. Some are the result of years of concentration, dedication, and hard work. Agricola labored for twenty years to produce one work, the *De re metallica.* William Smith spent a lifetime literally walking over every inch of England as he mapped it geologically. John Flamsteed's life was dedicated to counting the stars, and the number of autopsies Morgagni performed was exceeded only by the number of pounds of pitchblende Madame Curie and her husband, Pierre, refined. In contrast to these works of scientific accretion are the thunderbolts of scientific illumination experienced by the likes of Newton and Einstein. Their well-known laws and theories were formulated during a relatively short burst of creative energy and were the product of real inspiration. Such insights into nature are obviously rare and spectacular. A curious combination of long, hard work and inspiration are found in the case of Charles Darwin, who seemingly spent years in preparation for his special moment of insight—ratifying Pasteur's comment about "the prepared mind."

As to the recognition of the significance of a work at the time of its appearance, some met with instant acceptance while others were dismissed outright as worthless. The great human anatomist Andreas Vesalius saw his pioneering work arouse such strident opposition that he quit research altogether. And yet his work, *De humani corporis fabrica,* contained some of the most startlingly accurate drawings of the human body ever made. It is Isaac Newton who provides the ultimate example of the contemporary recognition of a work's content. His *Principia* may also be the best example of the least understood but most accepted scientific work in history. Possibly because Newton's awesome intellect was already acknowledged, the recognition of his 1687 work was a foregone conclusion. Nonetheless, nature's laws became Newton's laws. Surprisingly, there were also a good number of pioneering works that remained totally unknown in their time, offering their contemporaries no chance to judge. The melancholic Dutch naturalist Jan Swammerdam is today regarded as the founder of modern entomology, yet his research was never published during his lifetime. It was not until his work was discovered by Hermann Boerhaave over fifty years later that his discoveries became known. Two hundred years later, the abbot of a monastery in Bohemia, Gregor Mendel, discovered the laws of hereditary characteristics and published his work in an obscure Bohemian journal, only to have his efforts go unnoticed. The world waited another generation until Hugo DeVries retraced the monk's footsteps. It is to DeVries's credit that he not only unearthed Mendel's long-neglected work (subsequent to his own independent discovery) but offered his own work as a confirmation of Mendel's.

Also unpredictable is the manner in which a work may have affected its author's personal life and fortunes. Many successful scientists were drawn or summoned to a monarch's court and experienced the hazards and headiness that attend proximity to power. Unable to withstand the winter regimen of instructing Queen Christina in philosophy three times a week at 5 A.M., René Descartes finally died at the Swedish court. The prodigiously talented and energetic Leibniz, on the other hand, flourished at the Prussian court and served it for forty years in nearly every imaginable capacity. The undoing of Francis Bacon was related to his court duties, but the prolific mathematician Leonhard Euler prospered with such duties. Euler spent nearly his entire working life serving two royal courts—of Catherine I and Catherine the Great at St. Petersburg and Frederick the Great at Berlin—and did everything from planning a national canal network in Berlin to supervising the Russian system of weights and measures in addition to his creative mathematical work. The prestige and renown that being a great authority bestows was lavished upon Cuvier and Linnaeus during their lifetimes and both men surely relished their authoritative roles. Conversely, all his life Anton van Leeuwenhoek held the sinecure of janitor at the Delft City Hall, despite having become world famous for his microscopial discoveries. Both the queen of England and Tsar Peter the Great came to visit this modest amateur scientist, who managed his fame by simply ignoring it and going about his business. Some individuals prospered financially from their discoveries or inventions. Others were not as fortunate and despite dedication and hard work, ended life as poor as they began it. William Smith spent all his slender earnings and more to produce his great geological map of England. Eventually he lost both his house and his rare collection of fossils and was financially ruined. Others paid a similar price to science. Claude Bernard's family life was ruined by his wife's revulsion at his animal experiments, and many a chemist shortened his life by sampling new compounds.

In sum, men and women of science have enjoyed and endured the same pain and happiness, failure and success, and doubts and hopes that impress themselves on the lives of everyone. One who experienced both the pleasure of accomplishment and fame and the despair of rejection in the same long lifetime was Galileo. The heights of his intellectual achievements were exceeded only by the breadth of his fame. Yet as a man of nearly seventy years, he was humiliated by the Inquisition and forced to recant the essence of his scientific work. Surely few have experienced such extremes.

Another interesting contrast lies in the scientific nature of the works themselves. Most of the works discussed here have significantly enlarged the body of knowledge in a particular field. Only a few did something more for science, and these are generally described as "revolutionary" works. Such works mark scientific turning points—they redefine our scientific reality—yet they are not all alike. Some are revolutionary in effect and consequence rather than content. Some explode into controversy from the moment of publication; others are quietly released and slowly percolate until they become part of the accepted wisdom. Some unify and some refract. Some reassure us and some sow doubt. The *De revolutionibus* of Copernicus, a book recognized as marking the birth date of modern science, caused no real stir after its publication in 1543. Although the

European scientific intelligentsia became well acquainted with the book and its premise of heliocentrism, nearly a century passed before Galileo openly touted its ideas, only to discover that he had miscalculated the liberality of his own times. Given that Copernicus offered an idea that overturned everything mankind thought it was sure of, the actual and immediate effect was inconsequential. It was an idea which grew increasingly powerful over time.

In sharp distinction was the instantaneous uproar and debate that followed publication of Darwin's *On the Origin of Species* and resulted in essentially new concepts of how nature worked and what the place of man was in the natural order of things. Einstein's relativity theories had this same wrenching effect. These two shocking, unsettling, and provocative ideas probably had their impact heightened considerably by occurring in modern times.

One of the best examples of a revolutionary scientific idea that had the opposite effect is the synthetic, almost consoling theory of Maxwell's electromagnetism. Maxwell was able to prove mathematically Faraday's great intuitive idea, that light is an electromagnetic phenomenon. This brought about a productive unification of three main fields of physics—electricity, magnetism, and light. The irony of Maxwell's centrifugal accomplishments, however, is that they led to the Michelson-Morley experiments, which in turn set the stage for the shattering centripetal discoveries of Einstein. So it is with the expanding universe of scientific knowledge in which, it appears, answers lead always to more questions.

A survey of these and other major works of science provokes random thoughts and observations concerning both the individuals and the processes of science. One of the most curious is the startlingly high number of very prolific and creative "old men"—that is, individuals whose scientific contributions were made at a relatively advanced age. The most obvious and celebrated is perhaps Galileo, whose *Dialogo* was published when he was sixty-eight years old. Seemingly broken by the Inquisition, Galileo then produced, at seventy-four, what is now considered his greatest work, *Discorsi e dimostrazioni matematiche*. Anton van Leeuwenhoek continued his microscopial discoveries until he died in his ninety-first year, and Jean Lamarck, who founded modern invertebrate zoology, did not even begin his serious natural history studies until he was fifty. Giovanni Morgagni, the founder of pathology as a major branch of medicine, achieved the height of his fame in his eightieth year with the publication of his *De sedibus et causis morborum*. Only then did the great anatomist feel he had done a sufficient number of dissections to link the manifestation of a disease with a certain anatomical change. No doubt these anecdotes can be balanced by a similar litany of those whose best work came in their very early years, but they nonetheless cause one to at least question the modern conceit that creative science can only be done by the young.

A corollary to this is the obvious and almost total absence of women in the history of science. Despite the well-known exceptions, scientific activity has until fairly recently been virtually a male concern. The now-familiar statement, "of all the scientists who ever lived, 90 percent are alive today," might be even more appropriately applied to women scientists.

The extreme and often sincere religiousness of many a great scientist should also put in doubt our contemporary idea that science and religion have always been at odds. Given the undeniably significant role religion has played in mankind's history, there ought to be little surprise when it plays a similar part in the life of an individual—albeit a scientist. Thus Isaac Newton spent what might be considered an inordinate part of his life writing about matters of theology, and Christiaan Huygens dabbled in a sort of mysticism. Blaise Pascal became so obsessed with religion that he put all science out of his life forever. Even that skeptical chemist Robert Boyle carried an extremely devout form of Christianity with him throughout his life. Some scientists tried to keep separate their science and religion and some sought reconciliation; others saw no conflict. In the end, the significant role of religion in the history of science serves to underscore the idea that although science is primarily an intellectual activity, it is carried out by individuals and is therefore subject to the various influences and forces that affect the individual. On a conceptual level, this may in part explain why metaphorical thinking has always been such a useful tool for the scientist. Creative analogies taken from almost any aspect of our culture have played a significant role in the history of science. Darwin described the moment when his thoughts coalesced into a theory of natural selection—he was reading Thomas Malthus's *Essay on the Principle of Population.* Later, Darwin's own theories of survival of the fittest made the cultural leap the other way, offering a "scientific" rationale for laissez-faire economics.

Two final observations must be made about the nature of science—observations which we often forget or overlook, especially when writing such a book as this. The first has to do with the inability of science to be truly objective. Modern evidence of this abounds, from Planck's quantum theory, which posits the disturbing effect of observation, to Heisenberg's uncertainty principle and its mathematical counterpart, Gödel's "proof." Albert Einstein recognized the subjective aspect of science when he said, "It is the theory that decides what we can observe." His revealing statement testifies to the fact that each of us imposes our own set of preferences and preconceptions on the natural world. No scientist begins work without some assumptions or working hypotheses. In this regard, the words of William James apply in a general way:

> Pretend what we may, the whole man within us is at work when we form our philosophical opinions. Intellect, will, taste, and passion co-operate just as they do in practical affairs. . . . It is almost incredible that men who are themselves working philosophers should pretend that any philosophy can be, or ever has been, constructed without the help of personal preference, belief, or divination. . . . every philosopher, or man of science either, whose initiative counts for anything in the evolution of thought, has taken his stand on a sort of dumb conviction that the truth must lie in one direction rather than another.

It is part of the beautiful uniqueness of science, however, that its traditional methods provide boundaries to circumscribe these natural human tendencies. We may therefore speculate as wildly and investigate as creatively as we can, always secure that the self-correcting criterion of demonstrable fact and experimental evidence will not let us stray too far from the straight and true.

The second observation reminds us of the insufficiency of science to the whole human person. Although earlier I took license with Descartes, stretching his maxim to the point of saying "I seek to know, therefore I am human," such a statement overemphasizes a single dimension. If men seek truth, they also seek beauty—and love and justice and all of the so-called universals. Science alone is insufficient because it serves to nourish only one part of us—albeit a large part. On this thought George Sarton has the last word.

> Science is not distinct from religion or art in being more or less human than they are, but simply because it is the fruit of different needs or tendencies. Religions exist because men are hungry for goodness, for justice, for mercy; the arts exist because men are hungry for beauty; the sciences exist because men are hungry for truth.

The following sketch of the general course of scientific development is offered with no specific discipline in mind. Its purpose is to provide a brief framework within which the later, individual chapters may fit. Such a general sketch also provides an opportunity to discuss individuals, organizations, themes, or specific books that cut across a wide range of topics and scientific disciplines.

As with any human activity, science is affected by the economic, political, religious, and overall cultural climate of its times. Evidence of how science can flourish in propitious times is offered first by the Greek experience. During the golden age of Greek science, the fifth and fourth centuries before Christ, the great Greek philosophers established the elements of nearly all the basic disciplines. Even with the political changes that occurred and the emergence of an altered culture called the Hellenistic Age, Greek science continued apace. It is significant to note that the golden age of Greek science was similarly golden for Greek art and literature. The rise of Rome signaled the real decline of Greek science, although what could be called Roman science was done primarily by Greeks. With Rome's decline and the subsequent Moslem conquest of a great part of the Mediterranean world, the scientific centers as well as actual libraries shifted eastward.

As Western science became increasingly distant from its origins, the break with the past became nearly complete, causing first a standstill and then only a retrograde movement. After the Germanic invaders had done their work, little of the ancient knowledge remained in the West. It was left to the likes of the Latin encyclopedists to preserve what remained. None of these scribes was original, and each was basically a compiler, but given the paucity of any sort of organized knowledge, their works re-

Here Isidore of Seville reflects the simplicity of medieval geographical thinking by depicting the earth as a wheel encircled by the ocean. *Etymologiae*, 1472. Isidore of Seville.

mained influential throughout early medieval times. Three of the most popular were Boethius, who flourished from 480 to 524 and whose manual on mathematics and logic, *De institutione arithmetica,* became a standard medieval text along with his *De consolatione philosophiae;* Isidore of Seville (560–636), whose *Etymologiae* was enormously influential; and the Venerable Bede (673–735), whose *De rerum natura* reflected largely derivative cosmological ideas. During early medieval times, these works were the best the West could produce, and their main contribution was preserving what science had survived, not contributing anything original. With the invention of typography, their works went from manuscript to book form and all three writers are represented in the Library's collections.

During this time, Arabic and Persian schools, having translated every Greek author available, were beginning to go beyond those texts to make contributions of their own. From A.D. 800 to 1100, Arabic learning was in its prime, and its international culture spread from Spain to India. But as the empire began to disintegrate and the provinces reasserted themselves culturally and politically, it was in Spain that the medieval West rediscovered its lost scientific legacy. More than most occupied regions, Spain had undergone a remarkable and distinct change as the centuries of cross-fertilization of Muslim, Jewish, and Christian cultures took hold. The Spain that was slowly retaken from the invaders had changed profoundly over the centuries, and when the West finally reclaimed its lost sister it also embraced her Eastern progeny. Western scholars and intellectuals were stimulated equally by the great store of Greek-Arab learning left behind in schools and libraries and by the Arab habit of a tolerance of thought. Thus by the middle of the twelfth century, Western scholars already were translating into Latin, the common language of the West, the Arab versions of the Greek manuscripts. In the span of an additional generation, the most significant works had been translated—sometimes badly, but nonetheless producing a treasure-house of new wisdom. No individual more aptly symbolizes this twelfth-century phenomenon than does Gerard of Cremona. It was he who, in less than thirty years time, translated over seventy works into Latin—among them many of the core works of Aristotle, Euclid, Archimedes, Ptolemy, Galen, and the Persian Rhazes. Among other famous and prolific translators, Adelard of Bath and the much later William of Moerbecke certainly deserve mention. The significance of what happened at Toledo cannot be overemphasized, for seemingly overnight the lamp of knowledge was relighted for the West. The sound Greek foundation of modern science was unearthed nearly intact, preserved by its Arab caretakers, and upon that, construction began on the edifice of science.

This transitional period was marked too by flaws and mistakes. Some translations were so literal as to be almost useless. Some were done haphazardly, and others were deliberately and selectively edited. Still, the intellectual upshot of the entire movement was so overwhelmingly beneficial to the West that these and all other qualifiers do not really apply. With the fall of Toledo in 1085, the chasm that had separated the West from its cultural heritage was at once spanned and Europe was on the

The medieval universe was strictly hierarchical, with the changeable and imperfect earth set well below the immutable and perfect domain of God. *Buch der Natur*, 1481. Konrad von Megenberg. See p. 101.

road to a revival of learning. One irony that becomes apparent through hindsight is that Spain, whose sanguinary and successful "reconquista" may be said to have fertilized the roots of modern science, never came to know its intellectual pleasures or practical advantages. Sadly, its extreme authoritarianism and asceticism stifled any scientific advance in Spain.

These two extremes may be said to characterize the medieval mind-set and its general approach both to nature and to the place of mankind in the natural world. Although such a mind-set persisted longer in Spain, it eventually gave way in much of Europe. Its demise was essential to the growth of modern science, for as the Spanish experience would subsequently instruct, science cannot prosper in an atmosphere dominated by the theological and the moral. The medieval mind was a strange mixture of hardy pragmatism, manifested in impressive technical accomplishments, with a dreamy, otherwordliness demonstrated by its obsession with religion and magic. A society thus given to symbolic and metaphorical thinking may be devout and romantic but not truly scientific.

How people view the natural world and seek to explain the universe determines to a great extent the type of questions they ask, which in turn largely affects the answers that are given. The medieval adaptation of Platonism by the Church (principally by St. Augustine) and the pervasiveness of its transcendental world view, effectively forestalled any real penetrating questions about nature. Indeed, its indifference to the sensible world and its rejection of things physical and sensual made medieval times an era of almost quiet reflection. A pervasive sense of stoical acceptance is hardly the atmosphere in which scientific inquiry thrives. Although knowledge of the natural world was not entirely neglected during this time, it was regarded as secondary knowledge at best and certainly not the type of knowledge that by itself would lead to any actual illumination. Knowledge of nature was sought primarily as a symbolic, although imperfect, reflection of a higher and truer reality. This nearly absolute theological domination of all of medieval life and thought began to weaken as the fabled light came from the East. The exotic, worldly, and tolerant Arab ways made reconquered Spain a shimmering temptress to the rest of Europe. Among her many gifts, that of a new way of looking at the natural world might have been her most precious offering.

This reorientation toward natural knowledge for its own sake was, by today's standards, an excruciatingly slow process, and its full flowering did not occur until the seventeenth century. But the opening of the Eastern window had, by the beginning of the thirteenth century, stimulated an irrepressible intellectual yearning in the West. It soon became apparent that the range of the existing monastic and cathedral schools was severely limited, and the new "Universitas" arose to meet these new needs. In Bologna, Paris, and Oxford, and finally in most of the major European cities, there arose new, more secular schools of learning, although still run mostly by churchmen. Science as we think of it today was certainly not on the curricula of these great universities, dominated as they were by the presence of theology at the top of the academic ladder, with the seven liberal arts at the bottom. At the University of Paris, for instance, the

An illustration of grammar as part of a university's curriculum. *Margarita philosophica*, 1503. Gregor Reisch.

curriculum consisted of theology, philosophy, law, and the liberal arts. The liberal arts were divided into the "trivium," consisting of grammar, rhetoric, and logic, and the "quadrivium," made up of arithmetic, geography, astronomy, and music. In fact, the University of Paris, which by 1210 had replaced the famed School of Chartres as the center of learning in France, actually banned in that very year the teaching of any of Aristotle's scientific writings. Perhaps nothing better symbolized the recovery of Greek knowledge than did the works of Aristotle, and in a way this entire Scholastic period of scientific history can be described as medieval man's attempts to reconcile the new learning—of Aristotle—with the old—religious dogma.

This condemnation in 1210 of Aristotle's "natural philosophy" has become a minor historical footnote because of the Church's later happy reconcilation of Aristotelian thought with its own religious teachings. At

Oxford, Robert Grossteste translated and commented upon a number of Aristotle's works. It was he who initiated the great scientific tradition at Oxford and had as a pupil the legendary Roger Bacon. At Paris, Albertus Magnus and his pupil Thomas Aquinas were mainly responsible for making Aristotle palatable and finally complementary to religion. Indeed, Aquinas did his work so well that his more rigid Scholastic followers eventually held that there was no greater knowledge than that of Aristotle.

During this thirteenth-century revival of learning, the individual who captured the essence of the new science and saw clearly its unlimited potential was the great Franciscan thinker Roger Bacon. Like his teacher Grossteste, Bacon belonged to the Franciscan order—unlike his contemporaries Albertus Magnus and Thomas Aquinas, who were Dominicans. There were repeated and bitter controversies between the mendicant Franciscans and the studious Dominicans over Aristotelian philosophy, and the unswervingly orthodox Dominicans later proved to be the ideal order to carry out the purges of the Inquisition. Bacon was a Franciscan, however, and was delightfully nonconformist. His approach to natural knowledge anticipated his famous namesake of three hundred and fifty years later. The medieval Bacon offered what might be the earliest and certainly the clearest argument concerning the role of the experimental method. "Reason does not suffice," he said, "but experience does."

Basically, Bacon argued with some vehemence that certainty in science can come only by observation and experiment, for although reasoning alone about the natural world can lead to conclusions, it does not verify them. Ironically, Bacon's typically medieval passion for alchemy coexisted splendidly with his seemingly modern ideas. He is perhaps most famous for his startlingly prophetic vision of the fruits of the scientific method.

> Machines for navigation can be made without rowers so that the largest ships on rivers or seas will be moved by a single man in charge with greater velocity than if they were full of men. Also cars can be made so that without animals they will move with unbelievable rapidity. . . . Also flying machines can be constructed so that a man sits in the midst of the machine turning some engine by which artificial wings are made to beat the air like a flying bird. . . . Also machines can be made for walking in the sea and rivers, even to the bottom without danger. . . . And such things can be made almost without limit, for instance, bridges across rivers without piers or supports, and mechanisms, and unheard of engines.

Such originality and insight were met first with suspicion and later with imprisonment. And while Bacon's writings were condemned, his contemporary Thomas Aquinas was sainted.

Ideas such as Bacon's were the first ripple in what would become an ineluctable wave of science. Among Bacon's thirteenth-century kindred spirits were the Spanish alchemist Arnold of Villanova and the French scholar called Petrus Peregrinus, who experimented with magnetism. During the next century, such major establishment figures as William of Ockham, Jean Buridan, and Nicole Oresme chipped away at that medieval monolith, authority. All three supported the heretical "impetus theory"

EPISTOLAE
Gulielmi Budęi Regii Secretarii.

Venūdantur in officina Iodoci Badii cū gratia & priuilegio in triennium.

The press of Joducus Badius as represented in a woodcut used for his printer's mark. This is one of the earliest representations of the printer's trade. *Epistolae*, 1520. Guillaume Budé.

that opposed Thomistic, or Aristotelian, proofs for the existence of God. The fact that the heavenly spheres were in motion proved only that an original push had come from somewhere, not necessarily from God, they argued. They also disputed the dogma that the angels continued to propel these bodies on their course through the heavens. These ideas are not important as religious heresy but rather as examples of complex notions that began to search for the natural causes of natural phenomena. Some of the scientific commentaries of these pre-Renaissance greats were printed during the first century of mechanical printing but they are not part of the Library's collections.

During the fifteenth century, when the creative energy of the Renaissance was at its height, science began to benefit from the robust spirit of free inquiry. The second half of the century is especially significant for science, for it marks the births of Columbus, Leonardo, and the process of typography. Columbus, though certainly no scientist, was to offer a very real vision of new worlds and limitless horizons. Leonardo, the archetypal Renaissance scientist and artist, was to offer a universal, almost premodern approach to nature. And the invention of printing with movable type was to offer the means to reproduce and to circulate on a large scale ideas and information, both new and old. These and other forces, such as the Reformation, served to make the sixteenth century a time of transition—one in which traditional modes of thought were breaking down and being replaced by new ones.

By the beginning of the sixteenth century, the business of printing books had been an ongoing phenomenon for nearly fifty years and had produced an estimated forty thousand titles or recorded editions of books totaling somewhere between fifteen and twenty million individual copies. Much that was printed came from the deeply medieval past, including many old manuscript books. Certainly there were new and original scientific efforts, but nothing to approach the volume of older knowledge. This disequilibrium between new and old was characteristic of the transitional sixteenth century. It was a time of danger and of opportunity; a time when science had not entirely found its way; a time of confusion resulting from the push-pull of the old and the new. Compared to the spectacularly assured and successful scientific revolution of the seventeenth century, with its impressive list of individual accomplishments, the sixteenth century produced a decidedly modest number of major scientific works. That century's pioneers of science—Paracelsus, Copernicus, Vesalius, Fuchs, Cardano, Gesner, Agricola, and Brahe—were sufficiently small in number to fit comfortably in any room, whereas those of the next century would have required an auditorium. Nonetheless, in the quality of their revolutionary thought, the likes of a Copernicus or a Vesalius was the match for nearly all of their successors. Indeed, most agree that the one work most representative of the coming scientific revolution, both actually and symbolically, is the *De revolutionibus* of Copernicus, published in 1543, the year of his death.

In the first year of the historic seventeenth century, two starkly contrasting and portentous events occurred. In Rome an uncompromising heretic

NICOLAI CO-
PERNICI TORINENSIS
DE REVOLVTIONIBVS ORBI-
um cœleſtium, Libri VI.

Habes in hoc opere iam recens nato, & ædito, ſtudioſe lector, Motus ſtellarum, tam fixarum, quàm erraticarum, cum ex ueteribus, tum etiam ex recentibus obſeruationibus reſtitutos: & nouis inſuper ac admirabilibus hypotheſibus ornatos. Habes etiam Tabulas expeditiſsimas, ex quibus eoſdem ad quoduis tempus quàm facillime calculare poteris. Igitur eme, lege, fruere.

Ἀγεωμέτρητος οὐδεὶς εἰσίτω.

Norimbergæ apud Ioh. Petreium,
Anno M. D. XLIII.

The symmetry and simplicity of this title page bespeak the essence of the Copernican idea. Stating that the sun and not the earth was at the center of our universe, Copernicus argued from common sense: "For who could set this luminary in another or better place in this most glorious temple, than whence he can at one and the same time brighten the whole?" *De revolutionibus orbium coelestium*, 1543. Nicolaus Copernicus.

was burned to death in a sadly spectacular show of institutional force, while in London a book was quietly published. The heretic was Giordano Bruno, a wandering visionary more philosopher than scientist, though he has been characterized by many as a martyr to science. The published book was William Gilbert's *De magnete,* a treatise as remarkable for its virtual omission of any reference to God or theology as for its dedication to the experimental method. No two individuals could have been more different in background, temperament, or social position than were Bruno and Gilbert, yet both shared the spirit of the coming scientific revolution. Bruno lived and taught his ideals, always laboring in the fields of philosophy, or what might today be called theory. Gilbert was very much the man of action and experiment whose trust was placed in the world of the senses. Bruno was an apostate monk, part genius and part magician, whose animist religious views caused his hasty departure from ten cities in ten years. His was a uniquely inflammatory presence that managed to alienate not only the Catholic establishments of France and Italy but their Lutheran and Calvinist counterparts as well. His significance to science is found in his incautious espousal of the heliocentrism of Copernicus and his quite serious concepts of infinite space and the existence of other worlds.

William Gilbert seems a milder type, as anyone would compared to Bruno, and he was a respected member of the Royal College of Physicians. Shortly before his death, he became Queen Elizabeth's personal physician.

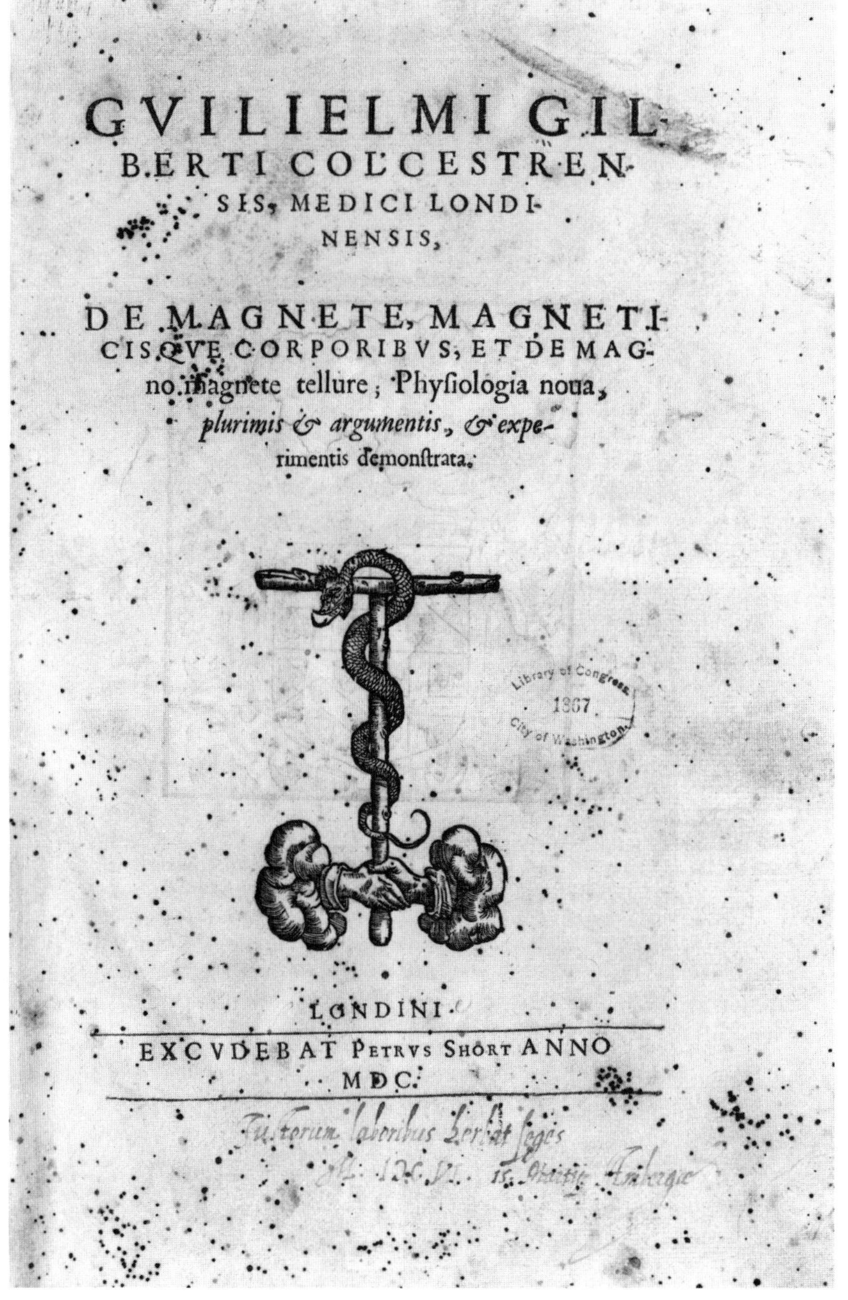
GVILIELMI GIL-
BERTI COLCESTREN-
SIS, MEDICI LONDI-
NENSIS,

DE MAGNETE, MAGNETI-
CISQVE CORPORIBVS, ET DE MAG-
no magnete tellure; Physiologia noua,
plurimis & argumentis, & expe-
rimentis demonstrata.

LONDINI
EXCVDEBAT PETRVS SHORT ANNO
MDC.

Gilbert's work was an influence and an inspiration to seventeenth-century scientists—especially so to Galileo. Gilbert offered the coming generation both an empirical and a theoretical paradigm. His systematic approach of repeated experiment and observation showed off the emerging scientific method at its best, and his inductive translation of these findings into a magnetic theory of the earth demonstrated the power of an organizing idea. *De magnete*, 1600. William Gilbert.

His famous work, *De magnete,* aside from its substantive scientific contributions, stands as one of the first works of science based on the modern methods of research and experiment.

The distinguished Gilbert and the "insufferant" Bruno are linked by more than the happenstance of chronology. The two men certainly knew of each other and may even have met. Bruno spent his most productive years in England, 1583–85, under the ambassadorial sponsorship and protection of the French embassy. He was a worldly man who had met with popes and kings alike and while in England accompanied the ambassador to Queen Elizabeth's court. By 1583, Gilbert had been a member of the Royal College of Physicians for two years and was well on his way to becoming one of the most prominent men of Elizabethan science. Each of the two men in his own way was a harbinger of the new century's scientific spirit. Each struggled against the authority of traditional learning, which Bruno called "the three-headed hellhound of Aristotle, Ptolemy, and dogma" and which Gilbert said produced "oceans of books whereby the minds of the studious are bemuddled and vexed." Through hindsight, we see that both men were revolutionaries. Bruno preached the idea of a scientific revolution and Gilbert showed the way to achieve it. Gilbert died a natural death and was honored both in life and death, but Bruno spent the last eight years of his life in a Roman prison before being burned alive on the Campo de' Fiori in Rome as a dangerous and unrepentant heretic. His religious heresies were indeed severe and extreme and it was probably for being a magician that he was silenced—his heliocentrism and ideas of infinity only making matters worse for him. But it was exactly those radical ideas that, when linked to Gilbert's empiricism, would serve to guide European science out of the closed maze of medieval times into an open vista of scientific adventure and learning. The two men are the ideal figures of transition.

In 1559 William Cuningham, a contemporary of Giordano Bruno, published under the patronage of Queen Elizabeth *The Cosmographical Glasse.* This illustration from that work shows the Ptolemaic conception of a finite universe—a tidy idea that had no place for Bruno's notions of other worlds and infinite space. In this picture, Atlas, dressed like an ancient king, holds the earth-centered universe on his shoulders. *The Cosmographical Glasse*, 1559. William Cuningham.

The first decade of the seventeenth century was notable for other auspicious beginnings. Sometime between 1601 and 1603 a society was formed in Rome—the Accademia dei Lincei—that became the principal forerunner of the first true scientific society. This academy had its own forerunner in the Accademia Secretorum Naturae, founded in Naples in 1560 by the physician Giambattista della Porta. Called an "academy of curious men," it met at Della Porta's home for scientific discussion and experimentation. It was soon closed by the pope on suspicion that its members were meddling with the black arts. The Academy of the Lynx had as its primary purpose the scientific study of nature conducted by members who were linked through a bond of brotherhood. Its early history was irregular, and it was closed on suspicion of poisoning and incantation. But in 1609 it was successfully reorganized and numbered among its loyal members both Galileo and Della Porta. The academy never really flourished, however, and when the death of its founder, Duke Federico Cesi, was followed by the official condemnation of Galileo in 1633, the society lost both its patron and its members. Some of the pride and loyalty felt by its members were demonstrated by Galileo who, on the title page of both his *Dialogo* and *Discorsi,* called himself "Galileo Galilei Linceo."

DIALOGO
DI
GALILEO GALILEI LINCEO
MATEMATICO SOPRAORDINARIO
DELLO STVDIO DI PISA.
E Filoſofo, e Matematico primario del
SERENISSIMO
GR.DVCA DI TOSCANA.
Doue ne i congreſſi di quattro giornate ſi diſcorre ſopra i due
MASSIMI SISTEMI DEL MONDO
TOLEMAICO, E COPERNICANO;
Proponendo indeterminatamente le ragioni Filoſofiche, e Naturali tanto per l'vna, quanto per l'altra parte.

CON PRI VILEGI.

IN FIORENZA, Per Gio:Batiſta Landini MDCXXXII.

CON LICENZA DE' SVPERIORI.

The title page of Galileo's most famous work identifies him proudly as "Galileo Galilei Linceo," a member of the Academy of the Lynx. *Dialogo*, 1632. Galileo Galilei.

The experimental spirit of Galileo connects the failed Lyncean Academy to what is regarded as the first organized scientific society, the Accademia del Cimento. This Academy of Experiment was founded at Florence in 1657 by Galileo's most famous pupils, Evangelista Torricelli and Vincenzo Viviani, who garnered the patronage and protection of the Medici brothers, the Grand Duke Ferdinand II and Leopold. Besides these two former pupils of Galileo, the best known names on the academy's rolls were Giovanni Borelli and the Dane Nicolaus Steno. For ten years, these and other scientists labored methodically to live up to the academy's purposive name. They worked together at experimentation of all sorts and eschewed speculative thinking. Together they worked out new research methods, invented new instruments, and devised better standards of measurement. The results of this uniquely cooperative experimental venture were summarized in a publication, *Saggi di naturali esperienze fatte nell' Accademia del cimento,* published in Florence in 1666. The Library of Congress has in its collections this original edition as well as those published in 1667 and 1691. The elegant folio was an expensive book to produce and contained many precisely engraved plates depicting the instruments used in the experiments. Its overall focus was that of experimental physics, and when finally translated into Latin in 1731, it became the "laboratory manual" of the eighteenth century. Many regard this collaborative work as the beginning of modern physics. The academy was discontinued in 1667 and many believe that with its termination, Italian leadership in physics came to an end.

Unlike the short-lived Academy of Experiment, the Royal Society of London endures to this day. Incorporated formally in 1662 by charter of Charles II, the society had its origins in the informal associations and meetings of devotees of the new experimental science. Meeting almost weekly from 1645, the group discussed a wide range of topics, with politics and theology deliberately excluded from consideration. Although devoted to the experimental method, as was the Accademia del Cimento, the Royal Society had three significantly different characteristics. First, it was a more open, almost egalitarian society, for, despite its affiliation with the Crown, its existence was not contingent on the vagaries of political fortune. Second, its members proceeded with their work in a much more individual manner, unlike the cooperative, almost anonymous work of the Italian academy. Third, the society endeavored not only to communicate the results of its work to all of Europe by letter but to be receptive to non-English work.

Four years after the Royal Society was formally incorporated, a corresponding French institution was established. Founded by Louis XIV in 1666, the Académie des Sciences burst upon the scientific scene in a spectacularly grand manner. Unlike the informal, individualistic Royal Society, which was in constant financial trouble, the French academy was created as an actual institution of the government, both benefiting and suffering from all that affiliation implied. Thus, its members received fixed pensions and enjoyed the best of experimental facilities, and they labored together as state employees—members of a bureaucracy. The trade-off of some freedom for security could be regarded as worthwhile in this instance, since the patronage of Louis XIV offered many a scientist—like Huygens—the luxury of not having to pause from his researches to worry about earning a living.

The basic soundness of these British and French models of scientific academies was demonstrated not only by their own longevity and productivity but by how often they were imitated. Beginning with the Akademie der Wissenschaften in Berlin in 1700, the formation of other national scientific academies was usually patterned after one or the other. Both in spirit and in deed, they all reflected the exuberance and the sobriety of the new experimental science. The academies were a tangible example of the unifying power of truth and the recognition that science and its secrets belonged to no one individual—that at the core of the new science was one social imperative, to communicate one's knowledge to another. The scientific academies that first grew up in the seventeenth century came to be, in part at least, because of the failure of other institutions—mainly the universities. Ironically, the same universities that in medieval times had been so progressive had become scientifically sterile, dedicated as they were to dogma. Learned men needed to talk with others, to exchange ideas and methodology, to sound out their fellows on their theories, and to compare notes—it has always been so.

An awareness of a sort of international community of science became increasingly apparent in the role played by the famous "postman of Europe," Marin Mersenne. This priest who seldom left his monastery in Paris became the living center of an informal network of European intellectuals. As a former pupil and close friend of Descartes, he was very well connected and either met or corresponded with all the scientific savants of his time. Mersenne died in 1648 and in his own way was a precursor of the great scientific academies. The academies that sprang up during the seventeenth century were characteristic of an enthusiastic, self-aware, and at times visionary new age of science, one whose voice would grow significantly louder.

It was the academies that transformed the voice of science into the word of science, for they stimulated the creation of the first scientific journals. The periodical appearance of a journal to publicize the work of these groups was a logical outcome of the attitude and approach that gave rise to the academies themselves. Thus at the very beginnings of real organized scientific effort, it was obvious to its practitioners how important reliable information was. The first independent scientific periodical was the *Journal des sçavans* founded in Paris by Denis de Sallo. Its first number appeared on January 5, 1665, shortly before the French academy was officially created. The journal was suppressed on March 30, 1665, as a result of Jesuit criticism but it resumed publication the next year and became a widely popular and successful journal, of which the Library of Congress has a nearly complete set of the Amsterdam edition.

Two months after the first issue of *Journal des sçavans* appeared, the Royal Society endorsed its central concept by issuing its own journal, *Philosophical Transactions,* on March 6, 1665. Although affiliated with the Royal Society, the publication itself was actually the private property of Henry Oldenburg, the society's secretary, and it was not until sometime in the middle of the eighteenth century that the society officially embraced it. Although the publication date of *Philosophical Transactions* follows that of the *Journal des sçavans,* the former may be said to be the genuine inventor of the journal as a means of disseminating new scientific information. In this regard, the French model concentrated more on reviews and appealed to a wider audience, whereas the British version offered for the most part serious (and signed) articles of substance, usually detailing a new scientific observation or theory. The creation of several other scientific journals followed shortly after 1665, with either the French or British

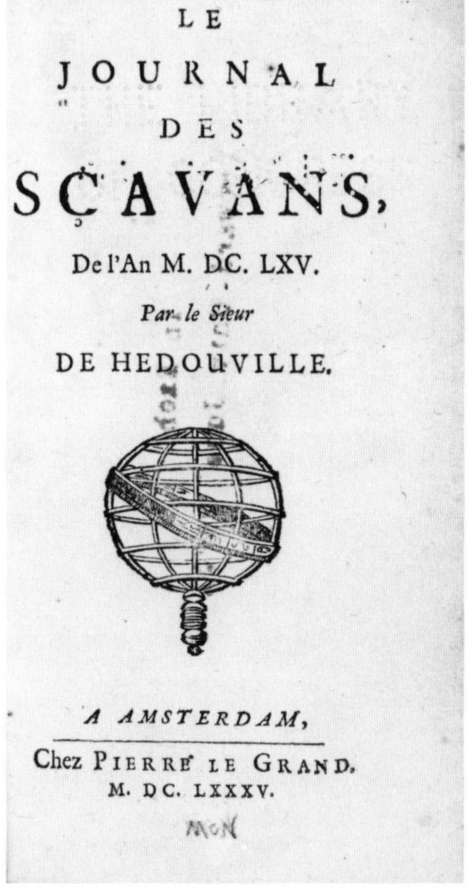
LE
JOURNAL
DES
SÇAVANS,
De l'An M. DC. LXV.
Par le Sieur
DE HEDOUVILLE.
A AMSTERDAM,
Chez PIERRE LE GRAND.
M. DC. LXXXV.

Journal des sçavans was the first independent scientific periodical. From the beginning of its publication in Paris, January 5, 1665, an edition was also published in Amsterdam and appeared at the same intervals. The Library of Congress has the Amsterdam edition. *Journal des sçavans,* 1665–1756, 1764–69.

version serving as the model. One of the more significant journals was the German *Acta eruditorum* first published in Leipzig in 1682. The Library does not have the early years of this journal but does have nearly 320 years worth of the Royal Society's monthly publications in its collections.

The emergence of scientific academies and the success of the regular scientific journal as the standard mode and instrument of scientific communication are perhaps the clearest illustrations of the institutionalization of science so characteristic of the late seventeenth and eighteenth centuries. Through their meetings and journals, prizes and medals, the scientific academies contributed at least as much to the spirit of the scientific revolution as to its substance. Science was becoming a successful, vital, and, equally important, acceptable field. Modern science was beginning to emerge, and its practitioners were evolving quickly from philosophers to scientists. This seventeenth century breakaway from the traditions of the past was the direct result of the enthusiastic embrace of the experimental method propounded and prophesied by Francis Bacon at the beginning of this incredible century.

Francis Bacon has been called by one twentieth-century scientist the "father figure of western science," and it is Bacon who links the members and founders of the great scientific academies and journals to the next century's encyclopedists. Indeed, Bacon's philosophy connected the simple empirical stirrings of the medievalists to the haughty extremes of the Age of Reason. Not that Bacon was moderate in his espousal of the scientific method, however. No man who boasted that "I have taken all knowledge to be my province," and who proposed "the enlarging of the bounds of Human Empire, to the effecting of all things possible," can be described in terms of moderation. Bacon came from a very prominent family. His father, Sir Nicholas Bacon, was lord keeper of the great seal. Francis was well educated and rose through the courts of Elizabeth and James I to the position of lord chancellor and then viscount of St. Albans. His character has been described by some as mean and obsequious, and a conviction of bribe-taking eventually ended his career. But Bacon's importance to science transcends any personality traits, for it was he more than any other individual who both understood the revolutionary essence of the "new science" and became its articulate spokesman and popularizer.

Bacon began his work with the assumption that scholastic philosophy, with its Aristotelian emphasis on final causes, was a complete failure. Since the Greeks, no real progress in understanding nature had been made, he said, and he dismissed contemptuously the authority of the Scholastics. Science, he argued, must be based not upon authority but upon observation and experiment. Bacon was acutely aware that a new era was about to dawn, and the titles he chose for his works exemplify this consciousness (*Novum organum* means "new tool or instrument" and *New Atlantis* is self-explanatory). His magnum opus, *Instauratio magna,* has perhaps the most obviously deliberate message, meaning literally the "great restoration after decay." That Bacon's understanding of the scientific method was overly simplistic and over-ambitious is not surprising, since he was no scientist himself. Yet his dogged insistence that only by adhering to the inductive method would science ever amount to anything gave to a new generation of seventeenth-century scientists the requisite enthusiasm and vision to persevere. Bacon was by no means the originator of the inductive method by which the general laws of science are drawn out of a mass of specific data by hypothesis, observation, and experiment, but it was he who gave the new science an articulate and respectable philosophy of method as well as a sense of mission and a scholarly grace.

Several ironies are evident in Bacon's life and work. First among them is the fact that this influential proponent of experimentation conducted few experiments of his own and in fact was not a scientist at all. This leads to the further observation that he wrote theoretically about the factual—that is, he gave to method a guiding philosophy and a vision. One final irony is Bacon's refusal to accept the views of Copernicus—proof that some "facts" are not self-evident to everyone.

Among the many original works of Bacon's in the Library's collections are the two considered most significant. His *Instauratio magna,* published in London in 1620, and better known as *Novum organum,* offers a new method of reasoning. *Instauratio magna* was planned as Bacon's largest and greatest work. Intended to be composed of six parts—much of which was either not written or not published—it presented what Bacon called "a total reconstruction of sciences, arts, and all human knowledge." The 1620 edition contains mostly the second part, *Novum organum*—explaining the title the work is known by. The other major work by Bacon in the Library's collections is his *New Atlantis,* published posthumously in London in 1628. This utopian fable carried Bacon's vision of the possibilities of science far and wide. Bacon told a glowingly optimistic story of philosopher-princes who founded the "House of Salomon"—a community given over entirely to science and totally dedicated to it. It is no accident of history that within a generation after Bacon's death, the idealism of Salomon's House saw partial fruition in the founding of the Accademia del Cimento and its successors.

Perhaps no individuals followed Bacon's words so literally and so earnestly as did the French encyclopedists of the eighteenth century. At the vanguard of this group was the brilliant and indefatigable Denis Diderot, whose brainchild became the monumental seventeen-volume work called *Encyclopédie ou Dictionnaire raisonné des sciences, des arts, et des métiers.* Published in Paris between 1751 and 1765 and augmented by eleven volumes of plates published between 1762 and 1772, this huge work can be considered the centerpiece of the French Enlightenment, espousing as it did the idea that intellectual progress would ensure the general progress of mankind. The conception of an encyclopedia was not new with Diderot—the great medieval compilers followed the same impulse—but Diderot improved upon both style and substance by ambitiously opting not only to include every aspect of the "new science" in comprehensive and sometimes definitive detail but also to commission the best scholars in France to write these articles. Organized alphabetically, the work typifies eighteenth-century formalism and its urge to organize and to codify—an understandable and necessary undertaking after the

ENCYCLOPÉDIE,

OU

DICTIONNAIRE RAISONNÉ DES SCIENCES, DES ARTS ET DES MÉTIERS,

PAR UNE SOCIÉTÉ DE GENS DE LETTRES.

Mis en ordre & publié par M. *DIDEROT*, de l'Académie Royale des Sciences & des Belles-Lettres de Pruſſe; & quant à la PARTIE MATHÉMATIQUE, par M. *D'ALEMBERT*, de l'Académie Royale des Sciences de Paris, de celle de Pruſſe, & de la Société Royale de Londres.

Tantùm ſeries juncturaque pollet,
Tantùm de medio ſumptis accedit honoris! HORAT.

TOME PREMIER.

A PARIS,

Chez
BRIASSON, *rue Saint Jacques, à la Science.*
DAVID l'aîné, *rue Saint Jacques, à la Plume d'or.*
LE BRETON, Imprimeur ordinaire du Roy, *rue de la Harpe.*
DURAND, *rue Saint Jacques, à Saint Landry, & au Griffon.*

M. DCC. LI.

AVEC APPROBATION ET PRIVILEGE DU ROY.

Title page of the first volume of the *Encyclopédie*, brainchild of Diderot and centerpiece of the French Enlightenment. *Encyclopédie*, 1751–65.

creative outburst of the previous century. But underlying each article, no matter how technical, could be found the Baconian concept of scientific progress pushed to the limit—the philosophy that reason will unveil truth and that all of nature is ultimately knowable. Progress in science and technology was viewed as a race toward the perfection of mankind—an admirable but philosophically precarious and sometimes dangerous position to hold. Its easy transferral to the sphere of politics did not go unobserved by the French monarchy, and publication of the *Encyclopédie* was plagued by real and threatened censorship.

Diderot had considerable help producing the *Encyclopédie,* and his collaborators numbered over two hundred. Among them were such notable writers as Jean Jacques Rousseau, Voltaire, Montesquieu, and Buffon. The great mathematician Jean le Rond d'Alembert was Diderot's assistant editor from the beginning and wrote not only the articles on mathematics but the general introduction to the entire work. D'Alembert left the project in 1759, the same year Pope Clement XIII described it as "containing false, pernicious, and scandalous doctrines and propositions, inducing unbelief and scorn for religion" and the King's Council legally suppressed it. Both the council's and the clergy's condemnation of the work as subversive was predictable since the *Encyclopédie* was indeed a work of propaganda. Most of its writers were committed intellectuals who fervently hoped to better the lot of mankind through the enlightened use of knowledge. By today's standards, the work is by no means objective, nor was it intended to be so. Perhaps no individual so exemplifies the writer as activist and prophet of the scientific movement as does the marquis de Condorcet. In 1792 when the *Encyclopédie* was long completed and the French Revolution was in full flower, Condorcet articulated the essence of its guiding philosophy. "All errors in government and society," he stated, "are based on philosophic errors which, in turn are derived from errors in natural science." The simplistic and usually dogmatic ideas of these genuine believers are cloaked with such sincerity and optimism—a sort of secular faith in mankind itself—that they engender grudging admiration in even their most severe critics. This monumental work now is better known for its politics than for its content. In actual substance, it is an exhaustive testament of all aspects of eighteenth-century thought and accomplishment with a decided emphasis on the sciences. The Library has the complete first edition of this monumental work, seventeen folio volumes of text and eleven volumes of elegantly precise engravings that graphically and accurately illustrate the technology of the day.

As the nineteenth century began, the exaggerated scientific claims of the French encyclopedists led some to the conviction that mankind was not far from a final explanation of all of nature. But as the century neared its close, the fertile seeds of skepticism had already begun to sprout. The nineteenth century was an age not only of great technological advances but of the popularization and professionalization of science itself. It was during this century that entire branches or disciplines of science were not only blocked out as to scope and method but substantively defined and described. Contemporary science as we know it today came to be.

Der·scaepherders

Kalengier

1. *Astronomy: Observing the Cosmos*

The history of astronomy, more than most scientific disciplines, has a rather pleasing progression about it. Over the centuries, the pattern of discovery has been like a series of connected events, each making the next possible or, sometimes, inevitable. One discovery has usually led to another. But if astronomy overall as a discipline is a rigorous science, its scientific subjects, its themes, and the individuals who played a role in its development have had an almost storybook quality about them.

Astronomy was always an easy collaborator to the romantic—drama is inherent in the stars, in the cosmos, and in the story of Galileo. Perhaps no science attracts so many dreamers. What other science offers such a grand and infinite workshop as the universe? And what other science offers its practitioners the luxury, indeed the pleasure, of combining unbridled otherworldly speculation with the most practical and concrete experiments and methods? Besides these stimulating practical and theoretical qualities, astronomy has always easily spilled over onto wider issues—usually of a philosophical nature. It is a science with ancient roots whose themes are grand and double-edged. This equivocal quality makes it all the more appealing.

When we ponder man's place in the heavens, we could be the fifteenth-century Nicolaus of Cusa, mystically positing ideas of infinity and relativity, or we could be John Flamsteed, three hundred years later, using a lifetime to count and catalog the stars. Asking such a question might put us in the realm of the technical or the philosophical—or both. All the best works of astronomy have this fascinating dualism.

Consequently, a trip through the major landmarks in astronomy represented in the collections of the Library of Congress reveals both mystery and orderliness. The mystery is found in the nature of the science itself. The very words that have been used to describe it—cosmos, heavens, luminaries, galaxy, orbs—suggest a magical realm of wonder and surprise. Some of its most famous investigators have been more mystic than scientist. Yet there is also an orderliness, almost a tidiness, in this science. From our modern perspective, we perceive the historic inevitability of a single true idea; one core theme that links each classic work with its predecessor and provides a bridge over which we might gaze at Copernicus while talking to Newton.

This unifying theme is summed up in the single word *heliocentrism* and all that it implies scientifically and historically. This idea, now so obvious, that the sun and not the earth is at the core of our system of moving planets, links the third-century B.C. Greek thinker Aristarchus with any

Opposite page:
This fifteenth-century illustration depicting a shepherd gazing at the heavens is taken from *The Shepherd's Calendar*, a compendium of practical advice relating to the whole existence of the common man. *Der scaepherders kalengier*, 1516. See p. 101.

twentieth-century astronomer. The scientific, political, and religious story contained in mankind's discovery, long-time rejection, and final embrace of that idea, makes it one of the grander themes of science and history. The ultimate rejection of the notion that the universe is man-centered, and the recognition and acceptance of the truth, no matter how deflating, is the story of science and the scientific method in a microcosm.

A final irony concerning this science that most consider to be the oldest is that it had as its subject not what immediately confronted those early thinkers but exactly what was farthest away from them. Despite an abundance of more accessible and easily observable earthly wonders before them, these early scientists instead looked skyward and beyond to seek the causes of terrestrial events in the heavens. No doubt the influence of religious thinking was a strong factor for many who hoped to discern some of the divine scheme in the pattern of the stars. Yet there must also have been the naturally curious who, lacking any motivation save the desire to know something for its own sake, took full advantage of their vertical stance and gazed upward in natural wonderment and delight. Motivation aside, astronomy has always attracted the best minds, as the following will attest.

Nearly all the great early Greek philosophers took an interest in some aspect of astronomy, contemplating the skies from a relatively rational perspective. As early as the fourth century B.C. the Greek astronomer Eudoxus knew that a year was not exactly 365 days long. During the next century, Aristarchus argued that all the planets, including the earth, revolved around the sun and that the reason the stars appeared motionless was that they were immensely distant from the earth. In the following century, Eratosthenes used his knowledge of the summer solstice to calculate almost correctly the circumference of the entire globe. The greatest astronomer of this pre-Christian era was Hipparchus, who numbered among his accomplishments the discovery of the precession of the equinoxes, the construction of the first systematic star catalog, and the founding of classical positional astronomy.

After Hipparchus, Greek astronomy languished for nearly three centuries until Claudius Ptolemy became his real successor. Although more is known of Ptolemy's works than of the second century Greek himself, few individuals have so dominated a discipline for so long a time. The title of his greatest and most influential work, *Almagest*—an Arabic title derived from the Greek—means "The Greatest." Originally titled *The Mathematical Collection,* it was later known as *The Great Astronomer.* The title's transmutation to its final, superlative case indicates the regard with which his work was held.

Over the centuries, Ptolemy's ideas dominated Greek, Arabic, and medieval thought. Among several of his printed works in the Library's collections is a 1515 vellum-bound book titled *Almagestū.* Printed in Venice, it

Opposite page:
The scene within the initial Q depicts a man pointing with one hand to the sun, moon, and stars and with the other to the things of this world. *Almagestū,* 1515. Claudius Ptolemaeus.

¶ L. Ptolemei Alexandrini Astronomorū principis Almagesti seu Magne ꝯstructionis liber: omniū celestiū motuum rationem clarissimis sententijs enucleans: fausto sydere incipit. Et primo in eūdē p̄fatio.

Vidam princeps nomine Albuguafe in libro suo (quem Sciētiarum electionem: et verborum nominauit pulchritudinem) dixit: ꝙ hic Ptolemeus fuit vir in disciplinarū scientia prepotens: preeminēs alijs. In duabus artibus subtilis: idest Geometria ⁊ Astrologia. Et fecit libros multos. de quorum numero iste est: qui Megasiti dicitur. cuius significatio est Maior perfectus. Quem ad linguam volentes conuertere Arabicā: nominauerunt Almagesti. ¶ Hic autem ortus et educatus fuit in Alexandria maiori terra egypti. Cuius tamē ꝓpago de terra Sem: et de prouincia que dicitur Pheuludia. Qui in Alexandria cursus syderum considerauit instrumentis tempore regis Adriani et aliorum. Et super considerationes quas Abrachis in Rhodo expertus est: opus suum edidit. ¶ Ptolemeus vō hic nō fuit vnus regum egypti: qui Ptolemei vocati sunt: sicut quidam estimant: sed Ptolemeus fuit eius nomen: ac si aliquis vocaretur Cosdrohe aut Cesar. ¶ Hic autem in statu moderatus fuit: colore albus: incessu largus: subtiles habens pedes. in maxilla dextra signum habens rubeum. barba eius spissa ⁊ nigra: dentes anteriores habens discoopertos et apertos. Os eius paruum: loquele bone et dulcis: fortis ire: tarde sedabatur: multum spaciabatur et equitabat: parum comedebat: multum ieiunabat: redolentem habens anhelitum: et indumenta nitida. Mortuus est anno vite sue septuagesimooctauo. ¶ Hec sunt de disciplinis et sapientijs Ptolemei huius. ¶ Conueniens est intelligenti pro deo verecundari: cū ea que ei sunt grata cogitat. ¶ Intelligens est qui semper linguam suam refrenat: nisi ad hoc vt de deo loquatur. ¶ Insipiens est qui suiipsius ignorat quantitatem. ¶ Cum aliquis sibi placet: ad hoc deductus est: vt ira dei sit super ipsum. ¶ In bono quod deus operatur: quasi bonitatem largi datoris attendere debes: et in malis aduersis quasi purgationis et eterne remunerationis bonitatem. ¶ Quanto plus fini appropinquas: bonum cum augmento operare. ¶ Hominis disciplina sui intellectus socius est: et apud homines intercessor. ¶ Non fuit mortuus qui scientiam viuificauit. nec fuit pauper qui intellectui dominatus est. ¶ Qui inter sapiētes humilior est: sapientior existit: sicut locus profundior magis abundat aquis alijs lacunis ¶ Non disseras nisi cum eo qui veritatem concedit: nec respondeas nisi a te querenti consilium: ⁊ cupide recipienti. ¶ Tuum consilium non committas nisi qui ipsum celauerit. ¶ Qui in mundo permanere voluerit: cor patiens aduersitatibus preparet. ¶ Parua domus est dolor minor. ¶ Plus gaudeas ꝙ non dixisti errorem: q̄ꝫ ꝙ bene dicendo non tacuisti. ¶ Cum irasceris non extendas manū ad peccandū. ⁊ cum dimissio vindicte non fuerit debilitas: parce. ¶ Ultime hominis promissiones cane sunt. ¶ Iustorum corda secretorum sunt monumenta. ¶ Qui per alios non corrigitur: nec alij per ipsum corrigentur. ¶ Manus intellectuum: animarum tenent habenas. ¶ Uulgi habenas regere melius est q̄ꝫ multos habere milites. ¶ Fiducia est socius ꝯsolans: quā licet non consequaris: eam tñ angariasti. ¶ Securitas solitudinis dolorē remouet: ⁊ pauor multitudinis cōsolationē aufert. ¶ Inter hoies altior existit mūdo: qui nō curat i cuius manu sit mūdus. ¶ Inuido videtur ꝙ ablatio boni alterius sit sibi bonum. ¶ Hoies lucrantur census: ⁊ census lucrant̄ hoies. ¶ Qui scientiā suam vltra astutiā que in ipso est extendit: est sicut pastor debilis cum multis ouibus. ¶ Qui in dignitate sua multum extollitur: in amissione eius multū deprimitur. ¶ Qui male operando vult celari: satis discoopertus est. ¶ Qui in mendacio cōfidit: tempestiue deficiet ei. ¶ Meditatio veritatis existit clauis. ¶ Intercessor est petentis ala. Anima non egredietur a fiducia vsqꝫ ad mortem. ¶ Anima ignorās suo socio magis inimicatur. ¶ Quidam rex inuitauit Ptolemeū ad prandium. qui rogans fore se excusatū: dixit Regibus contingit fere quod contingit considerantibus picturas. que cum a longe videntur placent: propinque vō non dulcescunt.

a

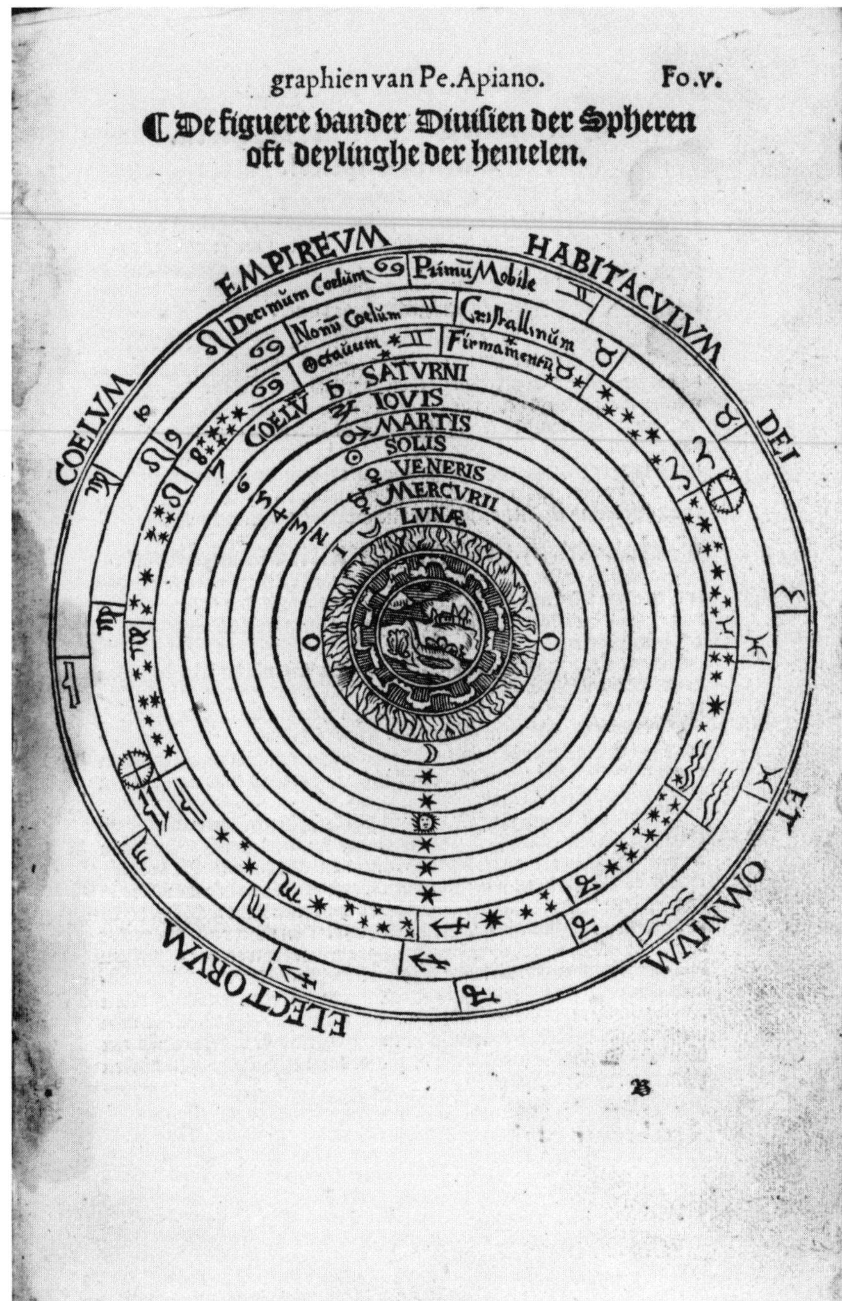

Ptolemy's cosmology was basically that of Aristotle. In this depiction of Aristotelian cosmology by Petrus Apianus, a motionless earth lies at the center of the universe. It is surrounded by seven concentric spheres, "crystalline spheres," which carry the sun, the moon, and the planets. An outermost ring of fixed stars rotates about the entirety. *Cosmographia,* 1545. Petrus Apianus.

is dense with tables, and its wide margins contain many diagrams. Ptolemy's *Almagest* is also represented in a massive 1541 collection of his works, the *Opera* printed in Basel. Overall, its tone is serious and authoritative.

At the core of his astronomical theory, so revered for so long, is the concept of a fixed spherical earth placed at the center of the universe. Not so much an original thinker as a synthesizer, Ptolemy took his geocentric idea directly from Aristotle. Though erroneous, the excruciatingly complex system of an unmoving earth at the hub of planets whose movements were explained by epicycles, deferents, and equants dominated cosmological thinking until the time of Copernicus—some fifteen hundred years after Ptolemy's death.

One man who greatly contributed to Ptolemy's authoritative sway was a fifteenth-century East Prussian, Johannes Müller, who called himself Regiomontanus (king's mountain) after the Latin name of his birthplace. Ptolemy's *Almagest* is represented in the Library's collection also by the Regiomontanus version *Epytoma in Almagestum* published in Venice in 1496. This weathered and much-used volume has a full-page woodcut at the beginning of the book that portrays the author and Ptolemy side by side.

As a student of the German astronomer Georg Peurbach, Regiomontanus was a confirmed Ptolemean who learned Greek so as to go beyond the faulty Arabic translations of Ptolemy's works. It was at the request of Cardinal Bessarion that Regiomontanus completed the project begun by the dead Peurbach and produced a completely revised and corrected version of *Almagest.* Regiomontanus also saw the immense potential of the mechanical printing process, then in its infancy. His corrected version of Ptolemy's astronomical work was published in Latin, the language of the learned Western world. It was in this volume, then, that the ancient astronomical knowledge, so long lost to the Western world and then misshapen by translators, was restored and presented to the eager minds of the Renaissance.

In the history of astronomy there is a direct line from Ptolemy's *Almagest* to the *De revolutionibus* of Nicolas Copernicus. It is not exaggerating to say that in the nearly fourteen intervening centuries, no major astronomical discoveries were made. Yet with the publication of his *De revolutionibus,* Copernicus shattered the false certitude of centuries of astronomical tradition and initiated the inevitable abandonment of Ptolemaic dogma.

With the words "In the midst of all dwells the Sun," Copernicus rejected the earth-centered Ptolemaic concept of the universe and offered instead a heliocentric mode (or, strictly speaking, a "heliostatic" one). His systematic, mathematical universe was attractive in its simplicity: "For who could set this luminary in another or better place in this most glorious temple, than whence he can at one and the same time brighten the whole?" To him, the earth's movement or rotation on its axis and around the sun was a physical reality that could be understood mathematically. This concept of the natural world as simple and knowable lay at the core of the scientific revolution in all other fields as well as astronomy. Such ideas were truly radical in the early fifteenth century.

Copernicus chose a lyrical and ultimately equivocal title for his work, calling it *On the Revolutions of the Heavenly Spheres.* Indeed it was a work destined to bring about fundamental changes in the way men thought, but no immediate changes, as it lacked strong empirical evidence. Nor was its publication always assured. The core principles of the book were worked out and written by Copernicus more than thirty years before the 1543 publication date, circulated privately, and then put away. The

The revival of Greek studies was an important aspect of the Italian Renaissance. With the fall of Constantinople to the Turks in 1453, Greek scholars and manuscripts flooded the West. The papal legate to the Holy Roman Empire at this time was an enlightened Byzantine, Cardinal John Bessarion, whose aim it was to bring the writings of ancient Greece to the attention of the West, by translation. Bessarion persuaded the astronomer Georg Peurbach to translate Ptolemy's *Almagest,* but Peurbach died after completing Book Six and his student, Regiomontanus, completed the project. Regiomontanus owed much of his fluency in Greek to Bessarion's teaching. Regiomontanus died in Rome in 1476 at age forty, having been summoned to that city by Pope Sixtus IV to emend the incorrect Julian calendar. This frontispiece shows a crowned Ptolemy and a gesturing Regiomontanus sitting beneath an armillary sphere. *Epytoma in Almagestum,* 1496. Regiomontanus.

NICOLAI COPERNICI

net, in quo terram cum orbe lunari tanquam epicyclo contineri
diximus. Quinto loco Venus nono menſe reducitur. Sextum
deniq; locum Mercurius tenet, octuaginta dierum ſpacio circũ
currens. In medio uero omnium reſidet Sol. Quis enim in hoc

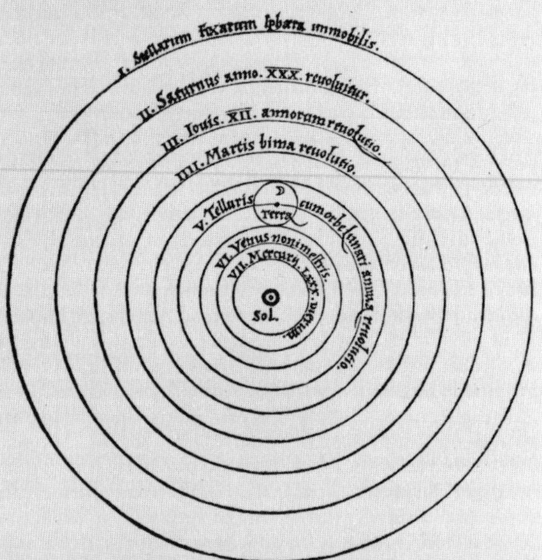

pulcherimo templo lampadem hanc in alio uel meliori loco po
neret, quàm unde totum ſimul poſsit illuminare? Siquidem non
inepte quidam lucernam mundi, alij mentem, alij rectorem uo-
cant. Trimegiſtus uiſibilem Deum, Sophoclis Electra intuentẽ
omnia. Ita profecto tanquam in ſolio regali Sol reſidens circum
agentem gubernat Aſtrorum familiam. Tellus quoq; minime
fraudatur lunari miniſterio, ſed ut Ariſtoteles de animalibus
ait, maximã Luna cũ terra cognatio nẽ habet. Concipit interea à
Sole terra, & impregnatur annuo partu. Inuenimus igitur ſub
hac

"In the midst of all dwells the Sun," said Copernicus. To him, the universe was systematic, mathematical, knowable, and above all simple. In this diagram from *De revolutionibus,* the earth is number V, "Telluris," counting from the outermost ring of stars. Copernicus was not entirely liberated from Aristotle and Ptolemy, however, for he too believed the orbits to be circular and uniform. *De revolutionibus orbium coelestium,* 1543. Nicolaus Copernicus.

work might never have been published at all had it not been for the intervention and urging of Rheticus, whose story is told later.

Its scientific merits and status aside, the book has a number of oddities that are historically interesting. It is estimated that four hundred first edition copies were published in the spring of 1543, but few sold and no real stir was created. The book's preface, written by a Lutheran theologian, seems to contradict the author's dedication. The latter asked for intellectual freedom "to follow the truth wherever it might lead." The former disclaimed the book's premise, saying it was not necessarily "or even probably true." The book was written by an astronomer who was certainly no natural observer of the heavens—in the book's 400 pages, only twenty-seven actual observations are noted. And although the author's name did appear on the book's cover, the text never included the name Copernicus but referred to its author as *domine praeceptor,* or master teacher. Despite the crucial role played by the young Rheticus in persuading Copernicus to publish it and in supervising its printing, the book nowhere contains his name either. Copernicus named several others in his dedication, but not Rheticus. And, finally, legend has it that the first copy of his book reached Copernicus on his deathbed, although this is probably apocryphal. Although prudently dedicated to Pope Paul III, the book was placed on the *Index librorum prohibitorum* by the Catholic Church in 1616 (not to be removed until 1822). Despite the Church's condemnation, the Copernican view came to be accepted within a century by the leaders of science. The Library's first edition copy was printed in Nuremberg in 1543 and was rebound during the nineteenth century in handsome Moroccan leather. The pages are gilt-edged.

The quotation beneath this vivid frontispiece scene is taken from the Bible's Acts of the Apostles, which says that, "Many of them also which used curious arts brought their books together, and burned them before all men." This scriptural account refers to the voluntary burning of superstitious books at Ephesis by St. Paul's new converts. Although the church's censorship of books dates to Roman times, the first catalog of forbidden books to include the word *index* in its title was published in 1559 by the Sacred Congregation of the Roman Inquisition. Copernicus's book was placed on the *Index of Forbidden Books* in 1616. *Index librorum prohibitorum*, 1758.

A reasonably important mathematician in his own right, Rheticus is one of those individuals in history whose fate and fame became linked and subordinate to another. Named Georg Joachim von Lauchen, he took the Latin name of the province (Rhaetia) in which he was born. He is best known today not for preparing the best trigonometric tables of his time but for his *Narratio prima,* a succinct abstract of Copernican theory. Rheticus first met Nicolas Copernicus when he was twenty-five years old and the master was sixty-six. The young man read the never-published manuscript that Copernicus had written before Rheticus was born and became the master teacher's first disciple. His *Narratio prima,* published in 1540, is the result of a compromise with Copernicus whereby Rheticus was allowed to publish this brief account of the hidden manuscript as a means of testing the scientific and political climate. No great protest emerged and Copernicus became convinced the times were right for his revolutionary manuscript. The young man's counsel proved correct. Despite having written the book in which the Copernican theory first appeared, the *Narratio prima,* and having guided *De revolutionibus* through the long and tedious printing process, Rheticus was ignored by Copernicus in his dedication. It has been suggested that in a volume dedicated to the pope, the Catholic Copernicus would not bring himself to thank the Protestant Rheticus. After all, Martin Luther had been excommunicated and outlawed only twenty-two years earlier. Although the 1540 edition is not in the Library, the text appeared again in the second edition of the *De revolutionibus* (Basel, 1566), which the Library does have. The Library also has a facsimile edition of the 1540 original owned by Copernicus's friend Johann Schöner.

Thirty years after the publication of *De revolutionibus,* a book written by Tycho Brahe appeared that supported Copernicus's challenge of the immutability of the heavens but did not fully endorse his heliocentric idea. Tÿcho Brahe, the irrepressible, argumentative Dane, is one of astronomy's most colorful and interesting figures. From his false metal nose (which replaced the real one he lost in a duel) to his three-foot tall court jester, Tycho was out of the ordinary. He did nothing in a quiet or modest manner. Yet this hard-drinking aristocrat was also a precise, methodical scientist who became astronomy's greatest naked-eye observer. His accurate observations put the first crack in the foundation of Aristotelian cosmology.

Trained in the law, Tycho was fortunate to have his astronomical appetite whetted early by an eclipse of the sun he observed in 1560. This event led him to further studies in astronomy and mathematics and prepared him for the great event in his life—the supernova of 1572. This exploding star, which seemed to appear out of nowhere, eventually grew brighter than Venus before dwindling away in a year and a half. The appearance of a new body in the heavens astounded, impressed, and shocked nearly everyone who saw it, peasants and learned men alike. It was deemed especially significant by all of sixteenth-century Europe, which placed so much import on astrology and the role the heavens played in men's lives. So it is not surprising that Tycho's *De nova stella* appeared in Copenhagen in 1573 along with scores of other books interpreting the star's meaning. Tycho's little book of fifty-two pages not only offered an explanation that this new body was indeed an unmoving new star but, more significantly, presented a theory that was the logical result of the author's precise celestial observations—the modern scientific method in a nutshell. He went where the truth led him. In this case, it led him to an implicit rejection of centuries of traditional cosmological thinking. Although Copernicus had published *De revolutionibus* thirty years earlier, the learned world still embraced the Aristotelian/Ptolemaic concept of an unchanging and perfect universe. Tycho not only claimed that this universe had changed, he offered scientific evidence, based on rigorous proof, what he called "hard, obstinate facts," that this phenomenon was indeed a new star, a recent addition to an obviously not-so-immutable or perfect universe.

Tycho proved himself human, however, by failing to embrace the full implications of the new star. A disciple of Ptolemy, Tycho never did fully support the heliocentric theory of Copernicus but rather attempted a great compromise between the two rival systems. He was on shaky ground, not being any great theoretician, and his faulty system of celestial machinery came to be disregarded. Accurate observation was his forte, however, and it was through his painstakingly precise measurements that the revolutionary implications of *De nova stella* could not be ignored. The Library's copy of *De nova stella* is a facsimile printed in Brussels from the original. Two later works by Tycho are in the Library's collections: *Astronomiae instauratae mechanicae,* published in 1602 in Nuremburg, which contains descriptions of his astronomical instruments, and *Opera omnia; sive, Astronomiae instauratae progymnasnata* (Frankfurt, 1648), which contains his *De nova stella.*

Some say that Tycho's greatest discovery was his young German assistant, Johannes Kepler. Kepler's creative genius asserted itself despite tragedy, war, plague, and religious persecution. That one man could accomplish such scientific breakthroughs despite the personal traumas inflicted by a chaotic Europe is amazing. That such a giant of science was, in

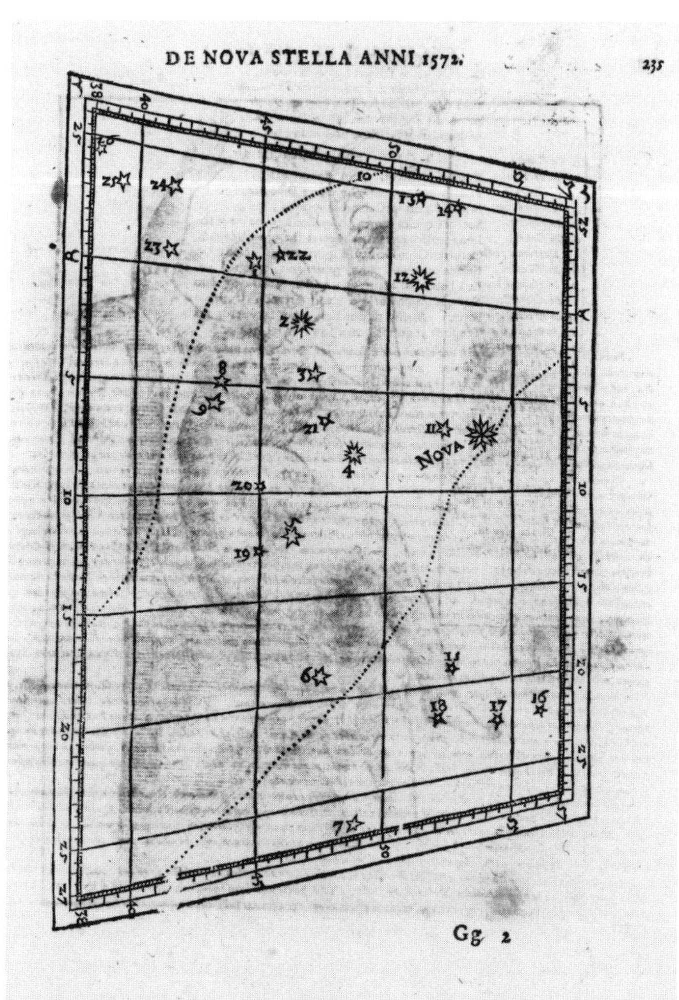

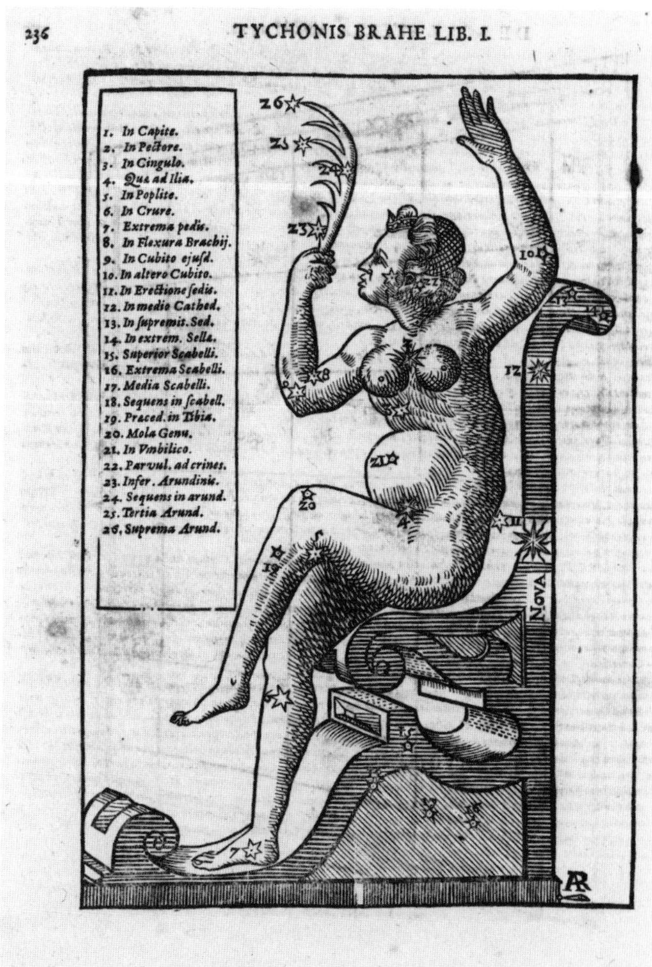

As history's greatest naked-eye astronomer, Tycho Brahe located the new star of 1572 (which we now know to be a supernova or exploding star) in its proper celestial place in the constellation Cassiopeia. The illustration on page 236 shows why the constellation, located between Andromeda and Cepheus, is also called Cassiopeia's Chair. *Opera omnia*, 1648. Tycho Brahe.

the flesh, a lonely, insecure man with weak eyes and a penchant for mysticism, gives some insight into the varieties of the individual scientific experience.

As was the case with so many astronomers, it was Kepler's studies in mathematics that led him to astronomy. But unlike many astronomers of his time, he became a thoroughgoing Copernican and chose for himself the intellectual goal of determining what the mathematical order of the universe was. With a single-mindedness he himself described as a "sacred madness," he discovered not only the true movements of the planets but the mathematical and physical laws that control them. His greatest work, aptly titled *Astronomia nova,* for indeed it laid the foundation for the new science of physical astronomy, was published in 1609 in Linz. This was eight years after the death of the great Tycho Brahe, with whom Kepler had lived and worked for nearly two years. Tycho bequeathed to the young Kepler not only his mass of precise astronomical data but also, and perhaps even more significantly, a respect for what has become the scientific method. That is, Kepler learned the essential value of applying theory to observable facts and testing his ideas against the real and the verifiable.

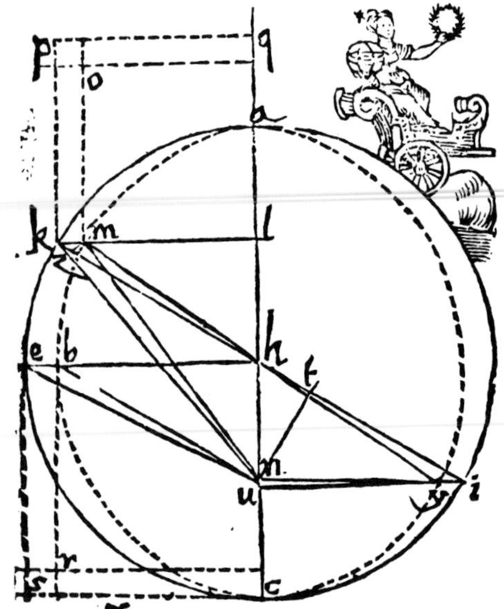

In this fairly simple diagram (topped by a drawing of "Victorious Astronomy"), Kepler demonstrated his two laws of planetary motion. First, he showed that Mars rotates in an almost circular elliptical orbit around the sun (the broken line is Mars's orbit around the sun, *n*). This at last broke the grip of Greek astronomy with its sacred, perfect circles. Second, he showed that if a radial line were drawn from the sun to Mars (*n* to *b*, *n* to *m*) it would cover equal areas in equal time. Kepler thus explained mathematically why planets change velocity, since they speed up as they near the sun and slow down as they move away. *Astronomia nova*, 1609 [1968]. Johann Kepler.

Combining his native genius with Tycho's observations, Kepler labored for six years and produced his massive *Astronomia nova.* Always a modest man, Kepler was nonetheless well aware of the significance of his masterpiece, for in his introduction he compares his celestial goal with the earthly adventures and aspirations of Columbus and Magellan. In that claim he was not boasting, for *Astronomia nova* certainly broke new ground. It contained what came to be known as Kepler's first two laws of planetary motion. The first stated that planets move in ellipses with the sun in one focus. The second explained that a line joining the sun and a planet sweeps out equal areas of the ellipse in equal times. These two laws of planetary motion at last broke the grip of Greek astronomy and its sacred circles. With them, the new science of physical astronomy was born, which henceforth would enable astronomers to compute a planet's orbit by a given mathematical formula. No more circles on circles, epicycles and eccentrics, or deferents and equants. Astronomy had begun to grow up.

Ten years later, Kepler completed his *Harmonices mundi,* a mystical work which, despite its Pythagorean musical analogies, contained his third law of planetary motion: that the squares of the periods of the planets around the sun are proportional to the cubes of their distances from it. With this final major discovery, Kepler had put the capstone on a mathematical system that would allow future astronomers to predict the movements of those wanderers the Greeks called "planets."

Kepler was indeed a genius, and an open-minded one at that. His exposure to the recently invented telescope resulted in a correct theory of vision. Also, he was the first to employ Napier's new logarithms, and he popularized their use in Germany. He was a voluminous writer, and among his volumes in the Library's collection is his *Somnium* or *Dream,* published in 1634 in Frankfurt, four years after his death. A mythical account of two people transported to the moon, it details what the moon would look like to visitors from earth and is regarded as the first work of modern science fiction concerning a trip to the moon. The Library's *Harmonices mundi,* published in Linz in 1619, is a first edition, with writing covering its vellum covers front and back. The 255-page book contains many musical symbols and abundant illustrations. Its fifth book, titled *De harmonia perfectissima motuum coelestium,* describes Kepler's third law. Although the Library's *Astronomia nova* is a recent facsimile, the collections include first editions of Kepler's *Ephemerides novae motuum coelestium,* published 1617–30 in Linz, *Tabulae Rudolphinae,* published in Ulm, 1627–30, and his mathematical work, *Chilias logarithmorum,* published in Marburg in 1624.

With the publication of *Sidereus nuncius,* Kepler's contemporary Galileo Galilei turned himself from a little-known mathematician into one of the most famous and celebrated men in all of Europe. More importantly, he triggered both the beginning of modern observational astronomy and the ultimate end of the Aristotelian-Ptolemaic world view. Galileo was a genius and an opportunist. Upon hearing of the Dutch invention of a "spyglass," he immediately designed and constructed several of his own—each time improving the magnification. Although he presented his new

HARMONICIS LIB. V. 207

mnia (infinita in potentiâ) permeantes actu: id quod aliter à me non potuit exprimi, quam per continuam seriem Notarum intermedia- CAP. VI

rum. Venus ferè manet in unisono non æquans tensionis amplitudine vel minimum ex concinnis intervallis.

Atqui signatura duarum in communi Systemate Clavium, & formatio sceleti Octavæ, per comprehensionem certi intervalli concinni, est rudimentum quoddam distinctionis Tonorum seu Modorum: sunt ergò Modi Musici inter Planetas dispertiti. Scio equidem, ad formationem & definitionem distinctorum Modorum requiri plura, quæ cantus humani, quippe intervallati, sunt propria: itaque voce quodammodò sum usus.

Liberum autem erit Harmonistæ, sententiam depromere suam: quem quisque planeta Modum exprimat propiùs, extremis hîc ipsi assignatis. Ego Saturno darem ex usitatis Septimum vel Octavum, quia si radicalem ejus clavem ponas *G*, perihelius motus ascendit ad ♮: Jovi Primum vel Secundum; quia aphelio ejus motu ad *G* accommodato, perihelius ad *b* pervenit; Marti Quintum vel Sextum; non eò tantùm, quia ferè Diapente assequitur, quod intervallum commune est omnibus modis: sed ideò potissimùm, quia redactus cum cæteris ad commune systema, perihelio motu *c* assequitur, aphelio ad *f* alludit: quæ radix est Toni seu Modi Quinti vel Sexti: Telluri darem Tertium vel Quartum: quia intra semitonium ejus motus vertuntur; & verò primum illorum Tonorum intervallum est semitonium; Mercurio verò ob amplitudinem intervalli, promiscuè omnes Modi vel Toni convenient: Veneri ob angustiam intervalli, planè nullus; at ob commune Systema, etiam Tertius & Quartus; quia ipsa respectu cæterorum obtinet *e*.

Tellus canit MI FA MI ut vel ex syllaba conjicias, in hoc nostro domicilio MIseriam & FAmem obtinere.

CAPUT VII.

Harmonias universales omnium sex Planetarum, veluti communia Contrapuncta, quadriformia dari.

NUnc opus, Urania, sonitu majore: dum per scalam Harmonicam cœlestium motuum, ad altiora conscendo; quâ genuinus

So hoch als Keppler stieg, ist noch kein Sterblicher gestiegen, – Und Keppler starb in – Hungersnoth!

In this page from his *Harmonies of the World*, Kepler gives the musical scales or range of sounds that he ascribed to each planet, including the earth. He arrived at these harmonies by calculating the velocities of each planet when nearest and furthest from the sun. Kepler's mystical forays into the music of the spheres was more than a revival of Pythagorean and Platonic ideas, for it was exactly these investigations that led him to the discovery of the principles of planetary motion. *Harmonices mundi*, 1619. Johann Kepler.

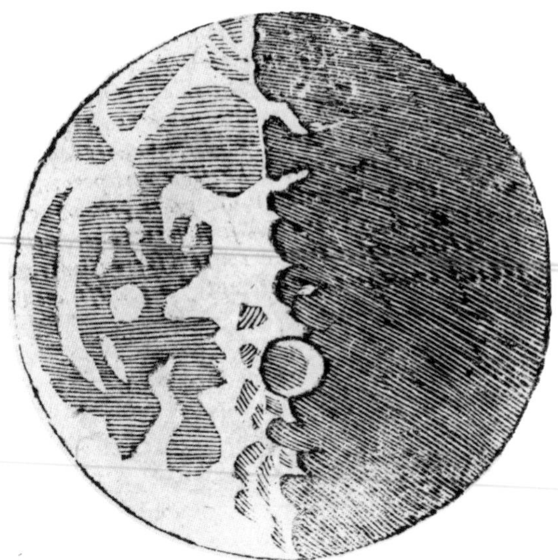

Galileo's *Starry Messenger* contained the first telescopic drawings of the moon to be published. Galileo showed the moon to be a solid body with irregular surface features. This drawing correctly shows mountaintops catching the sunlight and casting shadows, the length of which Galileo used to estimate the mountain's height. *Opere di Galileo Galilei*, 1655. Galileo Galilei.

Opposite page:
A frontispiece by Stefano della Bella to Galileo's *Dialogue Concerning the Two Chief World Systems*, 1632, shows Aristotle, Ptolemy, and Copernicus discussing matters of astronomy. Galileo was able to pursue his research in Florence during the years 1616–32 despite papal disapproval, since he worked under the aegis of the Medici family (symbolized here by the Grand Ducal crown and the banner). The family banner did not protect him from the wrath of the Roman Inquisition, however, and it was the *Dialogo* and its open defense of Copernican heliocentrism that occasioned his trial and his abjuration of the work itself. *Dialogo . . . sopra i due massimi sistemi del mondo tolemaico, e copernicano*, 1632. Galileo Galilei.

device to the burghers of Venice and dazzled them with its commercial potential, Galileo saw beyond that mundane application and turned his telescope upward to spy on the heavens. On each cloudless night from September 1609 to March 1610 he studied the moon and planets, noting feverishly the new wonders revealed to his extended eyes. His discoveries were dazzling, and he rushed them into print. His little twenty-eight-page *Sidereus nuncius* told what he saw. The moon was not smooth and polished but "is in fact rough and uneven, covered everywhere, just like the earth's surface, with huge prominences, deep valleys, and chasms." Also, the sun has spots, Venus has phases, and Jupiter has several moons of its own. The Library has a facsimile of the 1610, Venice, first edition of this work as well as a 1655 edition in *Opere di Galileo Galilei* published in Bologna.

To Galileo, these startling observations had enormous implications. Above all, they seemed to totally affirm the Copernican idea that the sun and not the earth was the center of the universe. Certainly this new window on the universe showed him a decidedly different place than the immutable, ethereal realm of Aristotle. There appeared to be an entire universe out there which took little note of man, nor did it function for the sake of man alone. The publication of his slim book in 1610 led Galileo into an inevitable defense of the Copernican system, which he published twenty-two years later. At this point in time, however, the forty-five-year-old Galileo was still circumspect and did not proclaim the implications of *Sidereus nuncius*. Rather, he rode discreetly on a wave of acclaim and popularity, so that only four months after the book's publication, he was appointed court mathematician to the Grand Duke of Tuscany. Galileo had the foresight to dedicate his book to the duke, Cosmo II de Medici, after whose family he named Jupiter's four moons "Medicean stars."

Over twenty years later, another dedication, this time to the pope, met with much less success. During the intervening years of war, plague, and religious dogmatism, Galileo had engaged in many small-scale skirmishes

DIALOGO
di
GALILEO GALILEI LINCEO
AL SER.mo FERD. II. GRAN. DVCA DI
TOSCANA
Stefan. Della Bella. F.

because of his Copernican advocacy. But by 1630, after laboring for six years, he was prepared for an open fight, having completed his *Dialogo . . . sopra i due massimi sistemi del mondo.*

Assuming that a friendly Pope Urban VIII was, if not on his side, at least not openly against him, Galileo went to Rome in May 1630 to obtain an imprimatur for *Dialogo.* Although the pope did not read the book, he regarded the work as a strictly hypothetical exercise and urged Galileo to change titles. Galileo had proposed to call the book *Dialogue on the Ebb and Flow of the Seas*—a title the pope thought overemphasized the issue of physical proof. Galileo agreed to the change and pressed for the imprimatur. Although the blessing to publish was eventually given, it was granted as the result of rather questionable lobbying practices by Galileo's influential friends. Further, the book was published in Florence and not in Rome under the pope's eyes. Only a few weeks passed before the pope realized he had been duped. In August 1632, he ordered all unsold copies of *Dialogo* confiscated. Those already sold would be bought back. In less than a year's time, Galileo would sign his guilt, admitting the falseness of his writings. The machinery of the Inquisition proved a fearsome and harrowing experience for the seventy-year-old man.

The censor's fears of *Dialogo* were justified, for the book was a masterly polemic against the old and for the new science. Written in the form of a discussion among three friends—a Copernican, an Aristotelian, and a supposedly impartial listener—*Dialogo* inveighs against a thousand years of tradition. Aside from its open defense of Copernican heliocentrism, the book had more subtle implications. To Galileo, the ultimate test of a scientific theory was found in nature. Man, he said, was capable of dealing objectively with the world and of knowing it on rational terms. Today, we call this way of thinking and operating the scientific method.

This particular first edition of Galileo's *Dialogo* is rare, many of the volumes having been burned by the Inquisition. It contains the famous Stefano della Bella frontispiece which shows Aristotle, Ptolemy, and Copernicus in discussion. Written in the vernacular, the book is 485 pages long, has five imprimaturs listed (three in Florence alone), and has an index. A first edition was smuggled to Strasbourg, translated into Latin in 1635, and circulated widely throughout Europe. The Library has two copies of the 1632 first edition published in Florence.

As Galileo first made his mark with the telescope, so young Christiaan Huygens did also with his own much improved version of the same instrument. Huygens was a mathematical prodigy who in his twenties devised a better method for grinding lenses, with the result that his homemade telescopes provided a significant increase in magnifying power. Still following in Galileo's footsteps, Huygens used his telescope to study the changes in the appearance of Saturn—changes that were first noted by the great Galileo. Huygens, however, was able to see what others could not—that Saturn was surrounded by a thin ring that did not touch the planet and whose inclination to the ecliptic varied slightly, which thus accounted for the changes in the planet's appearance. He also discovered the first satellite of Saturn, which he later named Titan. Huygens first

[12]

AXIOMATA SIVE LEGES MOTUS

Lex. I.

Corpus omne perſeverare in ſtatu ſuo quieſcendi vel movendi uniformiter in directum, niſi quatenus a viribus impreſſis cogitur ſtatum illum mutare.

PRojectilia perſeverant in motibus ſuis niſi quatenus a reſiſtentia aeris retardantur & vi gravitatis impelluntur deorſum. Trochus, cujus partes cohærendo perpetuo retrahunt ſeſe a motibus rectilineis, non ceſſat rotari niſi quatenus ab aere retardatur. Majora autem Planetarum & Cometarum corpora motus ſuos & progreſſivos & circulares in ſpatiis minus reſiſtentibus factos conſervant diutius.

Lex. II.

Mutationem motus proportionalem eſſe vi motrici impreſſæ, & fieri ſecundum lineam rectam qua vis illa imprimitur.

Si vis aliqua motum quemvis generet, dupla duplum, tripla triplum generabit, ſive ſimul & ſemel, ſive gradatim & ſucceſſive impreſſa fuerit. Et hic motus quoniam in eandem ſemper plagam cum vi generatrice determinatur, ſi corpus antea movebatur, motui ejus vel conſpiranti additur, vel contrario ſubducitur, vel obliquo oblique adjicitur, & cum eo ſecundum utriuſq; determinationem componitur.

Lex. III.

[13]

Lex. III.

Actioni contrariam ſemper & æqualem eſſe reactionem: ſive corporum duorum actiones in ſe mutuo ſemper eſſe æquales & in partes contrarias dirigi.

Quicquid premit vel trahit alterum, tantundem ab eo premitur vel trahitur. Siquis lapidem digito premit, premitur & hujus digitus a lapide. Si equus lapidem funi allegatum trahit, retrahetur etiam & equus æqualiter in lapidem: nam funis utrinq; diſtentus eodem relaxandi ſe conatu urgebit Equum verſus lapidem, ac lapidem verſus equum, tantumq; impediet progreſſum unius quantum promovet progreſſum alterius. Si corpus aliquod in corpus aliud impingens, motum ejus vi ſua quomodocunq: mutaverit, idem quoque viciſſim in motu proprio eandem mutationem in partem contrariam vi alterius (ob æqualitatem preſſionis mutuæ) ſubibit. His actionibus æquales fiunt mutationes non velocitatum ſed motuum, (ſcilicet in corporibus non aliunde impeditis:) Mutationes enim velocitatum, in contrarias itidem partes factæ, quia motus æqualiter mutantur, ſunt corporibus reciproce proportionales.

Corol. I.

Corpus viribus conjunctis diagonalem parallelogrammi eodem tempore deſcribere, quo latera ſeparatis.

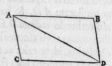

Si corpus dato tempore, vi ſola *M*, ferretur ab *A* ad *B*, & vi ſola *N*, ab *A* ad *C*, compleatur parallelogrammum *ABDC*, & vi utraq; feretur id eodem tempore ab *A* ad *D*. Nam quoniam vis *N* agit ſecundum lineam *AC* ipſi *BD* parallelam, hæc vis nihil mutabit velocitatem accedendi ad lineam illam *BD* a vi altera genitam. Accedet igitur corpus eodem tempore ad lineam *BD* ſive vis *N* imprimatur, ſive non, atq; adeo in fine illius temporis reperietur alicubi in linea illa

Newton's three laws of motion laid the groundwork for his law of universal gravitation. They state: (1) that a body remains at rest unless it is compelled to change by a force impressed upon it; (2) that the change of motion (the change of velocity times the mass of the body) is proportional to the forces impressed; and (3) that to every action there is an equal and opposite reaction. *Philosophiae naturalis principia mathematica*, 1687. Isaac Newton.

wrote of Saturn's rings for a small, two-page tract by Giovanni Borelli called *De vero telescopii,* but his writing was in the guise of an anagram, keeping his discovery disguised until he could further confirm his observations. This accomplished, he wrote his *Systema Saturnium,* which was first published in The Hague in 1659. The Library does not have this work in its collections. It does, however, have the complete, twenty-two-volume *Oeuvres complètes de Christiaan Huygens* (which includes his correspondence), published in The Hague between 1888 and 1950. After this and other astronomical successes, Huygens left astronomy and turned his genius to the fields of dynamics and optics.

The year 1642 began and ended symbolically for astronomy—Galileo died in January and Isaac Newton was born in December. Twenty-four years later in his annus mirabilis, the young Newton concluded in a bold, intuitive stroke that the physical laws of the heavens and those of the earth were one and the same, with both planets and apples being subject to the same natural forces. Although he had worked out the mathematical proof of his theory in 1666, Newton would wait another twenty-one years to publish it. Regarded as the greatest scientific work ever written, Isaac Newton's *Philosophiae naturalis principia mathematica* described the entire world as subsumed under a single set of laws. The translated title, in fact, makes its purpose quite clear. *Mathematical Principles of Natural Philosophy* is based on the Platonic belief and tradition that the study of nature should rest mainly on mathematical principles. The *Principia* consists of three books, the third of which, titled *System of the World,* affects most directly the history of astronomy. Book 3 deals specifically with the motions and mutual attractions of celestial bodies and contains Newton's famous law of universal gravitation. This discovery followed from his three laws of motion propounded in Book 1 and states that every particle of matter attracts every other particle with a force proportional to the product of the masses and inversely proportional to the square of the distances between them. By thus establishing that gravity is a universal property of all bodies, celestial and earthly alike, Newton is said to have "democratized" the universe, ending for all time the hierarchical and immutable concept of the cosmos. Thus, for the first time, a single mathematical law could explain movement both in the heavens or on earth. The entire cosmos was shown to be unified by knowable laws and predictable phenomena. Such a concept was truly a revelation in human thought and marked the glorious culmination of the scientific revolution begun by Copernicus.

Newton surely "stood on the shoulders" of others, as he modestly claimed, but where Copernicus, Kepler, and even Galileo could only describe, Newton explained. His ability to so penetrate to the core of a problem, to abstract and to speculate with precision, to so focus his magnificent intellectual ability, made Newton see and understand where others could not.

Both the origin and the publication of Newton's *Principia* are rather interesting stories to tell. Although written in 1686, some of its core calculations were worked out by Newton during the plague year of 1666. It was only in 1684 when Newton's friend, the astronomer Edmund Halley, posed the problem of planetary motion to him that Newton indicated that he had, indeed, worked out the calculations as a young man of twenty-four years. Halley reacted as had Rheticus to Copernicus, urging him to write and to publish. Newton became immersed in the work and fifteen months later produced the *Principia.* Yet publication was not assured. Robert Hooke, a lifelong adversary of Newton's, claimed priority of discovery. Although the Royal Society had intended to publish Newton's new work, it suddenly announced it was short of publishing funds rather than become embroiled in what promised to be a nasty dispute between two famous, influential scientists. Halley again offered his help, and, being a man of considerable means, financed all the publication expenses. The *Principia* was published in London in 1687 and the Age of Reason began. The Library's copy is a first edition.

Pierre Simon Laplace's monumental five-volume work on celestial mechanics was published in Paris over a span of twenty-seven years, from 1798 to 1825. Titled *Traité de mécanique céleste,* the work focuses on the perturbations of the members of the solar system and the overall stability of that system. Called the "Newton of France," Laplace not only codified the theories of his predecessors, specifically Isaac Newton, but he also developed them in a brilliant manner. On the questions of the irregularities of planetary movement, Newton could not explain the anomalies of Saturn and Jupiter, and resorted to God as the ultimate realigner. Laplace, on the other hand, maintained the universe was like a great self-correcting machine that was inherently stable. When asked why there was no mention of God in his lengthy work, he reportedly stated, "I had no need of that hypothesis." This extremely difficult treatise marked the culmination of Newton's application of mathematical theories to the science of astronomy. Furthermore, many of its more abstract theories were indeed confirmed by later mathematicians. Laplace's brilliant explanations and proofs make his *Treatise* a singular achievement, yet he is most famous for a footnote he added to a more popular book on astronomy, *Exposition du système du monde,* first published in Paris in 1796. This note, which is one of five at the end of the book, became known as his famous "nebular hypothesis." In it, Laplace offered the speculation that the sun originated as a giant cloud of gas or nebula that was in rotation. By centrifugal force, the core of the nebula became the sun and the rim of gas, the planets. This throwaway speculation on the part of Laplace took hold and is still popular today. Although the famous note is not found in the first edition, it is present in later editions, one of which the Library possesses (the third edition, published in Paris in 1808). The Library also has his complete five-volume *Traité de mécanique céleste* in first edition (Paris, 1798–1825).

Between these great works already described and the large mass of workmanlike but certainly minor astronomical efforts, there exists a small but significant corpus of first-rate, historically significant volumes in the Library's collections in astronomy. First among these, if only because its author predated even Ptolemy, is *De magnitudinibus, et distantiis solis, et lunae* of Aristarchus of Samos. Called the Copernicus of antiquity, Aristarchus may have been the first to propose the heliocentric hypothesis. Although no copies of his seminal work on the heliocentric hypothesis remain, his *Distantiis solis* did survive medieval times and was published in Pisa in 1572. His calculations as to the size and distance of the sun in this slim, thirty-eight-page book, which the Library has in first edition, were off by a factor of twenty, but his theory proved right.

Calculations were what *Almanach perpetuum* was all about. This famous astronomical work of over three hundred pages of tables contributed greatly to the age of maritime explorations and discoveries. Using its astronomical tables, ship captains were better able to navigate, using the sun and stars as a guide. Vasco de Gama was aided by its calculations on his expedition to India, and Christopher Columbus owned a copy, now in the Bibliotheca Columbina in Seville. The original text was written in Hebrew and translated later into Latin. The Library's *Almanach,* a wooden-bound text with a metal clasp, was published in Leiria, Portugal,

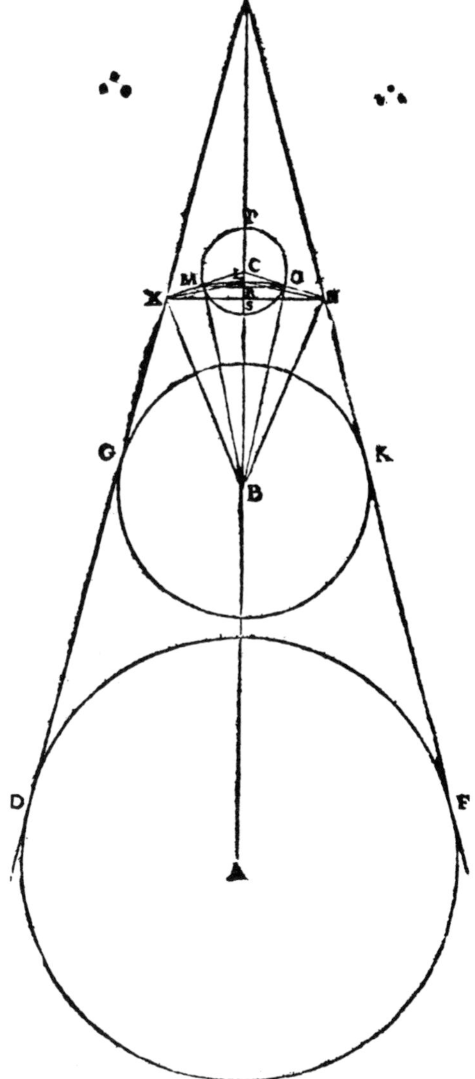

Aristarchus was the first to approximate the real scale of the solar system. He determined the size and distance of the sun and moon by using the earth's shadow on the moon and his knowledge of geometry. In theory, his method was correct though his final measurements were wrong. His calculation of the sun's distance from the earth was only four or five million miles, but his figure was sufficiently large to indicate generally the magnitude of the solar system. *De magnitudinibus, et distantiis solis, et lunae,* 1572. Aristarchus of Samos.

in 1496. The almanach's original tables are attributed to the Jewish mathematician Abraham Zacuto of Salamanca and Saragossa.

By far the most popular book on astronomy during the Middle Ages was the *Sphaera mundi* of Johannes de Sacrobosco (known also as John of Holywood). Little is known with certainty about him, but most agree Sacrobosco was an English clergyman from either Holywood or Halifax who died around 1256. He did spend the greater part of his life in Paris, where he taught mathematics and astrology. His *Sphaera mundi* was based primarily on Ptolemy and his Arabic commentators and was first published around 1220. Thereafter, it was eagerly copied and commented upon and became the standard textbook on astronomy in all the schools of Europe. With the new technology of printing in the fifteenth century, the *Sphaera mundi* was one of the first books on astronomy to be printed (Ferrara, 1472) and it was so generally adopted as a fundamental text that twenty-four more editions appeared during the next twenty-eight years. Indeed, it continued to be universally accepted until late into the seventeenth century. The text's obvious clarity and simplicity explain its continued use. It is a fairly small work, arranged in four chapters that treat the terrestrial globe, circular theories of the universe, the rising and setting of the stars, and the orbits and movements of the planets. Judging by this essentially derivative work, Sacrobosco had little to contribute to astronomy, but the clarity of his self-explanatory text and the later addition of an extraordinary series of woodcuts illustrating the constellations and the planets made his work enormously successful. The Library does not have the very rare 1472 Ferrara edition, but it does have several later editions of *Sphaera mundi* in its collections—the earliest being the 1485 Ratdolt edition printed in Venice. This contains some woodcuts printed partly in color and has, as accompanying text, the *Disputationes* of Regiomontanus and Georg Puerbach's *Theoricae novae planetarum.*

The intriguing title of *De docta ignorantia,* or *Learned Ignorance,* sets the tone for the mystical Nicolaus of Cusa. Described as the last great philosopher of the passing Middle Ages, Cusa held some very modern notions—such as a spinning earth moving about the sun, infinite space, and stars that were suns for other worlds. His *Learned Ignorance* was but a part, volume 1, of a larger work entitled *Opuscula theologica et mathematica,* published in Strasbourg around 1500. The Library's first edition copy is a most medieval-looking book with a dense, unrelieved text. The paradox of Cusa is that a medieval, almost intuitive philosopher could posit ultimately correct scientific ideas with no calculations, no observations, and seemingly no overall theory of his own. Yet in this cardinal's suggestions are found the seeds of modernism and revolution—ideas of relativity and infinity.

Still within the narrow realm of medieval astronomy is *Homocentrica* by Girolamo Fracastoro. This Veronese astronomer taught at Padua and may have known Copernicus, who was a student there. In this work published in Venice in 1538, Fracastoro labored long and hard to bring Ptolemy's geocentrism more in line with actual observations. Also in this work is found the suggestion of a telescope, where Fracastoro describes his experiments superimposing two lenses in order to magnify objects. The Library's first edition copy contains a portrait of the bearded author.

Petrus Apianus, whose real name was Peter Bienewitz, was a contemporary of Fracastoro. Both lived during the "great conjunction" of all the planets in the constellation of Pisces that took place on February 11, 1524. This event as well as the 1531 comet (which later became known as Halley's comet) provoked a great deal of astronomical and astrological speculation. Apian's principal work, *Astronomicum caesareum,* contains important observations on comets, most notably the fact that the tails of comets always pointed away from the sun. The Library has this large and handsome volume of *Caesar's Astronomy* in first edition. Published in Ingolstadt in 1540, it contains thirty-seven full-page colored volvelles that relate to the calculation of planetary movement and position. Apian believed these revolving disks to be of greater help in solving astronomical problems than the conventional mathematical tables. This volume was printed under the supervision of Apian and his brother at their private press in Ingolstadt, and it is thought to have taken eight years to produce. It is not an exaggeration to describe it as one of the most beautiful books ever made.

From *Astronomicum caesareum,* 1540. See p. 102.

Although both Fracastoro and Apian were contemporaries of Copernicus, it was Erasmus Reinhold, a German mathematics professor, who became one of the master's first converts. His main astronomical work, *Prutenicae tabulae coelestium motuum,* published eight years after *De revolutionibus,* presented the first set of planetary tables to be based on the Copernican theory. The work was sponsored by Reinhold's patron, Duke Albert of Prussia, and therefore was titled the *Prussian Tables.* Although these tables were eventually superseded by Kepler's more accurate tables, their mere existence contributed to the spread of the Copernican theory. Although there is some text to this book, it contains over 156 pages of tables. The Library's first edition copy was printed in Tübingen (Württemberg) in 1551.

One celebrated astronomer who publicly endorsed the Ptolemaic system but who privately embraced the Copernican idea was Michael Maestlin, the mentor and lifelong friend of Johannes Kepler. Since the Lutheran faculty at the University of Tübingen prohibited discussion of the Copernican theory, Maestlin was obliged, as an instructor there, to teach the geocentric system. This policy did not prevent him, however, from privately advocating the Copernican theory; and as Kepler's teacher, he instructed the young man in what he believed to be true—orthodoxy aside. His work is represented in the Library's collection by *Epitome astronomiae,* published at Tübingen in 1624. The book is of small dimensions but thick, with 543 stained pages. It is full of diagrams and tables and even has a foldout table. Maestlin outlived Kepler by one year and freely admitted that much of his pupil's work was beyond him.

The posthumously published works by the English astronomer Jeremiah Horrocks are noteworthy not only for their scientific accomplishments but because they are singular products of a genius who died too soon. In his short life of twenty-two years, Horrocks (whose name is sometimes spelled

Horrox) corrected Kepler's Rudolphine Tables and accurately predicted the transit of Venus between the earth and the sun. Following this prediction, he set up his telescope, aperture, and paper disk and became the first person ever to observe this phenomenon. From this single observation, he made what can only be described as an inspired (and nearly correct) guess as to the distance of the earth from the sun. Before his early death, Horrocks suggested the parallax effect, demonstrated that the moon's orbit was elliptical, and even gave a foretaste of Newton's law of universal gravitation. His 496-page *Opera posthuma* was published in London in 1673, thirty-two years after his death. The Library has this work in first edition.

The pioneering work of Johannes Hevelius (who Latinized his German name, Hevel) exemplifies how, soon after the invention of the telescope, the emphasis in astronomy was heavily on observation. Only one generation after Galileo first used a telescope to make his rough sketches of the moon for his *Sidereus nuncius,* Hevelius had built at Danzig Europe's best observatory and was probing the heavens with a 150-foot-long telescope. In 1647 he published in Danzig *Selenographia,* a magnificent atlas of 563 pages with 110 plates, detailing the surface of the moon. His lunar features are so accurate that we can recognize them today. In addition to this masterpiece, the Library has seven of his other titles dealing with his observations of comets and planets, most notably his *Cometographia,* published in Danzig in 1668.

Thirty-seven years after Galileo's moon drawings were made, Hevelius published *Selenographia*, his map of the moon. This magnificant work was the first real atlas of the moon and contains lunar features we can recognize today. Its level of detail reveals the rapid advances in telescope optics since Galileo's 1610 moon drawings. *Selenographia*, 1647. Johannes Hevelius.

Although Hevelius named many of the moon's features (he called the flat areas seas, or *maria,* a name we retain), his older friend and contemporary Giovanni Riccioli began the practice of naming lunar features after great scientists and prominent historical figures. Riccioli used this system in his *Almagestum novum,* published in Bologna in 1651, which, since he rejected the views of Copernicus, he titled in Ptolemy's honor. It should be noted that Riccioli named one crater after himself. The Library has a first edition copy of *Almagestum novum,* whose significance stems only in small part from Riccioli's contributions to naming and describing lunar features. A large portion of the two-volume folio concerns the Copernican controversy—a problem that fascinated Riccioli. As both a Jesuit priest and a man of science, he had difficulty accepting either the Aristotelian-Ptolemaic theory or the Copernican idea—each for different reasons. In addition to offering his own theory of the universe, *Almagestum novum* contains an account of Riccioli's pendulum experiments and his theories of lunar libration and planetary parallaxes, as well as the complete Latin text of Galileo's abjuration and the papal judgment against him.

The Cassini family of astronomers is represented in the Library by two works of the father Giovanni Domenico Cassini and two works of his son Jacques. Cassini père was an Italian who founded a dynasty of five generations of astronomers, all connected with the Paris Royal Observatory. The first Cassini made his mark by measuring the periods of rotation of Jupiter and Mars. He then discovered that Saturn's rings were divided in two and cooperated in an experiment to determine the solar parallax. The Library has his *Abregé des observations,* Paris, 1681, and his *La meridiana del tempio,* published in Bologna in 1695. Cassini has the dubious distinction of being the last of the great astronomers to refuse to accept the heliocentric theory of Copernicus. This stubbornness became a family trait, as each succeeding generation seemed to defy the wisdom of its times.

As the eighteenth century began, astronomy in its observational phase began to flower. And with observation came the testing, refining, and rebuilding of theory. Two British astronomers performed these functions admirably. John Flamsteed, who became the first Astronomer Royal in 1676, spent literally a lifetime systematically charting the heavens. Over a period of forty-three years he made an immense number of celestial observations and produced his *Historiae coelestis britannicae,* the first great star map of the telescopic age. The Library has in first edition this massive three-volume catalog of over twelve hundred pages and nearly three thousand stars, which was completed by his two assistants and published in London in 1725, six years after Flamsteed's death. Although his lifework was not marked by any great discoveries, his rigorous methods and penchant for accuracy made astronomy a more practical science.

Edmund Halley, Flamsteed's contemporary and sometime rival, is best known for the comet that bears his name. The erratic comings and goings of comets had long plagued and puzzled astronomers. Even Isaac Newton was uncertain about whether his law of universal gravitation applied to them. Halley's exhaustive researches indicated that the comet of 1682 had

Hevelius rivaled Tycho Brahe as the greatest naked-eye astronomer—owing much of his accuracy to the precision of his instruments. These instruments were designed, made, and engraved by Hevelius himself, and in volume 1 of this work he proudly shows them off in thirty exquisitely detailed plates. This one shows a six-foot brass sextant being used by Hevelius and an assistant to measure the angular distances between pairs of stars. The large sextant is cross-membered in iron to prevent flexing and is finely counterpoised with weights, ropes, and pulleys. Hevelius engraved this instrument in divisions of one-twelfth of a degree or five minutes of arc and he is shown using a micrometer to make fine adjustments to his movable sight rule. *Machinae coelestis,* 1673–79. Johannes Hevelius.

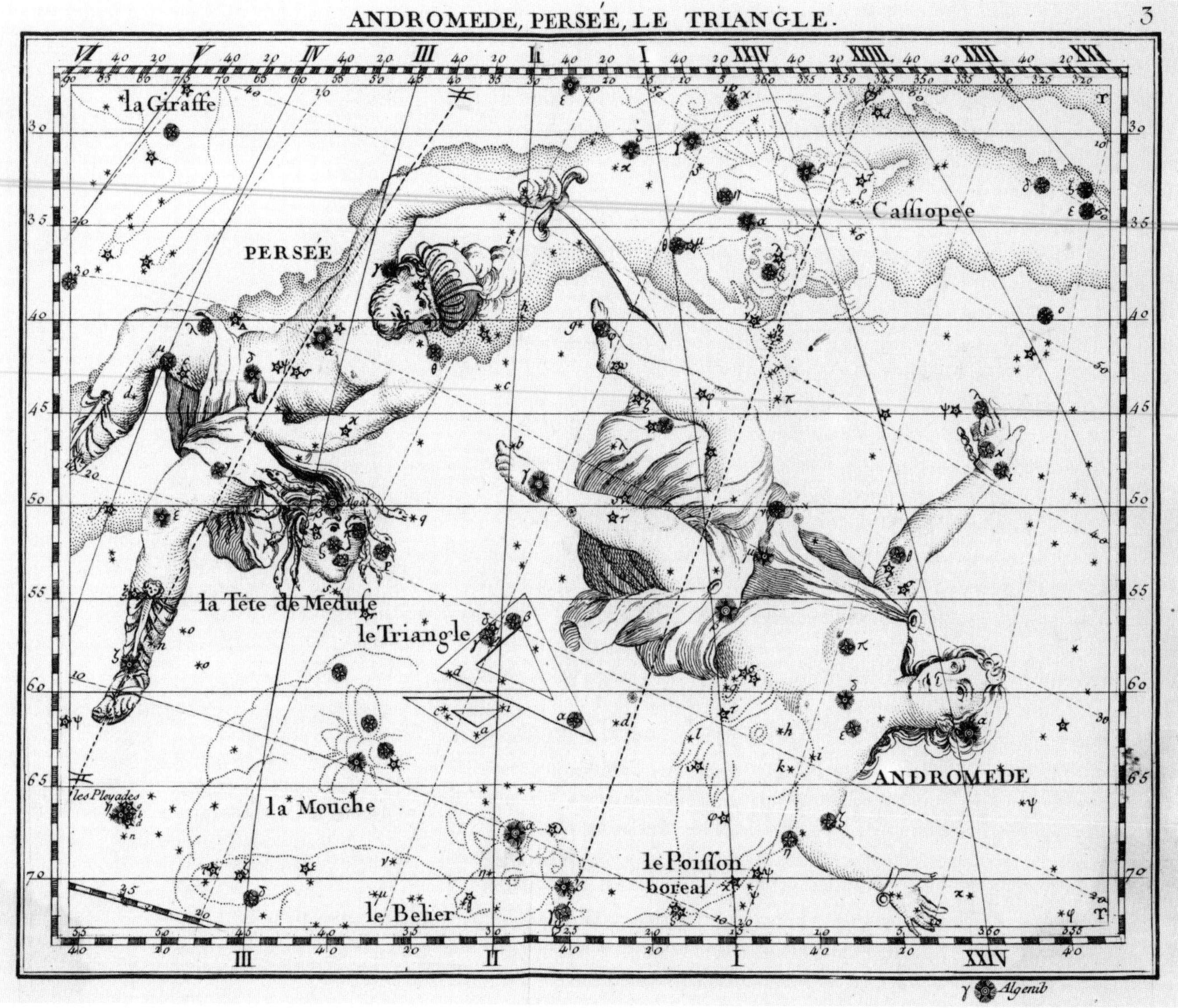

This decorative star map shows the adjacent constellations Andromeda, Perseus, and Triangulum as depicted in Flamsteed's mammoth star catalog, *British History of the Heavens*. Published in full after his death, Flamsteed's work was three times as large as Tycho's and six times more precise. *Atlas céleste de Flamstéed*, 1776. John Flamsteed.

orbits similar to those of 1456, 1531, and 1607. In his "Astronomiae cometicae synopsis," published in London in 1705, Halley expounded the highly original theory that comets are not erratic but periodic, that they belong to the solar system, and that they move around the sun in eccentric orbits. This theory was proved when, sixteen years after his death, the comet returned as he predicted, at the end of 1758. It has since returned in 1835 and 1910 and will be seen once more in 1986. His "Astronomiae cometicae synopsis" first appeared in the *Philosophical Transactions* of the Royal Society, a journal which is in the Library's collections. The Library also has his *Astronomical Tables,* published in London in 1752, ten years after his death.

By the beginning of the eighteenth century, science was starting to show its more practical side, and many individuals sought to apply theory to the problems of everyday life. Astronomy has always had its practical side—telescopes were first used for commercial rather than scientific purposes—and no practical problem was more pressing to the British than the determination of longitude at sea. If a navigator had a clock that would run accurately at sea, he could keep it set for Greenwich time, calculate the difference between it and local time, and astronomically establish his longitude. In 1714, the British Parliament offered a reward of £20,000 for a solution. The Library has a copy of this act—an early example of an enlightened and desperate government attempting to promote scientific progress. Results, however, were not immediately forthcoming. Nearly fifty years later, John Harrison, a self-educated Yorkshire carpenter, gave his "marine chronometer" to Nevil Maskelyne, Astronomer Royal, to test on a trip he would make to the island of Barbados.

Harrison's ingenious device proved able not only to withstand the sway of the ship but to compensate for changes in temperature as well, and his chronometer met the degree of accuracy specified for the award. Using the new instrument, Maskelyne determined the longitude of Barbados within one minute, a significant improvement over the conventional method of lunar distances, which entailed inaccuracies of four minutes. Four years later, in 1767, Harrison published in London the thirty-one page *Principles of Mr. Harrison's Time-Keeper,* which described the solution to this long-intractable problem. The Library's first edition copy has ten plates taken from the original drawings of the clock. Harrison received the full award in 1773, only after the direct intervention of George III.

During the last quarter of the eighteenth century, astronomers enjoyed the comfortable feeling that theirs was now a completed science, incapable of revealing anything really new. This complacency was shattered in 1781 with the revelation by William Herschel, an Anglicized German, that he had discovered a new planet—the first to be discovered in historic times. Herschel himself did not fully grasp the nature of his discovery when on March 13, 1781, using his own telescope, he first noticed an object ressembling a disk rather than a point of light. His assumption that he had discovered a comet was a natural one, and on April 26, 1781, the Royal Society published his "Account of a Comet." The Library has a complete set of the society's *Philosophical Transactions,* in which Herschel's findings appeared. Further observations revealed the object's orbit as circular, and by summer of 1781 Herschel was convinced that indeed he had discovered a new planet.

Herschel was correct and the new body was named Uranus. Herschel's sister Caroline shared, from the beginning, in all of his astronomical work and is known as the first important woman astronomer. Caroline Herschel is represented in the Library's collections by her *Catalogue of Stars,* published in London in 1798. It is interesting to note that although she prepared the lengthy tables for this volume and is credited as author, she wrote not one word of the text—both introductions being signed by her famous brother, William.

By mid-nineteenth century, Uranus was still the most distant known planet. However, studies had shown deviations of 1.5 minutes in the planet's arc—an anomaly that could not be accounted for. Two men set out to calculate the size and position of a hypothetical planet whose presence would account for the deviations of Uranus. Neither knew of the other's work. Urban J. J. Leverrier, a Frenchman, and John C. Adams, an Englishman, worked independently on this most difficult mathematical exercise, and both succeeded in determining the probable path and place in the sky of this hypothetical planet. Although Adams was first with his calculations, Leverrier was the more fortunate in publishing his results and then persuaded J. G. Galle of the Berlin Observatory to search for his new planet at a certain spot in the sky. Galle thereupon made the optical discovery of the planet later named Neptune.

Leverrier's article postulating the existence of Neptune is found in the June 1, 1846, issues of *Comptes rendus* (Paris). It was titled "Recherches sur les mouvements d'Uranus." Adams's paper, "An Explanation of the Observed Irregularities in the Motion of Uranus," appeared in 1847 in the *Memoirs* of the Royal Astronomical Society (London). The Library has both of these journals. The significance of Neptune's discovery is found in the ability of the two men to predict the existence of a giant planet solely by calculation. This was the most dramatic confirmation yet of Newtonian theory.

Early twentieth-century astronomy is perhaps best represented by Harlow Shapley's *Starlight,* a popular work printed in New York in 1926 that marked the beginning of present-day galactic astronomy. The Library has this small work in first edition. Shapley's studies of globular clusters of stars led him to postulate that the sun was not near the center of our galaxy. His studies were thus the first to present a picture of our galaxy that was close to its actual size. All previous estimates had been far too small. So, as Copernicus had moved the earth from the center of the universe nearly four hundred years before, now the sun itself was placed in a less central but more accurate position.

Shapley's matter-of-fact conclusion reminds us of the implications of discoveries in modern astronomy.

> The future history of the stellar system appears, indeed, thoroughly independent of our temporary terrestrial career. Man's station in this scheme is not too flattering—an animal among many, precariously situated on the crust of a planetary fragment that obeys the gravitational impulses of one of the millions of dwarf stars that wander in remote parts of a galactic system. His place in the universe, from the standpoint of dimensions, duration, or physical influence, is unimpressive; and his importance in some non-material way is a subject not suited to scientific research or speculation. We leave the subject here, noting that man's role as an investigator and would-be interpreter of the universe is surpassingly fascinating, whether or not it is cosmically significant.

Shapley's closing words bring us to the opposite pole of Aristotelian-Ptolemaic homocentrism. No longer is man the center of the cosmos and the reason it exists. Yet despite our most ordinary place in the physical scheme of things, we can take pride in the simple discovery and actual acceptance of that truth—and this is the essence of the scientific method.

Throughout its very long history, the tradition of astronomy has been primarily that of observation. From Aristarchus of Samos to today's astronomers, whose extended eyes are telescopes fixed in space, the tradition of observation has thrived in spirit and in practice. Until very recently, astronomers could neither touch nor even closely observe the objects of their science—their single option being in-place observation. All pre-space age astronomy therefore was conducted necessarily by the naked eye or the aided eye, and it was always one giant step removed from what it studied. This physical limitation reduced the multifarious methods and techniques of science to a single, straightforward act. It is a testimony to this grand astronomical tradition that its greatest single idea—heliocentrism—was put forth during the era of naked-eye astronomy. This, in turn, explains the essential astronomer—he sees what we do, but with insight and understanding. He not only observes, he perceives.

PICTORES OPERIS,

Heinricus Füllmaurer. Albertus Meyer.

SCVLPTOR

Vitus Rodolph. Speckle.

2. *Botany: From Herbalism to Science*

Botany is surely the most gentle of sciences. The careful observation of a flower is a calm, unobtrusive action—the peaceful contemplation of a beautiful object. Reduced to its essentials, botany requires no laboratory but the natural world and few tools but the naked eye. Botany in its most scientific or purest form consists of seeking to know more about the plant simply for the sake of that knowledge. Plants have not always been regarded as worthy of knowing or studying in themselves, not on their merits as sources of food or drugs but as life forms. In fact, the history of botany can be viewed in terms of repeated rediscoveries of this one theme—that plants are worthy of study in and of themselves, quite apart from any use they might have for mankind.

The practical motives behind plant study should not be disparaged—the bulk of our medical history, for instance, is made up of accounts of herbal remedies. But the study of the medicinal properties of plants contained a self-limiting mechanism—if a plant seemed to have no utilitarian value, it was disregarded, and no further study of it was made. The Renaissance attitude to nature changed this overly practical bent and initiated the scientific study of plants.

Botany as a pure science has certain characteristics and makes certain assumptions that prove thought-provoking and interesting. One of its unspoken but basic assumptions is an implicit respect and regard for all living things. The botanist who studies a plant's structure or tries to understand its functions confronts nature on its own terms. Investigations of how a plant thrives or reproduces, or studies of the purposefulness of a flower's coloration and structure, are almost implicitly egalitarian and tautological. The botanist studies the flower because it exists, and because it exists it is worthy of study.

This unspoken assumption is both healthy and productive. It treats the natural world as at least an equal to man and does not subordinate it to his ends. It stresses the interrelationships of all living things and leads us to a better understanding of our role in nature's balance. It is also a productive assumption, for it spurs a scientific process in which there is always more to learn.

Botany also has its historical characters. Traditionally a one-to-one process, doing botany is more often than not the individual person confronting the individual plant—often in its natural environment. But the historical botanist was not always the local eccentric who would pad off to the nearest primrose. Often he was an explorer or a wealthy traveler who took up the science as a hobby and soon became intrigued by its

Opposite page:
To emphasize how different his work was ("according to the features and likeness of the living plant"), Fuchs showed how the plates were made and credited the artists. Here, upper right, is Albrecht Meyer, drawing a plant from nature, while across from him Heinrich Füllmaurer transfers the drawings from paper to woodblocks. Below is Veit Rudolf Speckle, who cut the final woodblocks. *De historia stirpium*, 1542. Leonhart Fuchs.

charm and its natural mysteries. The richness of botany is suggested by the personalities of these masters. It is a science that attracted both the rigorously methodical Linnaeus and the poetically inspired Goethe. It is a science that combines the dryness of a classification scheme with the artistry of the most delicate and beautiful botanical illustrations. It is a science that glorifies the humble garden pea.

Botany is marvelously represented in the collections of the Library of Congress. Besides possessing the majority of the botanical classics, the Library has a collection of illustrated botanical materials that is outstanding. The following pages offer a brief survey of some of the more significant items.

As a scientist who first regarded plants as worthy of study for their own sake and not solely for their utilitarian value, the Greek Theophrastus is rightly called the father of botany. Following his death about 287 B.C., eighteen centuries would pass before another outstanding pure botanist would appear. During the intervening period, plant study was almost completely dominated by practical considerations, regarding its objects as foodstuffs or as components of a pharmacopoeia but seldom studying plants for the sake of simply learning more about them. Theophrastus's scientific—or what was called philosophical—approach was unique even in third century Greece, where nature was studied primarily from a teleological standpoint, with man always its reference point.

Theophrastus is represented in the Library's collections by his classic *De historia et causis plantarum,* printed in Treviso in 1483. It is generally regarded as the earliest work of scientific botany. Originally written as two separate treatises, this incunabulum is devoted to detailed studies of a wide range of specific plants. In it, the botanist discussed seeds, grafting, budding, and the effects of disease and weather on plants. He also described their medicinal properties.

A pupil of Aristotle, Theophrastus went beyond the master's practical interest in plants. He classified plants, distinguished between and among them, described them, and enumerated and defined their many parts. His classification system was primitive, but it held until the mid-sixteenth century. Most significant perhaps was his description of the formation of the plant in the seed, which he likened to the fetus of an animal, something produced by it but not a part of it.

The place of Theophrastus is unique both as the author of the oldest distinctively botanical treatise extant and as the founder of botany. Though his *De historia et causis plantarum* is not a great theoretical work, it does contain all the essential principles of what is today scientific botany. Theophrastus observed, collected, and systematized his botanical information. His botanical work was clear and accurate and contained none of the fabulous and unscientific embellishments of plant lore. His gentle criticism and dismissal of such irrational but popular beliefs as

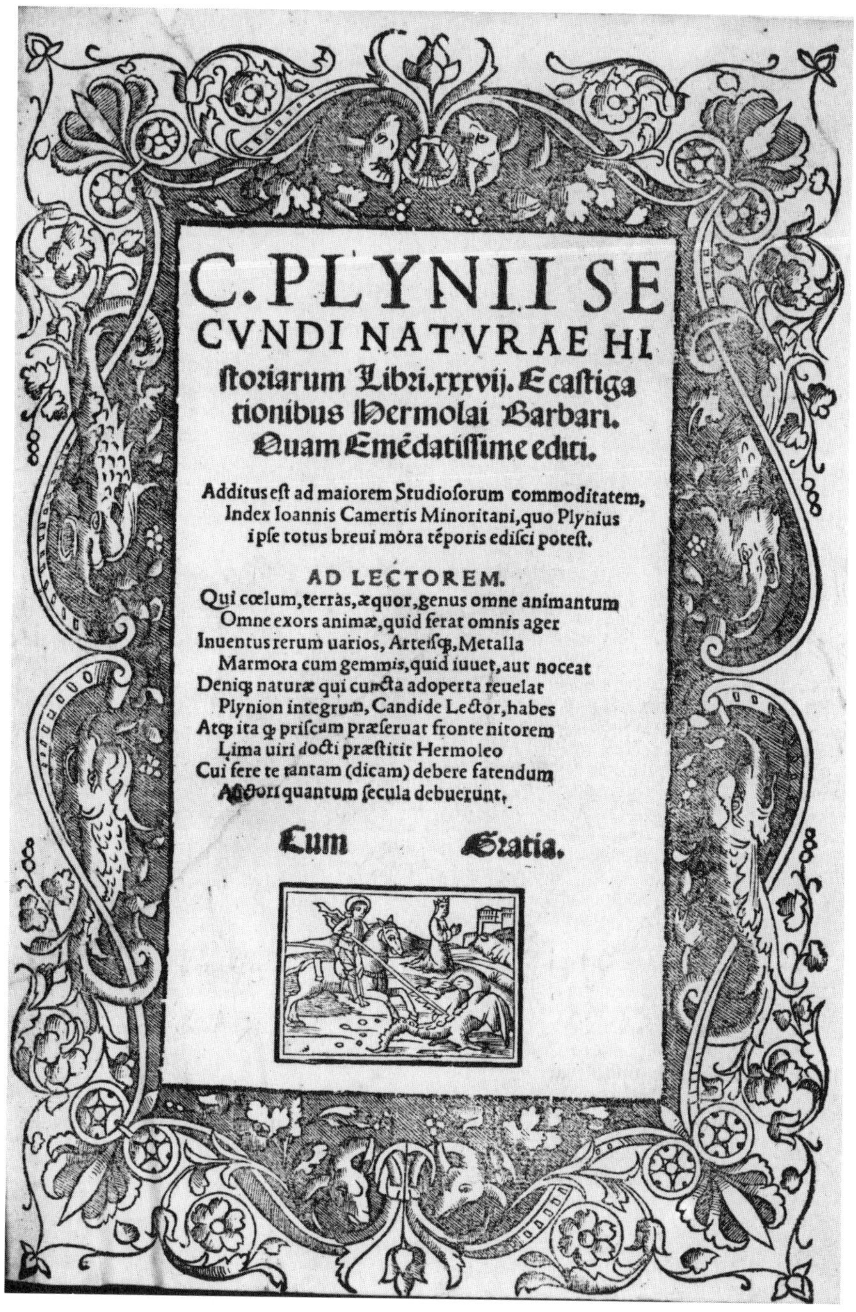

C.PLYNII SE
CVNDI NATVRAE HI
storiarum Libri.xxxvij. E castiga
tionibus Hermolai Barbari.
Quam Emédatissime editi.

Additus est ad maiorem Studiosorum commoditatem,
Index Ioannis Camertis Minoritani,quo Plynius
ipse totus breui mòra téporis edisci potest.

AD LECTOREM.

Qui cœlum,terràs,æquor,genus omne animantum
Omne exors animæ,quid ferat omnis ager
Inuentus rerum uarios, Artesq;,Metalla
Marmora cum gemmis,quid iuuet,aut noceat
Deniq; naturæ qui cuncta adoperta reuelat
Plynion integrum, Candide Lector,habes
Atq; ita q priscum præseruat fronte nitorem
Lima uiri docti præstitit Hermoleo
Cui fere te tantam (dicam) debere fatendum
Autori quantum secula debuerunt.

Cum Gratia.

The Library has several Plinys in its collections. This title page is from a 1519 edition. *Historia naturalis*, 1519. C. Plinius Secundus.

transmutation of plants presaged a modern scientific mind, but botany did not see his kind again for centuries.

Although Pliny's botanical work in no way approached that of Theophrastus, his thirty-seven-volume *Historia naturalis* is the earliest popular natural history book. Primarily an encyclopedic compilation, it was a secondary work derived, as Pliny himself said, from some two thousand volumes, most of which are now lost. Its significance, then, is as a summary of ancient knowledge of the natural world. For the medieval scholar, it became a standard reference work.

Pliny was not the most discriminating of Romans, and he included the fantastic as well as the factual. Throughout his work flowed the theme of nature in service to man. In this pragmatic perspective he was the most Roman of the Romans. Plants, however, do occupy a substantial portion of the work—sixteen of thirty-seven books. This work evinces little order and certainly nothing resembling the scientific method. Pliny's goal appears to have been to record voluminously, and that he did—albeit with an uncritical eye. Pliny's *Natural History,* which the Library has in first edition, dated 1469, is not only the first of the scientific classics to be printed, it was also one of the earliest books produced by the first press in Venice. Only 100 copies were printed.

The date of Pliny's death is known with certainty, for he died during the eruption of Vesuvius in A.D. 79. While in charge of the Roman naval fleet near Naples, Pliny went ashore to view the cataclysm more closely. He died from its poisonous vapors.

In Lib. quartum Dioſcoridis. 535

MANDRAGORA MAS.

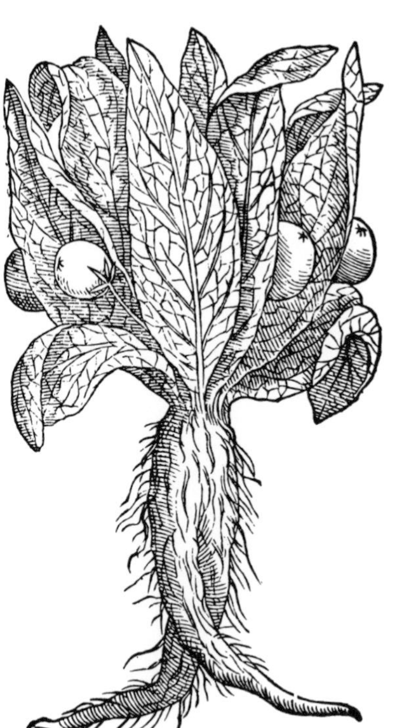

MANDRAGORA FOEMINA.

Dioscorides's most prominent editor was the late Renaissance Italian physician and botanist Pietro Andrea Mattioli. His masterwork, *Commentarii in libros sex Pedacii Dioscoridis*, was first published in 1544 in Italian, and it was phenomenally successful. As many as fifty editions in Latin, French, and German were published before the end of the sixteenth century. This illustration of the mandrake is taken from the Library's Latin edition, published in Venice in 1558. The legend of the mandrake, whose shape resembled a human figure, warned that the semihuman plant would shriek when pulled from the ground, causing madness or injury to the person who pulled it up. Since the mandrake was believed to have various narcotic, aphrodisiac, and cathartic properties, it was obtained by loosening the soil around the plant and attaching a rope to both it and an animal, usually a dog, who would then pull the plant from the ground. *Commentarii in libros sex Pedacii Dioscoridis*, 1558. Pietro Andrea Mattioli.

A contemporary of Pliny, Dioscorides is said to have written the most practically serviceable book that the world knew of for sixteen centuries. His *De materia medica,* published in Colle, Italy, in 1478, consists of five books. This incunabulum made available what had served as the authoritative source of herbal therapy for fourteen centuries. To the Sicilian Greek physician who wrote it should go the title of father of applied botany, for it was his task to describe the medicinal properties of plants. Dealing with some six hundred plants, *Materia medica* was held in high esteem even to the time of the Renaissance, and it was the first systematic pharmacopeia.

Dioscorides was widely traveled and became a much-experienced physician while serving the Roman armies. His advice on the medicinal properties of plants reveals him to have been an objective observer free of superstition. Concentrating mainly on the uses of plants, he nonetheless did attempt a rational classification and can be said to have anticipated the moderns in recognizing the natural affiliation of different species.

Soon after the invention of printing with movable type, many encyclopedic medieval manuscripts dealing with all aspects of health and natural history were put into print. Most had significant sections devoted to botanical matters, usually herbal remedies. Most were illustrated. The Library has three of the earliest and most important of these natural history encyclopedias.

Buch der Natur or *The Book of Nature* of Konrad von Megenberg, who translated the original Latin into German, had a wide circulation. The Library's copy was printed in Augsburg in 1481, though the original was compiled during the thirteenth century. Part of the book contains an account of the medicinal virtues of a small number of plants. Its historical significance is that it is the earliest printed work in which woodcuts representing plants were used purposefully to illustrate the text, rather than for decorative reasons.

De proprietatibus rerum was an encyclopedia written by the early thirteenth-century Franciscan friar Bartholomaeus Anglicus for the common people. The Library has the French version published in Lyons sometime after 1486, which is one of many later editions. Seventeen of its twenty books treat "de herbis et plantis," dealing chiefly with their medical uses.

The *Hortus sanitatis,* most notable as a work on medicine, was published in Mainz in 1491. This famous book also is wide in scope and deals

From *Buch der Natur,* 1481. Konrad von Megenberg. See p. 103.

In this time wheel, the large middle circle contains twelve vignettes, each showing an activity appropriate to a particular month of the year. Virtually every scene depicts an aspect of man's relationships to plants and planting. The outermost circle contains the twelve zodiacal signs, and the middle scene is one of opposites—woman and man, summer and winter, warm and cold, plenty and barrenness. The wheel or the circle was a ubiquitous medieval symbol for perfection—the perfection of a year, the perfection of a life, the perfection of the universe. *De proprietatibus rerum*, 1486. Bartholomaeus Anglicus.

From *Herbarium,* 1483–84. Apuleius Barbarus. See p. 103.

with animals, fish, and minerals as well as botanical matters. It is very rich in illustrations, with over one thousand small, crude woodcuts. Few of its drawings are original, with many of the illustrations as well as much of the text being derived from earlier manuscripts.

Two of the early herbal works in the Library's collections from which such books as *Hortus* were originally derived are the *Herbarium* (Rome, ca. 1483–84) and the *Latin Herbal* or *Herbarius Latinus* (Venice, 1499). As herbals, these books contained the names and descriptions of plants along with their particular properties and uses. The former, attributed to Apuleius Platonicus, sometimes called Apuleius Barbarus, may have had a manuscript career of a thousand years before it appeared in print. It has been suggested that its illustrations were made from metal plates and not wooden blocks. The *Herbarius Latinus* is also regarded among the doyens of the printed herbals. Upon printing, it met with immediate success and came to be known under many titles. As was frequently true for the early herbals, it was published anonymously and was a compilation of medieval, Arabic, and classical authors. Its many illustrations are typical of the early woodcut herbals—they are often formal and decorative, rather crude, and not very natural. As illustrated medical recipe books, they represent botany in its most practical or applied state.

These botanical works that numbered among the books produced during the first fifty years of printing were thus popular holdovers from medieval times. By the beginning of the sixteenth century, the reorientation toward nature brought about by the Renaissance had taken hold and botany was to enter a new, scientific era.

This woodcut of *acorus calamus* may be stylized, but the long, sword-shaped leaves and iris-like flower make it recognizable as a member of the *Arum* family, today called sweet flag or sweet rush. It grows in wet places (as the wading bird indicates) and has a purple flower. The plant's stout and aromatic rootstock was valuable to medieval herbalists and was used raw for toothaches, boiled for a physic, or made into tea for colic. *Herbarius Latinus*, 1499.

The so-called "German fathers of botany"—Brunfels, Bock, Fuchs, and Cordus—are represented in the Library's collections by one of the most famous, Leonhard Fuchs. His botanical masterpiece, *De historia stirpium,* was published in Basel in 1542 and is regarded by many as the most beautiful of all herbals. A large book, with over nine hundred pages and over five hundred full-page figures, the work typifies the Renaissance rediscovery of nature. Illustrated with "living portraits of plants" or "plants drawn from nature," this work is a far cry from the popular herbals of medieval Europe with their crude caricatures of plants. The illustrations were the result of the collaborative work of three men: Albrecht Meyer, who drew the plants from nature; Heinrich Füllmaurer, an artist who transferred the drawings to the blocks; and Veit Rudolf Speckle, who cut them in wood.

Fuchs was a physician and his botanical interests, as evidenced by his masterpiece, were certainly practical in nature. His plant subjects were the traditional herbs and his treatment was that of a pharmacopeia. He arranged his plants alphabetically and made no attempt at a natural grouping or classification. Besides setting a new standard for plant illustration, Fuch's work exemplified the new botany in its emphasis on firsthand observation and its attempt to rigorously describe the habits, locales, and characteristics of plants.

The Library also has a copy of his *New Kreüterbüch,* a German version of *De historia stirpium,* published in Basel in 1543. There are slight

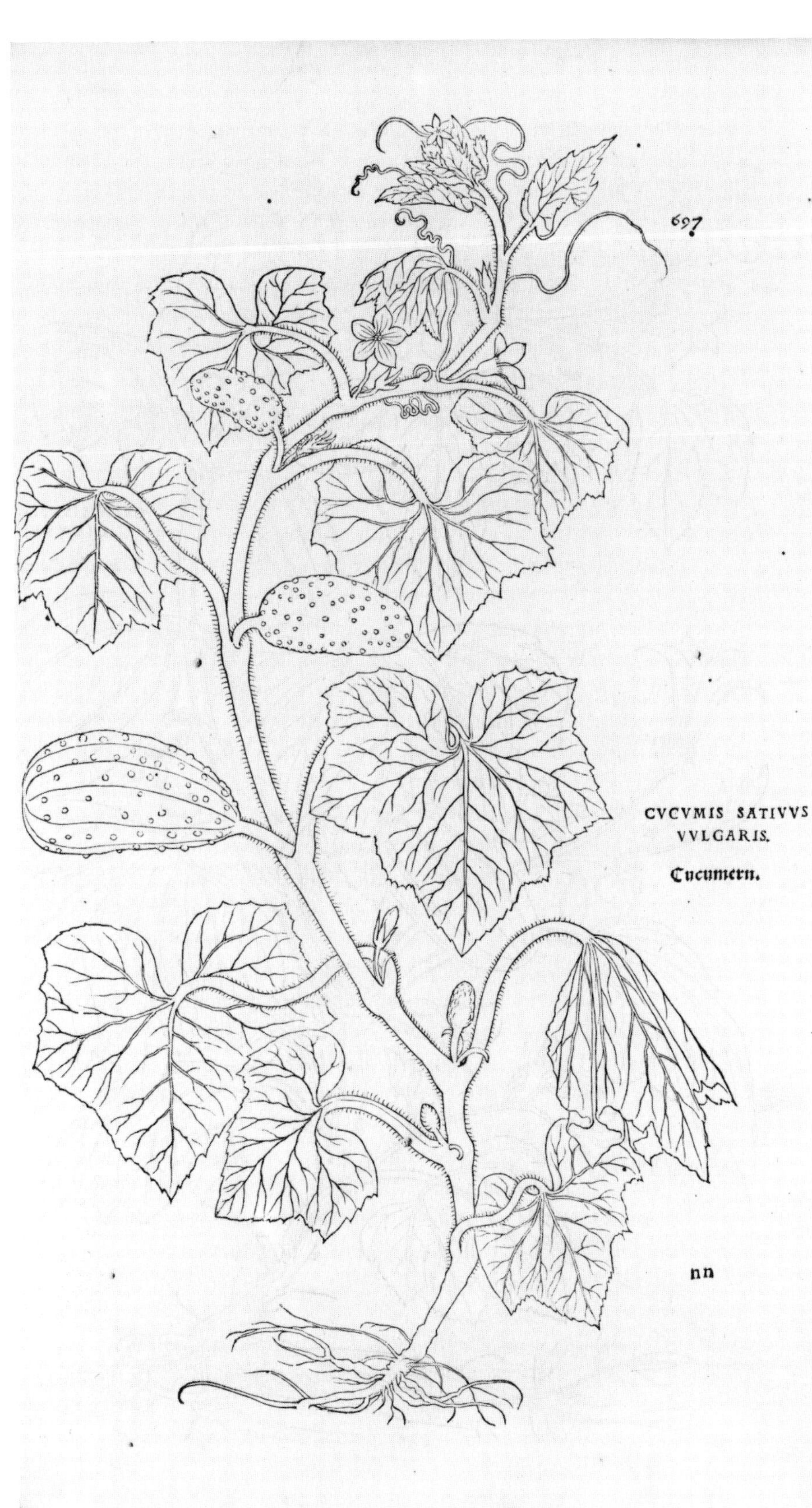

This is one of over five hundred full-page drawings of plants in *De historia stirpium*. Fuchs boasted that his work was illustrated with "living portraits of plants" and indeed, this cucumber plant is a far cry from the popular contemporary herbals with their crude caricatures of plants. *De historia stirpium*, 1542. Leonhart Fuchs.

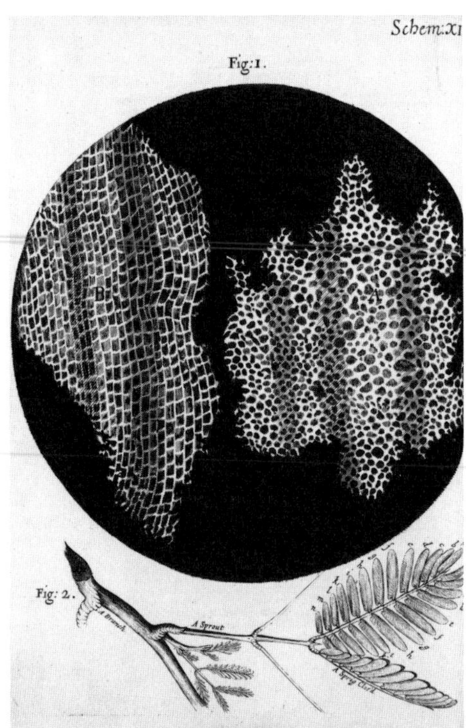

Robert Hooke examined a thin slice of a cork plant under his microscope and compared its composition to the cells of a honeycomb. Although Hooke had no conception of the nature of cells as it is understood today and was not really seeing living cells as such, he was the first to observe and represent the cellular structure of living tissue—a concept that would become one of botany's "big ideas." *Micrographia*, 1665. Robert Hooke.

differences between the two, with the new version containing six additional illustrations. Fuchs was later honored when a genus he described was named after him—fuchsia.

By the mid-seventeenth century, botanical investigation was being conducted not only in a more scientific manner but, in a sense, for its own sake. Botany was no longer tied strictly to the use of plants, and the emergence of lenses as aids to natural eyesight gave further impetus to more intimate investigations of plants.

Three men are linked to the rebirth (after Theophrastus) of plant anatomy and physiology, and all were microscopists. The most famous was the English mathematician and natural philosopher Robert Hooke. This versatile and argumentative scientist constructed a compound microscope, the first to resemble in any way a modern instrument, and promptly set out to examine nearly everything he touched. Unlike a simple microscope (one lens), a compound microscope has two lens systems—the eyepiece or ocular at the top and an objective at the bottom. The advantage of a compound microscope over a simple one is its greater magnification. However, the compound sacrificed much fine detail until the achromatic lens was introduced toward the end of the eighteenth century.

Hooke's *Micrographia* (London, 1665) is the result of his examinations. A first edition copy is part of the Library's collection. Written in English rather than Latin, this thick tome is copiously illustrated and its magnificent plates reveal some of the most handsome microscopic observations ever made. Most were from designs by the author himself, but some may have been made by Sir Christopher Wren. Hooke's botanical contribution in *Micrographia* and the discovery for which he is best remembered is that of the porous structure of cork, being composed of what he called "cells."

While Hooke was studying plant and animal alike under his microscope, two men, Malpighi and Grew, were systematically examining vegetable tissues under the microscope and laying the foundations of our knowledge of plant anatomy. Marcello Malpighi was an Italian physician and anatomist who produced a comprehensive study of plant anatomy called *Anatome plantarum* (1675–79). The Library's copy of this folio-sized work was published in London. Malpighi was a sharp observer with his microscope, as well as an ingenious experimenter. Among his many discoveries was that of plant stomata—the respiratory organs of plants. He saw them and sketched them, but their function was for others to discover.

Nehemiah Grew, working meanwhile in England, also had a medical degree but he, too, turned his microscope on plants. In *The Anatomy of Plants* (London, 1682), of which the Library has a first edition copy, Grew made available his extensive studies of plant structure, offering eighty-three full-page plates of microscopic sections of plant stems and roots. More important, it was here that the first statement of the sexual function of flowers was made. Both men, their particular scientific discoveries aside, demonstrated the scientific method at its best. Not accepting authority for its own sake, both went directly to nature and with their microscopes studied and literally probed the natural world deeply.

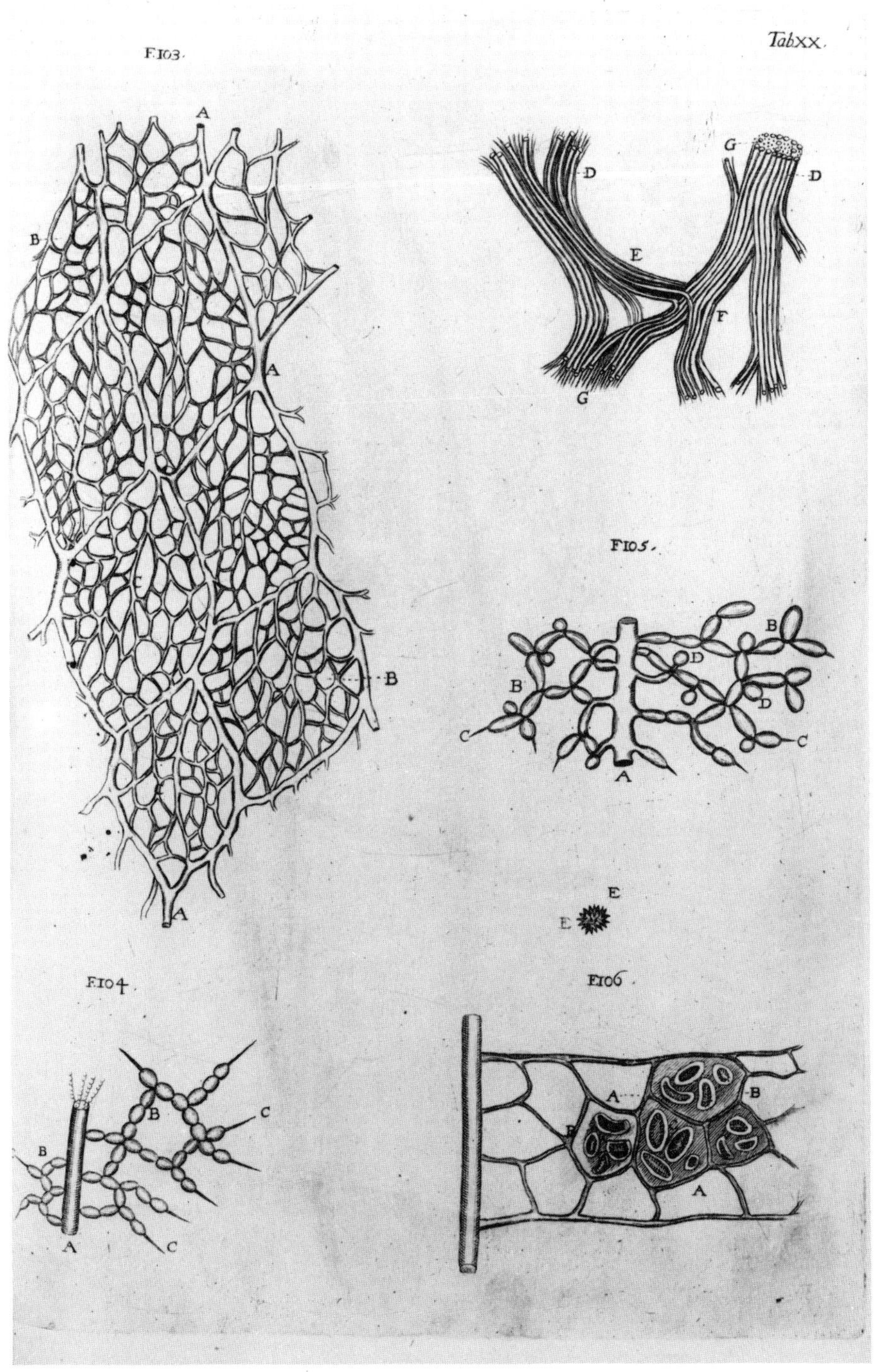

Marcello Malpighi was another of the very early microscopists. In figure 106, he shows what he discovered with his microscope on the underside of an oleander leaf. It was not until the nineteenth century that the oval-shaped openings within this network of fine veins were recognized as stomata—the respiratory organs of plants. *Anatome plantarum*, 1675–79. Marcello Malpighi.

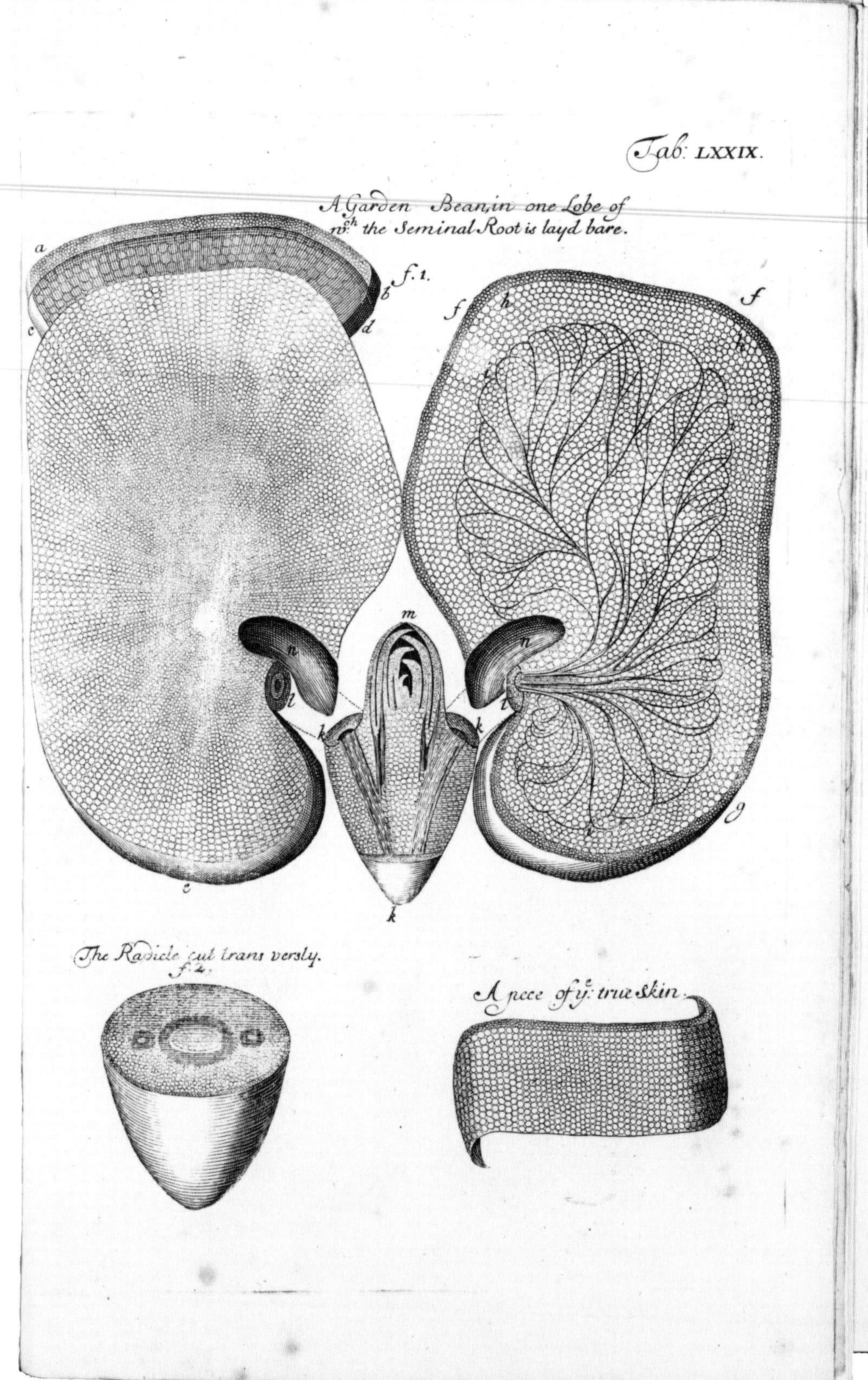
Tab: LXXIX.
A Garden Bean, in one Lobe of wch the Seminal Root is layd bare.
f. 1.
The Radicle cut trans versly.
f. 2.
A pece of ye true skin.

Before botanical science could begin to mature and explore the structure and workings of its living subjects, certain tasks had to be accomplished. Perhaps the most essential among these was the job of classification. As the seventeenth century ended, the state of botanical nomenclature was chaotic at best, and a uniform terminology was badly needed.

John Ray, the son of a blacksmith, did much to order botany's house. Working all of his life toward that end, Ray published in his later years a three-volume encyclopedia of plant life. This huge work described 18,600 different plant species, all arranged by a natural system of classification. The three volumes in the Library's collections were published in London between 1686 and 1704 and were titled *Historia plantarum.* Totaling 2,996 folio pages, these three volumes do not contain a single illustration. Ray's accomplishment is particularly significant since it laid the groundwork for the modern form of systematic classification of Carolus Linnaeus.

Linnaeus is the Latinized form of the name of Carl von Linné, the son of a Swedish clergyman. To say he had a passion for classification is an understatement, for he classified not only plants and animals but minerals and diseases as well. In his methodical way, he traveled through 4,600 miles of northern Scandinavia, then later through west Europe and England, observing plant and animal life (some results of which are covered in chapter 3). In 1735 he published *Systema naturae,* in which he established a method of classification or formal description of living things that is essentially the one we use today. He established the principles of class, order, genus, and species for all plants and animals. His second most important contribution to botany was his binary or binomial system, which is also still in use today.

Linnaeus was a purposeful individual who, early on, decided to devote his life to the unique task of classifying all plants and animals and working out a system that would remain valid and useful for all time. "I thought everything out by my twenty-eighth year," he later explained. He subordinated all botanical problems to that of classification and devised a system whose simplicity made it readily acceptable. For flowers, he used their sexual organs, especially the stamens, as a basis for classification. This sexual method of classification was far more revolutionary for plants than for animals and constituted a real breakthrough for botany. However, neither Linnaeus nor anyone who read him could remain entirely unaware of his system's prurient possibilities. It is to his credit that he sometimes dotted his work with a humorous touch, describing for example the class Polyandria (one which has many male organs, such as the poppy) as "Twenty males or more in the same bed with the female."

His *Systema naturae,* first published in 1735, was constantly revised and amplified as it went through many editions. The tenth edition was his final version and is the standard for today. In 1778 when Linnaeus died, a wealthy young Englishman purchased his books and collections. An apocryphal story says that the king of Sweden sent a warship after the British vessel that carried this treasure to England. The purchaser became the first president of the English biological association named the Linnean Society, which still owns the collections and library of its namesake. The Library of

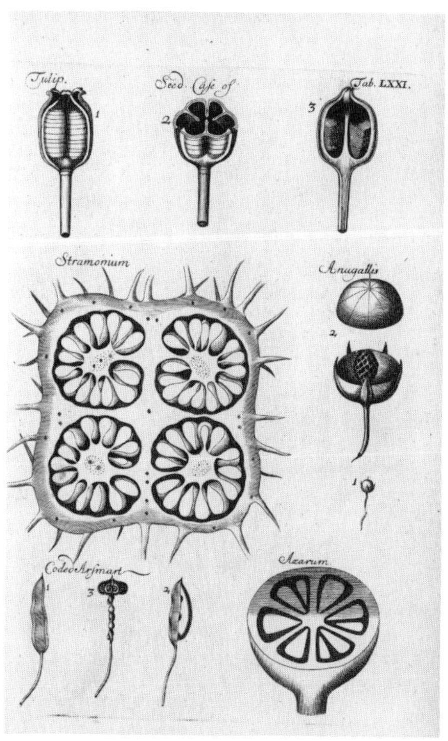

Here Grew reveals the structure of five different types of seeds. Although his chief contributions were in the field of plant anatomy and not physiology, it was Grew who first stated that the flower contained the sexual organ of the plant. *The Anatomy of Plants,* 1682. Nehemiah Grew.

Opposite page:
Nehemiah Grew was a Cambridge physician who turned his microscope on plants. In this highly detailed drawing of a cross-section of a bean, Grew was the first to examine and describe the delicate structure of seeds. *The Anatomy of Plants,* 1682. Nehemiah Grew.

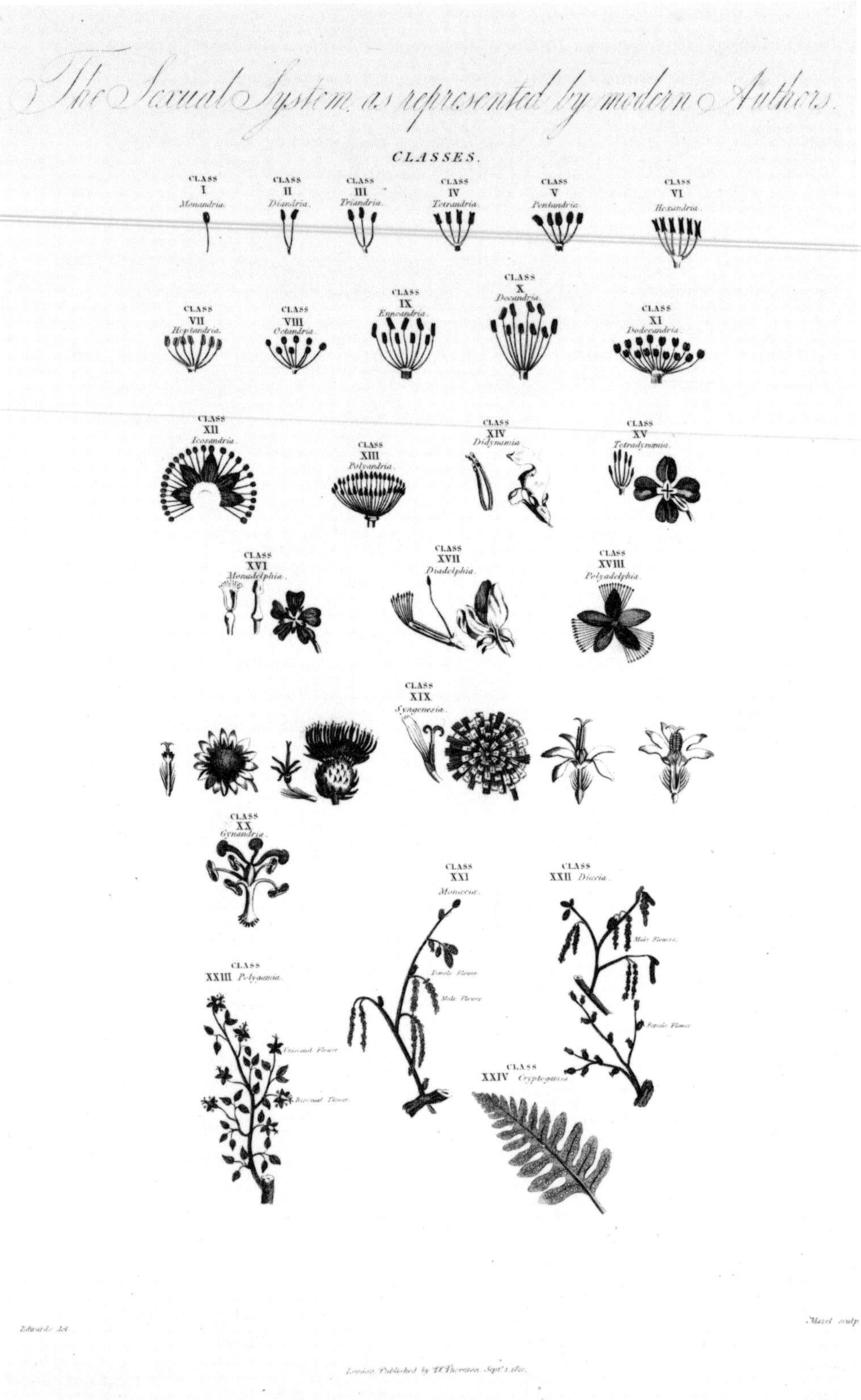

Between 1799 and 1807, Robert Thornton published *A New Illustration of the Sexual System of Linnaeus*. This illustration was taken from the book's third part, The Temple of Flora. It gives examples of some of the major aspects of the Linnaean system, which divided plants into "classes" according to the number and arrangement of their stamens or anthers. The twenty-four Linnaean classes were later subdivided into orders according to the number and arrangement of the female organs. Thornton was a somewhat eccentric English physician who spent his entire family fortune and more on this spectacular example of botanical romanticism. The Library has this large work in first edition. Thirty-one of its one hundred and forty-nine plates are in color. *New Illustration of the Sexual System of Carolus von Linnaeus*, 1807. Robert John Thornton.

Congress has a facsimile of the first edition of *Systema naturae* (1735) made from the first edition in the collections of the Amsterdam Zoological Library. This first edition is one of the more rare works in the history of science. It consists of only seven folio leaves, five of which are printed on both sides. The Library of Congress also has the enlarged second edition and subsequent editions as well as the much-amplified tenth edition published in 1758.

During the latter half of the eighteenth century botany and botanists became increasingly specialized and more scientific in their methods. The forerunner of the modern experimental method in botany was a country vicar, Stephen Hales. Hales has been called the first genuine plant physiologist and the founder of experimental physiology of plants. Physiology is tantamount to functions, and in Hales's time (he was born in 1677) little hard scientific knowledge was available as to how plants actually functioned. In a series of ingenious experiments, many of which are still repeated in the botanical laboratory, Hales measured rates of plant growth and sap pressure. At a time when nothing was known of the chemical composition of the air, Hales made the intuitive guess that "plants very probably draw through their leaves some part of their nourishment from the air."

The Library's first edition of his *Vegetable Staticks* (London, 1727) is a small, battered book and contains not only the record of Hales's clever experiments but twenty detailed drawings illustrating exactly how they were devised and conducted. Hales possessed a truly scientific mind and always sought to give exact mathematical expression to his experimental results.

This experimental method and tradition were continued by a Dutch physician, Jan Ingenhousz, whose *Experiments upon Vegetables* (London, 1779) contributed significantly to an understanding of the interrelatedness of all life on earth. Ingenhousz was stimulated by the discoveries of Joseph Priestley and Antoine Lavoisier concerning oxygen and carbon dioxide and the respiratory process of animals. During the summer months of 1779, Ingenhousz conducted five hundred experiments on plant breathing. He discovered and demonstrated that plants take up carbon dioxide and give off oxygen, but only in the presence of sunlight. In the dark, plants, like animals, absorb oxygen and give off carbon dioxide. This discovery, as revealed in his *Experiments upon Vegetables,* a compact work with no illustrations which the Library has in first edition, laid the foundation for our entire conception of the balance and economy of the living world. Ingenhousz's ecological insight was not as correct as it might have been, for he regarded the plant world as subservient to the animal world. He failed to grasp the degree to which the opposite is true. Although the significance of his work was not immediately appreciated, his discoveries elevated botany to a much-deserved higher plane.

Nineteenth-century botany is known for its work in the field of cell research. According to one of its practitioners, Matthias J. Schleiden, it was a science which "will be the sole and richest source of new discoveries and will remain so for many years." Schleiden was trained as a lawyer but

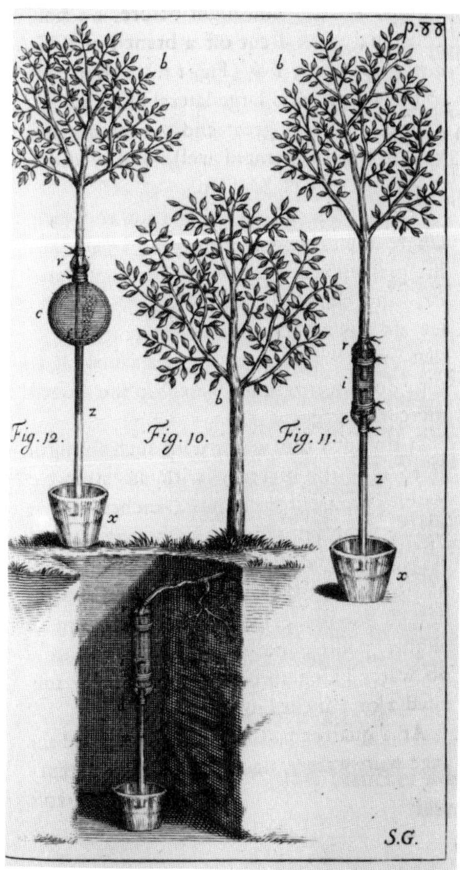

Stephen Hales's controlled experiments on plant physiology applied quantitative principles of weight and measure to the functions of a plant. In *Vegetable Staticks* he describes figure 10: "August 13. In the very dry year of 1723, I dug down 2 + ½ feet deep to the root of a thriving baking *Pear-tree*, and layed bare a root ½ inch diameter *n*." Hales wanted to measure the suction force of roots and so placed the cut-end of a root into a glass cylinder, filled it with water, and closed the top with an airtight seal. He then inserted a glass tube with an airtight seal into the lower end, turned it up while he filled it with water, held his thumb over the opening and quickly plunged it into a container of mercury. In only six minutes, the roots had sucked water from the tub with a force sufficient to draw the mercury up to level *z*. This and other experiments by Hales showed that the suction of roots can overcome the force of gravity. *Vegetable Staticks,* 1727. Stephen Hales.

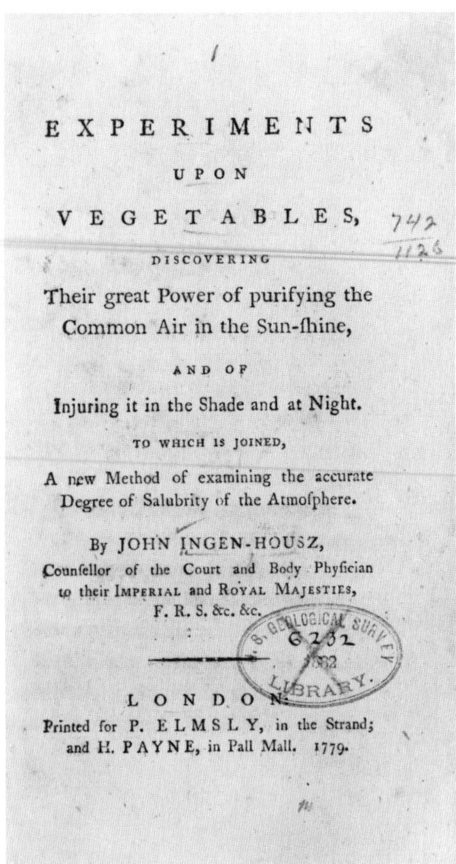
EXPERIMENTS

UPON

VEGETABLES,

DISCOVERING

Their great Power of purifying the
Common Air in the Sun-ſhine,

AND OF

Injuring it in the Shade and at Night.

TO WHICH IS JOINED,

A new Method of examining the accurate
Degree of Salubrity of the Atmoſphere.

By JOHN INGEN-HOUSZ,
Counſellor of the Court and Body Phyſician
to their IMPERIAL and ROYAL MAJESTIES,
F. R. S. &c. &c.

LONDON:
Printed for P. ELMSLY, in the Strand;
and H. PAYNE, in Pall Mall. 1779.

The elaborate title of Ingenhousz's main work pointedly describes what hundreds of experiments on plants had revealed to him—that sunlight was the driving mechanism or catalyst in plant breathing. "Their great power of purifying the common air in the sunshine" meant that in the presence of light, plants give off what Priestley called "vital air" (oxygen) and take in "vitiated air" (carbon dioxide). In the dark, the reverse occurs, thus "injuring it in the shade and at night." Ingenhousz had discovered an important key to the balance and economy of the natural world. *Experiments upon Vegetables*, 1779. Jan Ingenhousz.

soon took up botany professionally. Tired of the dry systematization of his time, he began studying plant tissue under a miscroscope and by 1838 offered the idea that the cell is the essential unit of the living plant. At the same time, Theodore Schwann was doing the same thing for animals. Schleiden identified and recognized the importance of the cell nucleus but erred as to its origin.

Schleiden is represented in the Library by his famous *Principles of Scientific Botany* (London, 1849), translated into English from the original German. Besides offering science the theory of the nucleated cell, this volume also performed another botanical service, for it replaced the botany textbooks of its time. A well-organized and thorough work, it also gives a resumé of the many authors who preceded Schleiden, along with his rather scathing comments and criticisms. Though critical of others, Schleiden was pathologically intolerant of any criticism of his own work—a trait which characterized his generally unstable emotional makeup. As a failed barrister in Hamburg, he attempted suicide by firing a gun at his forehead. Fortunately he failed in this, too, and gave himself only a superficial wound.

Between the final two major botanical works to be mentioned in this section, there exists a link of both substance and serendipity. One work is a forty-seven-page article published in an obscure journal by a modest Austrian monk who spent eight years breeding humble garden peas. The other work is a two-volume study produced by a distinguished Dutch professor of botany. Both men worked out independently what have come to be known as the laws of inheritance. Both worked with plants alone, yet their scientific legacy applies to all living things. Finally, one of them discovered and revealed the overlooked work of the other.

The Austrian monk, Gregor Mendel, who bred tall and short pea plants, also bred botany and mathematics. His years of breeding and cross-breeding led him to conclude that there exists a constant numerical ratio that determines the size and characteristics of succeeding generations. Furthermore, he concluded that such characteristics as tallness or dwarfness existed in sex cells or gametes and that tallness is dominant. Mendel published his paper "Versuche über Pflanzen—Hybriden" or "Investigations of Plant Hybrids" in the *Transactions of the Brünn Natural History Society* in 1865, but his work remained ignored and unnoticed until 1900 when De Vries discovered it.

Thirty-four years after Mendel's paper was published and sixteen years after the good monk's death, Hugo De Vries found Mendel's seminal work on hereditary characteristics. His discovery was not a simple literary find by the result of a deliberate literature search for confirmation of data he had already obtained. De Vries had sought to explain the one hole in Charles Darwin's theory—that of the unexplained manner in which individuals might vary. De Vries's theory of mutation offered the idea that sudden variations may occur in individuals, with a new species suddenly resulting. It was in search of evidence to confirm this theory that De Vries uncovered Mendel's laws of inheritance. Upon discovering that Mendel had worked out and proven his theory a generation earlier, De Vries announced Mendel's discovery and offered his own work as confirmation. The Library has De Vries's two-volume work entitled *Die Mutationstheorie* (Leipzig, 1901–3) as well as the volume of *Verhandlungen des Naturforschender Verein in Brünn* containing Mendel's 1866 paper. De Vries shares the distinction of this discovery with two others, Carl Correns and Erich Tschermak von Seysenegg. Amazingly, each man independently and unaware of the other followed the same research path and discovered Mendel's work. The separate publications of these three scientists appeared nearly simultaneously in 1900, and each offered his work solely as a confirmation of Mendel's.

In the Library's botanical collections there is a rich second tier of minor classics, oddities, and simply beautiful books. Among the earliest of these is the *Opera botanica* of Konrad Gesner. Gesner was a famous Swiss

D. CHRISTOPHORI IACOBI TREW
BREVIS HISTORIA FATORVM
EGREGII OPERIS BOTANICI
GESNERIANI.
EX VERNACVLA IN LATINAM TRANSLATA
PER
C. C. S.

LECTORI BENEVOLO.

Omnibus ad prela inſtructis, et Editori traditis, quae huic Volumini inſerenda erant, nuncius aduolat de repentino obitu *) Patroni et Promotoris Operum GESNERIANORVM, Illuſtris TREWII: luctuoſo pariter arti noſtrae, ac Reipublicae ſuae patriae, et amicis, dolendo mihi praecipue; cum abripuerit mihi Fautorem et amicum, quocum per quinque omnino luſtra coniunctiſſimus vixi animo, ſtudiis, et per ſexennium etiam officio. Is ipſe abitus ex hac vita, mortem enim non dixerim Viri, cuius memoria poſt fata durat et diutiſſime durabit, eo magis me perculſit, quo minus illum praeuideram; ſed amicum, ab aliquot etſi annis ſenectutis incommoda perſentiſcentem, animi viribus autem optime valentem credideram. Propoſuerat enim ſibi Vir Optimus, Hiſtoriam Fatorum Operis GESNERIANI, olim in ſubſidium laborum meorum mihi communicatam, ad incudem reuocare, et ſuo calamo ex vernacula in latinam conuertere, ſiquidem illa huic Volumini ex promiſſis addenda erat. Paullo autem poſt,**) quod adflictionem non parum auxit, mittuntur chartae manuſcriptae, in quibus praeter pauca aliqua verba, labor iſte ne quidem adfectus, multo minus abſolutus erat. Cum alius ſtatim non inueniretur, neque res moram diutius pateretur, in me translata eſt

A

verſio-

Gesner's botanical work was first published in 1751, nearly two hundred years after his death. Gesner is portrayed on the right, with his characteristic hat. Although Casimir C. Schmidel was the editor of this two-volume effort, he is not portrayed here. The prominent figure in the center is the Margrave of Bradenburg, Schmidel's patron, and the man on the left is Christoph Jacob Trew, whose library contained Gesner's unpublished botanical work. *Opera botanica,* 1751–71. Konrad Gesner.

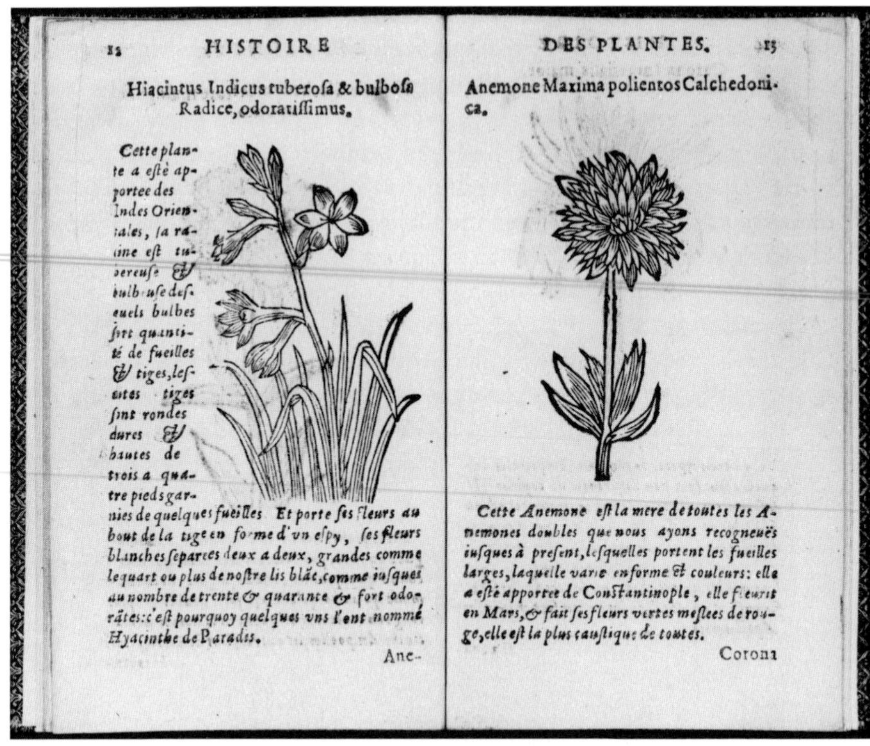

12 HISTOIRE

Hiacintus Indicus tuberosa & bulbosa Radice, odoratissimus.

Cette plante a esté apportee des Indes Orientales, sa racine est tubereuse & bulbeuse desquels bulbes sort quantité de fueilles & tiges, lesquelles tiges sont rondes dures & hautes de trois a quatre pieds garnies de quelques fueilles Et porte ses fleurs au bout de la tige en forme d'un espy, ses fleurs blanches separees deux a deux, grandes comme le quart ou plus de nostre lis blãc, comme iusques au nombre de trente & quarante & fort odorãtes: c'est pourquoy quelques vns l'ont nommé Hyacinthe de Paradis.

Ane-

DES PLANTES. 13

Anemone Maxima polientos Calchedonica.

Cette Anemone est la mere de toutes les Anemones doubles que nous ayons recogneuës iusques à present, lesquelles portent les fueilles larges, laquelle varie en forme & couleurs: elle a esté apportee de Constantinople, elle fleurit en Mars, & fait ses fleurs vertes meslees de rouge, elle est la plus caustique de toutes.

Corona

The hyacinth and the anemone are two of the flowering plants Robin described in this tiny book. *Histoire des plantes,* 1620. Jean Robin.

naturalist and is known today as one of the sixteenth-century "Encyclopedists"—scholars whose task it was to collect all known facts about living things. Gesner was a contemporary of the German fathers of botany and included botany among his interests. Although his best botanical work was not published until two hundred years after his death, he was a man of remarkable versatility and great learning. The Library's *Opera botanica* was published in Nuremberg (1751–71). It is a two-volume folio with many hand-painted color plates of plants and animals.

A little gem of a book usually overlooked and certainly seldom mentioned is *Histoires des plantes, nouvellement trouvées* (Paris, 1620) by Jean Robin. This diminutive, sixteen-page volume is generally considered to be an extract from a larger work relating to plants of the New World. Robin was a horticulturalist whose garden near the Louvre supplied the king's doctor and court with herbs. In 1601 he received the title of "arborist, herbalist, and botanist to the king, curator of the garden of the faculty." With help from his son Vespasien, Robin had many plants imported into his garden from many parts of the world. In particular, the black locust tree that he obtained from America and planted in 1601 in what is now the Square Viviani by Saint-Julien-le-Pauvre, was named after him ("Robinia pseudoacacia"). Most of the locusts now growing in Europe came from that single tree, which still blooms every spring. Robin's little book is extremely rare and contains several interesting wood engravings of his garden's contents.

Joseph Pitton de Tournefort, a professor of botany at the Jardin Royal in Paris, was considered in his time the leader of French botanical thought.

In his *Élemens de botanique* (1694) he inventoried and described 8,846 plants known at the time and offered an artificial system of classification that was widely accepted until Linnaeus. This book underwent an essential change in translation and emerged in English twenty-five years later as *The Compleat Herbal.* The Library has the English version published in London in two volumes dated 1719 and 1730. The translator was wise enough to retain the original illustrations and even mentions them on the title page as "about Five Hundred Copper Plates . . . all curiously Engraven."

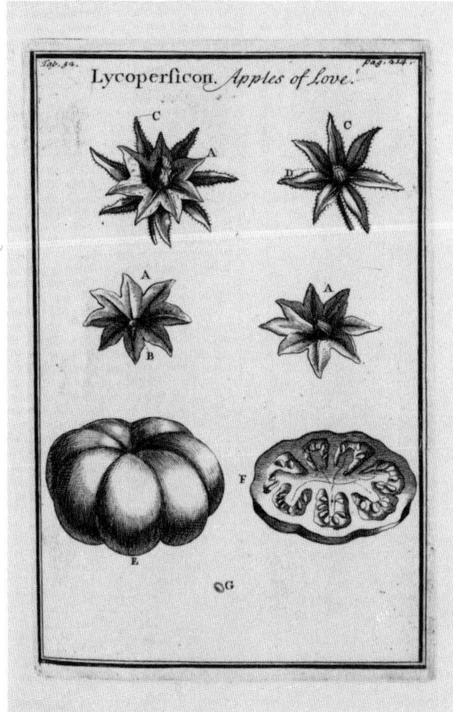

The English version of Tournefort's book was called *The Compleat Herbal.* It was published twenty-five years after *Elémens de botanique* and was an essentially different book. This illustration of a love apple, or tomato, is accompanied by the note: "The Italians eat the apples as we do cucumbers, with pepper, oil and salt; some eat them boiled: but considering their great moisture and coldness, the nourishment they afford must be bad." *The Compleat Herbal,* 1719–30. Joseph Pitton de Tournefort.

Joseph Pitton de Tournefort as portrayed in Robert Thornton's book *A New Illustration of the Sexual System of Linnaeus.* In addition to providing botanical illustrations, Thornton paid homage to the botanical heroes of his time with such honorific portraits. As a pre-Linnaean systematizer and professor of botany at the Jardin Royal in Paris, Tournefort was one of the leading botanists of his time. *New Illustration of the Sexual System of Carolus von Linnaeus,* 1799–1810. Robert John Thornton.

From *La botanique de J. J. Rousseau,* 1805. Jean Jacques Rousseau. See p. 104.

Although the genius and energy of Albrecht von Haller were devoted to many other areas besides botany, his botanical work was significant. He ranged from botany to medicine and anatomy, and he achieved distinction as a poet, philosopher, and novelist as well. He is best known for his pioneering work on the action of the nervous system, which laid the groundwork for modern neurology. In botany, he contributed an outstanding and indispensable bibliography. The Library has one of his large works, *Historia stirpium indigenarum Helvetiae* (Bern, 1768), which details the flora of his native Switzerland in two folio volumes.

The name de Jussieu recurs throughout the history of botany, and this one family produced eminent botanists for over a century. Bernard de Jussieu, the most famous, created botanical gardens at the Jardin du Roi, later called the Jardin des Plantes, in Paris, and at the garden of Le Trianon at Versailles. At these gardens it was his task to arrange the plants in the Linnean manner, so as to elaborate in a living way the "natural system" of classification. Although Bernard published nothing, the Library has the work *Genera plantarum* (Paris, 1789) in which his nephew, Antoine Laurent de Jussieu, developed his uncle's classification system.

The inclusion here of Johann Wolfgang von Goethe, Germany's most famous poet and dramatist, who ranks with the giants of literature, may surprise some. But his forays into science were both genuine and productive. In botany, Goethe's *Versuch die Metamorphose der Pflanzen zu erklären (1790),* or *An Attempt to Explain the Metamorphosis of Plants,* advanced the theory that all plant structures are but modifications of one fundamental organ, the leaf. In his botanical studies, Goethe coined the word *morphology* and directed scientific attention to the significance of plant form and structure. In his belief in organic evolution, he was a forerunner of Charles Darwin. Goethe's scientific methods were at times more deductive than inductive and often involved prolonged meditation in place of rigorous examination. It is said that when Goethe explained his botanical doctrine to Schiller, the latter commented, "This is not an observation, it is an idea." The Library's copy of Goethe's "idea" is a first edition original published in Gotha in 1790. It is a small, thin, unassuming book with no illustrations.

The work of Erasmus Darwin, grandfather (by his first wife) to the famous Charles Darwin, has been eclipsed by the latter's great accomplishments. Yet Erasmus was a noted scientist in his own right and became the first to state the thesis of the inheritance of acquired characters. His espousal of this incorrect theory preceded Lamarck. The Library has a copy of his *Zoonomia: or, The Laws of Organic Life* (Dublin, 1794–96) as well as *Phytologia* (Dublin, 1800), a botanical work published two years before his death. Interestingly, Erasmus Darwin was the grandfather (by his second wife) of another famous scientist, the English anthropologist Francis Galton.

Although Jean Jacques Rousseau is best known for his political and philosophical writings, he also wrote a general botanical textbook, published in 1771. This popular work, along with his other botanical writings, helped to further the general appeal that botany was experiencing during the late eighteenth century. The Library has a collection of his botanical work, *La botanique de J. J. Rousseau* (Paris, 1805), published twenty-seven years after his death. This lovely volume contains sixty-five color plates, made after the paintings of Pierre Joseph Redouté. As perhaps the greatest of all the French flower painters, Redouté contributed to over fifty books during his long career. The Library has in its collections the best known of all his works, the spectacular illustrated volumes entitled *Les roses.* Published in Paris in three volumes between 1817 and 1824, the Library's copy is the rare folio edition—one of only five copies made and printed on vellum paper with a double set of plates, black and white and color. Not only is it a most beautiful book, *Les roses* is also significant as a record of botanical knowledge of the genus *Rosa.*

From *Les roses,* 1817–24. Pierre Joseph Redouté. See p. 105.

In his botanical writings, Rousseau called for a purer science of botany and noted that when one is "used to looking at plants only as drugs or remedies . . . one does not imagine that the structure of the plant is worthy of attention in itself." Christian Conrad Sprengel took Rousseau's advice about the worthiness of botany to an extreme. As do some geniuses, Sprengel became so single-minded in his devotion to his studies that he neglected all else. It was his particular scientific passion to observe plants and flowers in their natural state—a consuming task which led him to neglect his duties as rector of Spandau College. Sprengel was eventually dismissed and led a solitary life in Berlin. His contribution to botany is contained in his *Das entdeckte Geheimniss der Natur im Bau und in der Befruchtung der Blumen,* translated as *The Discovered Secret of Nature in the Structure and Fertilization of Flowers* (Berlin, 1793).

In this work, Sprengel hit upon the real connection between the flower and the bee—specifically that the whole structure of the flower of nectar-bearing plants is directed to fertilization by insects and can be interpreted only by considering the function of each part in relation to the insect visits. Sprengel had indeed revealed one of nature's wondrous secrets; yet when his work received only scant attention he abandoned botany forever, and with some bitterness took up philology instead. The Library's copy of Sprengel's work was published in Leipzig in 1894, one hundred years after the first neglected edition.

Botany was one of the sciences to benefit most and to grow from the eighteenth-century practice of carrying naturalists on exploratory sea voyages. The earliest major ocean voyage of this kind was the Pacific trip of the *Endeavour,* which began in 1768. Under the auspices of the Royal Society and captained by James Cook, the *Endeavour* set off with scientific goals that were primarily astronomical and geographical but the voyage resulted instead in a natural history windfall. On board the *Endeavour* during its first Pacific trip was Joseph Banks, who later became president of the Royal Society, Daniel Solander, botanical student of the famed Linnaeus, and Sydney Parkinson, natural history artist. On the first voyage alone, Banks and Solander collected an estimated one hundred new families and a thousand new species of plants.

The idea of publishing illustrated accounts of Cook's voyages was in Bank's mind even before the *Endeavour* departed from England. The very

Captain Cook's voyage around the world took 1,051 days, leaving England on August 25, 1768, and returning on July 12, 1771. During this time, the *Endeavour* made landings at Madeira, Rio de Janeiro, Tierra del Fuego, Society Islands, New Zealand, Australia, Java, Cape of Good Hope, and Saint Helena. Sydney Parkinson was the sole botanical artist on board and he made his plant drawings from living specimens collected by Joseph Banks and Daniel Solander. Parkinson died at sea on January 26, 1771, after contracting dysentery at Java. His plant drawings constitute one of the most important scientific, historical, and artistic products of the voyage. Here, a black-and-white engraving based on Parkinson's color drawing of a morning glory is shown. It is from a recent publication, *Captain Cook's Florilegium,* 1973. Plates copyright © 1973 The Trustees of the British Museum (Natural History), London.

first voyage had its own artist, the talented young Quaker Sydney Parkinson, whose purpose it was to illustrate and thus document newly discovered animals and plants. William Hodges, a blacksmith's son who became a royal academician, served on the second of the *Endeavour's* voyages and John Webber, the son of a Swiss sculptor, on the third. A series of unforeseen and unlucky circumstances seemed to plague all efforts to publish the work of these artists, however, with the result that the bulk of Parkinson's work, now in manuscript in the British Museum of Natural History, was never published. Hodges's and Webber's work, too, remains in other institutions and in private collections.

Parkinson died during his voyage and the Library has his book *A Journal of a Voyage to the South Seas,* published posthumously in 1773. It also has a significant 1973 publication, *Captain Cook's Florilegium,* published in a limited edition of 100. This recent work contains a selection of black-and-white engravings based on Parkinson's color drawings. All of its engravings are botanical.

A later example of botany benefiting from a nautical connection is the work of Alexander von Humboldt, a most extraordinary man of universal talents and interests. In 1799, he began what was to be a five-year visit to the Americas—three years of which he spent in South America. Humboldt studied the natural world in all its variety—from fertilizer and ocean currents to volcanoes. The primary botanical result of his excursion was his *Essai sur la géographie des plantes* (Paris, 1805), written with his botanical assistant A. J. A. Bonpland.

Plant geography, which may be described as the study of the general principles that determine the regional distribution of plants throughout the world, hardly existed before Humboldt. It was he who had the vision to view the plant world on a gigantic scale and to search for laws that might govern the nature of these full landscapes. This universal, all-embracing approach was essential to botany becoming a mature science.

The Library has a facsimile of the 1805 edition of *Geography of Plants.* It also has a Latin edition of *De distributione geographica plantarum* published in Paris in 1817. This copy belonged to Thomas Jefferson. Humboldt met with then-president Jefferson on his way back to Europe from South America and presented him with this inscribed copy. The Library also has a German edition published in 1807, *Ideen zu Einer Geographie der Pflanzen.*

Last to be mentioned among these second-tier masterpieces is the work of Julius von Sachs, professor of botany at Freiburg-im-Breisgau and later at Würzburg. During the mid-nineteenth century, Sachs's researches in plant physiology contributed to a new, comprehensive view of plant metabolism. His rigorous studies of plant nutrition revealed the catalytic role of chlorophyll in the presence of light and gave botanists a real understanding of photosynthesis. Sachs was also the finest textbook writer of his time, and his *Lehrbuch der Botanik,* first published in 1868, was a major influence in the eventual emergence of botany as a comprehensive discipline. Hugo De Vries attributed his interest in plant physiology to his reading of this book. The Library has this work, which also contains

Sachs's espousal and application of Darwin's theory of evolution to plants, in its 1873 Leipzig edition.

After Sachs, the focus of botanical research and interest was in the area of genetics, where recent advances have taken the botanist to the molecular level. Progress there has been such that the propagation of new life forms seems possible. Thus, as often happens in science, old discarded ideas assume renewed vigor and relevance with the passage of time and the dawn of greater understanding. Though this chapter began by praising a third-century B.C. Greek for his dismissal of the popular but "unscientific" notion of the transmutation of plants, such a notion could very well be at the forefront of botanical research before the end of this century.

The tradition of botany is found deep within its herbal past. Long before there was any awareness of science as a human activity there was a purposive, albeit practical and exploitive approach to the natural world that was, in its own way, scientific. At some early point in man's existence, all vegetative life must have seemed as one, and when distinctions among plants were made, it was usually on the basis of their supposed usefulness. Although this pragmatic approach led to many discoveries and the accumulation of much information, its one-dimensional focus prevented much from being learned. Scientific botany only really began to emerge with the Renaissance discovery of the natural world. Once nature came to be regarded as an intrinsically worthwhile domain—irrespective of man—modern botany was inevitable. With this new attitude, the ancient science of Theophrastus that had been forgotten was rediscovered and wedded to the rich but limited herbal tradition, thereby producing the hybrid of modern botany.

II

3. *Zoology: Our Shared Nature*

If a fascination with the workings of nature and a desire to understand them defines mankind as an intellectual or reasoning race, then a genuine regard for the animals of nature makes us a compassionate, even estimable species. Well over two millenia ago Aristotle warned against easy patronizing of the animal world: "We must not betake ourselves to the consideration of the meaner animals with a bad grace, as though we were children; since in all natural things there is somewhat of the marvelous."

Interestingly, Aristotle likens man's feelings of superiority over the animal world to a child's natural but incorrect assumption that the world exists for his pleasure. His simile is especially apt, for with both a child and a scientific discipline, the process of maturing involves the recognition, understanding, and acceptance of things as they are. As the branch of biology dealing with the animal world, zoology took a long time to grow up. And unlike some disciplines, it had no real unifying theory until little over a century ago. Until Darwin, the history of zoology was an incomplete tale—a story rich in detail with a great deal of character but one with no recurring theme or unifying plot.

As the Greeks did with nearly all the sciences, they set zoology off on the right road, but with their decline, the scientific pathways became overgrown and were finally lost. Zoology then set upon a sometimes fantastic journey from the mythical, moralizing bestiaries of medieval times through the rigid classifying methods of the eighteenth century and the charm and naiveté of its popular natural history phase to the broader, sounder science of more modern times. Despite these scientific ups and downs and the vagaries and trends of scientific fashion, the core phenomenon of zoology has remained constant throughout—the phenomenon of life. Life and its perpetuation was what the animal world seemed to be about—indeed, the only thing it seemed to be about—and until Darwin, there was no one large encompassing idea under which the lush variety of animal life could be gathered and related to one another.

The sensitive subject of man's own physical nature and origin had a great deal to do with zoology's late development, since religious dogma dictated an essential and necessary distinction between animal and man, placing animals in a separate and decidedly inferior position. The conceptual revolution implicit in Darwinism was its renunciation of this distinction and its espousal of the opposite notion that all the varied forms and phenomena of life are part of a connected whole. The implications of Darwin's theory outside of science are broad and deeply felt—morally, socially, and politically. His removal of mankind from its self-appointed

Opposite page:
In what has been described as "the first comprehensive monograph on the anatomy of an animal," Carlo Ruini described each part of the horse's body and offered advice on the nature and treatment of its diseases. In style, detail, and beauty, the book's illustrations are obviously patterned after Vesalius's landmark work on human anatomy. Ruini was a Bolognese aristocrat and a high-ranking lawyer. He pointed out to his fellow Italians that their respected title "cavaliere," which denoted valor, courage, and nobility, was derived from this most worthy and noble animal. *Dell'anotomia et dell'infirmità del cavallo*, 1598. Carlo Ruini.

"most-favored-species" status offers a metaphor of seemingly unlimited application. But within the realm of science alone, his recognition of the oneness of nature transformed a static, satisfied discipline dominated by taxonomy into a dynamic and truly modern science.

Darwinism need not be regarded as the second fall of mankind. It can just as easily be viewed as the overdue elevation of the rest of the animal world to its rightful place. With this leveling out has come a slight conceptual shift in how we perceive our fellow beasts. Given the premise of a primordial kinship and a continual process of change, the study of animals not only becomes a worthy subject on its own merit but assumes a further, almost practical significance to mankind. Simply, we can learn more about ourselves by knowing better the great variety of animal life with whom we share this planet. A healthy respect, even reverence, for our fellow beasts need not be patronizing or anthropomorphic, but rather should result in a greater understanding of ourselves and our world.

Zoology is a wonderfully attractive science, as varied, surprising, and interesting as its living subjects. Encompassing the protozoan and the peacock, the mollusk and the man, it seeks to understand the mystery of life. Its subjects move, breathe, reproduce and are as bounteous, willful, and autonomous as nature itself. The following will trace, through the collections of the Library of Congress, mankind's intellectual efforts to comprehend this richness and to make a real science out of its work. The story ends with Darwin, who gave zoology its much-needed theory—a theory that has proven sufficiently strong and flexible to incorporate virtually all of the major twentieth-century biological discoveries and one that continues to endure.

From the beginning, mankind always has shared some sort of link with the animal world. The earliest and most primitive was surely that of hunter and prey—with man possibly playing the fatal role of victim, only later to reverse those roles as he became more skilled and intelligent. The later domestication of certain animals, along with the discovery of agriculture, made for a more settled and stable existence and was an essential step in the not-so-orderly process of becoming civilized. However, the intellectual distance between regarding an animal as the source of dinner or of material comfort and considering it a worthy subject for study, whose nature and habits are worth knowing and understanding for their own sake, is considerable.

Not until Aristotle did the animal world become a subject for serious scientific study. Although he wrote on seemingly every subject, Aristotle's work in zoology—studying animals as animals—is considered his most successful. He seems to have had a natural affinity and curiosity about the living creatures of the world and he took a special interest in marine life. His zoological writings reveal him to be a remarkably astute observer of the natural world, who wedded his observations to what might be called speculative reason. He was therefore a theorist as well. His overall theory was simple: nature was intrinsically purposive. "In the works of Nature," he said, "purpose and not accident is predominant." A thing is known then, when we know what it is for. He linked theory and practice by saying that interpretation of an observed phenomenon must always be made in light of its purpose. His zoological theory was thus a reflection of the essentially teleological nature of his overall philosophy.

Aristotle is called by many the first zoologist, and indeed he may be said to have started almost from scratch. "I found no basis prepared, no models to copy," he wrote. What he described modestly as "the first step and therefore a small one," resulted in a body of zoological work that was to remain dominant for well over a millenium. His major zoological treatises are three: *De historia animalium,* which outlines the observed facts, catalogs animals, and describes them mostly in terms of anatomy; *De partibus animalium,* which contains most of his physiology and attempts to explain the structure of animals in terms of function; and *De generatione animalium,* which discusses reproduction and development. Not surprisingly, it was these three works which, after the invention of printing, made up the first zoological compilation. Published in Venice in 1476 under the overall title *De animalibus,* this work in Latin contained a total of eighteen books and discussed over five hundred different animals. The Library has a copy of this first printed edition.

Although Aristotle's work contains many incorrect generalizations—falsehoods that were later slavishly accepted along with his true statements—his major emphasis on observation and his attempts to construct a systematic method of viewing the animal world established the science of zoology. Apart from being the first orderly storehouse of factual information about the animal world, his work in zoology accomplished something even greater. It posed the proper questions—questions of anatomy, physiology, morphology, and classification, among others, which were central to the science of zoology. Unfortunately, with the decline of Greece, this basic method of scientific inquiry was soon eclipsed in a world hostile to the rational approach.

Nearly four centuries after Aristotle, when the Romans ruled the known world, a Roman scholar of universal interests wrote a natural history that was in fact an encyclopedia of the knowledge of the ancient world. Gaius Plinius Cecilius Secundus, called Pliny, was a governmental official and a military leader, yet he found the time and energy to write not only the *Historia naturalis* in thirty-seven books, but sixty-five more books on other subjects. Of his thirty-seven volumes, five are concerned with zoology. Although little if any part of the *Natural History* is original, it is precisely its secondary nature that makes it so significant. Pliny referred to some two thousand previous works as sources and scrupulously acknowledged each one. His work has been described as unorganized and indiscriminate, including as it did both the factual and the fantastic with seemingly little distinction, but uncritical and derivative as it is, Pliny's *Natural History* is a mine of historical information. In it are preserved both the knowledge and the error of many an ancient and now-lost text. His zoological books

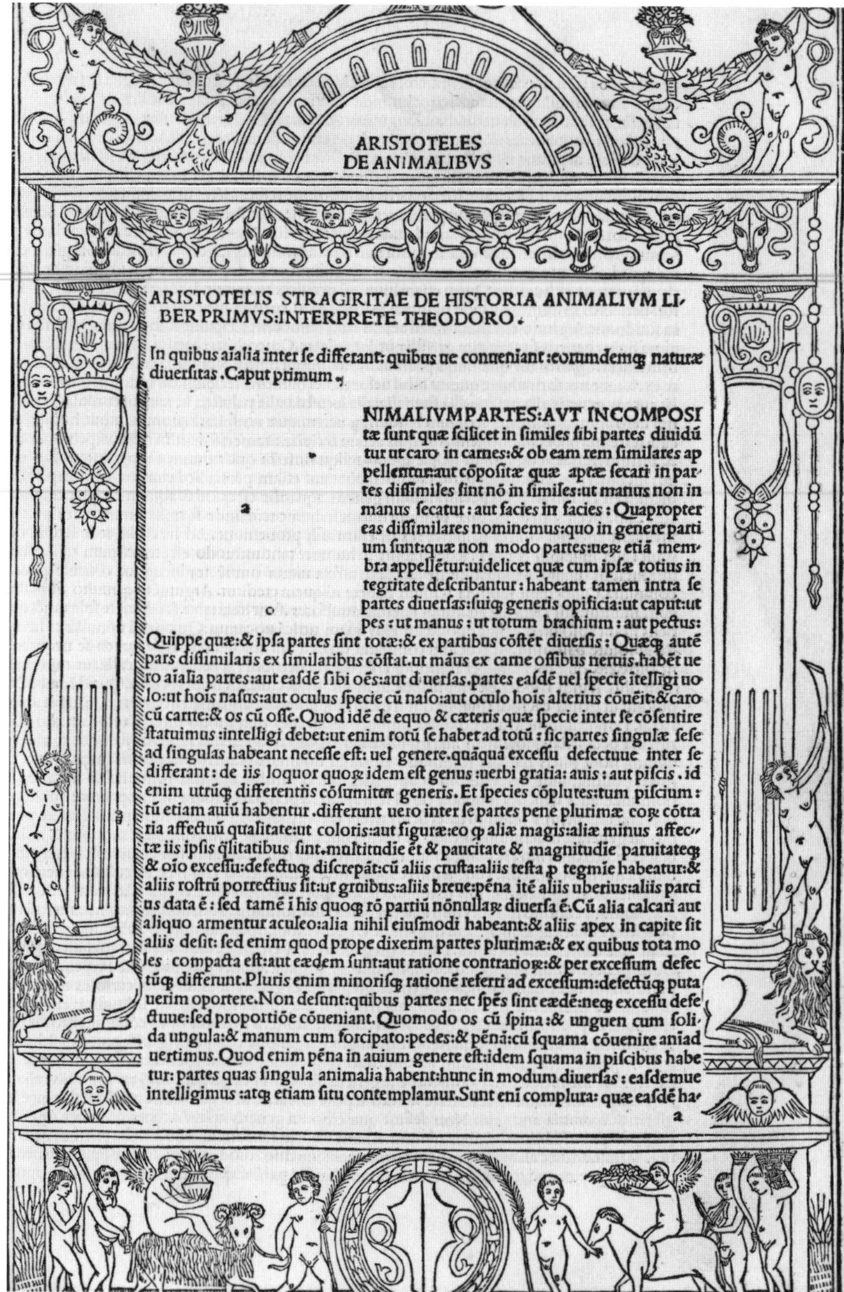
ARISTOTELES
DE ANIMALIBVS

ARISTOTELIS STRAGIRITAE DE HISTORIA ANIMALIVM LIBER PRIMVS:INTERPRETE THEODORO.

In quibus aĩalia inter se differant: quibus ue conueniant :eorumdemq́ naturæ diuersitas .Caput primum.

a

NIMALIVM PARTES:AVT INCOMPOSITæ sunt quæ scilicet in similes sibi partes diuidũtur ut caro in carnes:& ob eam rem similares appellentur:aut cõpositæ quæ aptæ secari in partes dissimiles sint nõ in similes:ut manus non in manus secatur : aut facies in facies : Quapropter eas dissimilares nominemus:quo in genere partium sunt:quæ non modo partes:uerũ etiã membra appellẽtur:uidelicet quæ cum ipsæ totius integritate describantur : habeant tamen intra se partes diuersas:suiq́ generis opificia:ut caput:ut pes : ut manus : ut totum brachium : aut pectus: Quippe quæ:& ipsa partes sint totæ:& ex partibus cõstẽt diuersis : Quæq́ autẽ pars dissimilaris ex similaribus cõstat.ut mãus ex carne ossibus neruis.habẽt uero aĩalia partes:aut easdẽ sibi oẽs:aut diuersas.partes easdẽ uel specie ĩtelligi uolo:ut hoĩs nasus:aut oculus specie cũ naso:aut oculo hoĩs alterius cõueĩt:& caro cũ carne:& os cũ osse.Quod idẽ de equo & cæteris quæ specie inter se cõsentire statuimus :intelligi debet:ut enim totũ se habet ad totũ : sic partes singulæ sese ad singulas habeant necesse est: uel genere.quãquã excessu defectuue inter se differant : de iis loquor quorũ idem est genus :uerbi gratia: auis : aut piscis . id enim utrũq́ differentiis cõsumitur generis. Et species cõplures:tum piscium : tũ etiam auiũ habentur .differunt uero inter se partes pene plurimæ corũ cõtraria affectuũ qualitate:ut coloris:aut figuræ:eo q́ aliæ magis:aliæ minus affectæ iis ipsis q̃litatibus sint.multitudie ẽt & paucitate & magnitudie paruitateq́ & oĩo excessu:defectuq́ discrepãt:cũ aliis crusta:aliis testa ꝓ tegmĩe habeatur:& aliis rostrũ porrectius sit:ut gruibus:aliis breue:pẽna itẽ aliis uberius:aliis parcius data ẽ : sed tamẽ ĩ his quoq́ rõ partiũ nõnullarũ diuersa ẽ.Cũ alia calcari aut aliquo armentur aculeo:alia nihil eiusmodi habeant:& aliis apex in capite sit aliis desit: sed enim quod prope dixerim partes plurimæ:& ex quibus tota moles compacta est:aut eædem sunt:aut ratione contrariorũ:& per excessum defectũq́ differunt.Pluris enim minorisq́ rationẽ referri ad excessum:defectũq́ putauerim oportere.Non desunt:quibus partes nec spẽs sint eædẽ:neq́ excessu defectuue:sed proportiõe cõueniant.Quomodo os cũ spina :& unguen cum solida ungula:& manum cum forcipato:pedes:& pẽnã:cũ squama cõuenire aniaduertimus.Quod enim pẽna in auium genere est:idem squama in piscibus habetur: partes quas singula animalia habent:hunc in modum diuersas : easdemue intelligimus :atq́ etiam situ contemplamur.Sunt enĩ complura: quæ easdẽ ha-

a

Aristotle has been described as a one-man university, given the breadth of his interests and abilities. That the study of animals was among his primary concerns is manifested in his zoological treatises, which make up a fourth of his entire corpus. His object and methods were simple and direct: to identify the formal groups of animals, to classify them, and to explain their functions. Not much escaped his eye or ear. He knew, for example, that the partridge would attempt to lure the hunter away from its young by limping and pretending to be wounded. He also correctly classified the dolphin with land animals rather than with fish.

The empty space near the top left of this page was left blank for a rubricated capital letter *A*, which obviously was never done. In the center of the space is a tiny guide letter *a*, telling the rubricator (who was not always literate) which letter to draw. *De animalibus,* 1476. Aristotle.

also contain as much myth as fact. Pliny had a natural sense of the marvelous and transmitted stories of many a strange race of men, later called "wonder people" by the medievalists. His sections on beasts include the unicorn and the phoenix as well as more commonly seen animals. He made no attempt to give any real description of an animal's internal or external structure, as Aristotle had done. Nonetheless, this seemingly unscientific treatise became one of the most widely used textbooks during the next fifteen hundred years and was the main source for all who sought to know more about the animal world. The popularity and stature of the

Historia naturalis was such that it was included in the first wave of classical scientific works to be printed. The Library has the first edition, which was printed in Venice in 1469. This is an exceedingly rare incunabulum, of which only 100 copies were printed.

The impulse to wonder and to marvel at the natural world so obvious and fresh in Pliny's work was indulged fully by the anonymous author of a treatise that appeared sometime between the second and fifth centuries. Titled the *Physiologus,* this work was to have a circulation and an appeal as wide as any save the Bible. The author made little or no attempt to compose a real natural history. Rather, he or she told fables using animals as actors in the stories. Quite often they were tales of fantastic creatures, invented so as to tell a better story, among them the siren, unicorn, phoenix, and ant-lion, a short-lived animal that resulted from an unnatural coupling of an ant and a lion. These stories were usually allegorical, and like all good edifying tales, each had a moral.

During the Middle Ages, as religion came to be a dominant force in everyday life, no preaching text proved more useful or more popular than the *Physiologus* with its animal symbolism. Its blend of allegory and magic appealed to the medieval mind. Editions appeared in both Latin and Greek as well as in many vernacular languages and their dialects. Throughout the centuries, the original text was modified and expanded, and by the twelfth and thirteenth centuries the *Physiologus* had become one of the leading picture books of its time. The Library's copy of this most famous natural (but unnatural) history book was published in Rome in 1587. The text of this version is in Latin and Greek, with its title transliterated as *Peri ton physiologon.*

Toward the middle of the thirteenth century, a work on falconry appeared that was in the tradition of Aristotle and therefore quite unlike the fantastic *Physiologus.* Titled *De arte venandi cum avibus,* it was a serious, scientific account of the habits and structure of birds. Based on observation, the book is strikingly modern to our eyes in that it is completely factual and rejects the medieval penchants for authority and magic. Still more remarkable is the fact that its author was Frederick II of Hohenstaufen. Frederick was the grandson of Frederick Barbarossa, and he became the Holy Roman Emperor and leader of the fifth crusade. He stands as a beacon of enlightenment in a confused age. Half-Sicilian, half-German, he grew up exposed to the heritage of Jewish, Moslem, and Christian cultures, living as he did in cosmopolitan Norman Sicily, and was nourished by all of them. Patron of the arts and sciences and founder of universities, this contentious and unpredictable iconoclast wrote what was in his time a zoological masterpiece. The Library has the first printed edition of his work, published in Augsburg in 1596.

Frederick's contemporary Albertus Magnus, the famed "Doctor universalis" of the thirteenth century, turned to natural history late in his life. Although the bulk of his writings deal with theology and philosophy, his *De animalibus* is significant to zoology in two ways. First, it redirected attention back to Aristotle—and therefore stimulated a greater interest in the natural world. Second, its descriptive approach to animals and its

S. EPIPHANII.

ΠΕΡΙ ΤΟΥ ΑΕΤΟΥ.

Κεφάλαιον ϛ'.

The popular bestiary *Physiologus* was one of the best known and most widely read books of the Middle Ages. As a work which drew upon animal legends and pseudoscience common to Indian, Hebrew, Egyptian, Greek, and Roman cultures, its moral tales always had a fascinating, exotic quality about them. Never intended as a zoological treatise, *Physiologus* offered animal allegories that were not only moralizing fables but stories that tried to reveal something about God's nature by interpreting how the Creator revealed himself symbolically in his creations. The story told about the eagle is one of renewal. *Physiologus* says that when the eagle gets very old, his wings become heavy and his eyes grow dim. He then flies toward the sun, burning away his wings and the dimness of his eyes, and descends to bathe himself in a fountain. He is restored and made new again. Like the eagle, we are told, mankind will be renewed by the fountain of baptism. *Peri ton physiologon,* 1587.

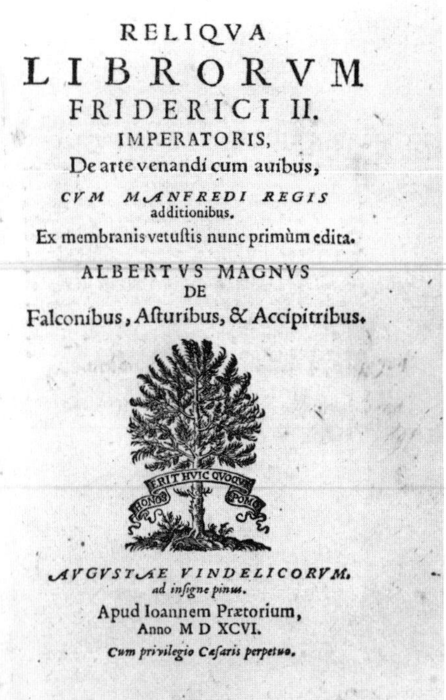

RELIQVA
LIBRORVM
FRIDERICI II.
IMPERATORIS,
De arte venandi cum auibus,
CVM MANFREDI REGIS
additionibus.
Ex membranis vetustis nunc primùm edita.
ALBERTVS MAGNVS
DE
Falconibus, Asturibus, & Accipitribus.

AVGVSTAE VINDELICORVM.
ad insigne pinus.
Apud Ioannem Prætorium,
Anno M D XCVI.
Cum priuilegio Cæsaris perpetuo.

Frederick II's book is far more than a treatise on hunting, for it displays a wide knowledge of avian anatomy and habits. The book symbolizes the character of Frederick himself, as an important transitional figure bridging medieval and modern culture. It was written with an almost modern spirit of investigation and experimentation and even went as far as correcting Aristotle. Frederick II was a versatile and interesting ruler who opened himself to all cultures and philosophies and took what he thought the best from each. As a Christian king who kept a harem, he was called the "baptized Sultan of Sicily." Not surprisingly, the arts as well as the sciences blossomed under the enlightened and liberal Frederick. Frederick II's son Manfred was king of Sicily from 1258 to 1266. Manfred revised his father's work, and it is Manfred, not Frederick, who is shown here with two falconers. *De arte venandi cum avibus,* 1596. Friedrich II.

incipient skepticism of such works as the *Physiologus* mark it as a transitional work of the era before the scientific revolution. His *De animalibus* was first printed in 1478 in twenty-six books. The Library's copy is the 1495 edition printed in Venice.

Since Albertus came to the study of animals rather late in his career, it is not surprising that one of his prime sources was the work of a former pupil, Thomas of Cantimpré. Like Albertus, Thomas was a Dominican friar, and as a true encyclopedist, he spent nearly fifteen years writing the *De natura rerum.* His major work was composed of nineteen books, well over a third of which dealt with animals. Although his books contained some moralizing, overall they attempted to consider most animals as natural phenomena to be accurately described. Thomas's work had a wide circulation in manuscript form and one hundred years later was translated into German by another cleric, Konrad von Megenberg. Konrad's work, entitled *Buch der Natur,* was much more than a direct translation, for he reworked the entire treatise—deleting what he thought incorrect, embellishing other parts, offering criticism, and generally revising. With the invention of printing in the next century, Konrad's *Buch der Natur* was printed in Augsburg in 1475. The Library has the 1481 Augsburg edition of this work, which stands as the first printed book to contain figures of animals, as well as the first major scientific book printed in German.

Another encyclopedist and contemporary of Albertus and Thomas was the monk Vincent of Beauvais. A man of immense industry about whom little is known, Vincent aptly called his great compilative work of eighty books the *Great Mirror*—or the *Speculum majus.* Of particular interest are its first thirty-two books, called *Speculum naturale.* There Vincent uncritically assembled a virtually complete collection of animal fact and lore—taken from Arabic and Hebrew as well as Greek and Latin sources. The Library's copy of *Speculum naturale* has no publication date, although it is known that the whole of the *Speculum* was printed in Strasbourg by Johannes Mentelin during the period 1473–76. The two huge volumes have vellum-bound wooden covers and many rubricated initials.

By the sixteenth century, the Renaissance spirit of inquiry had as much effect on zoology as on other disciplines, but zoology had no Copernicus or Galileo. No genius came forth to redefine and to revolutionize the science, to form it, to map out its scope, or to offer sweeping new unifying ideas. Instead, zoology was formed as a modern science in the sixteenth century by well-read universal types whose dedication and workmanlike performance laid a solid if undramatic base for the new science of animals. Progress in zoology had been hampered by the discipline's apparent lack of immediate utility, but with the great geographical discoveries of the late fifteenth century, fascination for new life forms caused the more familiar forms to be regarded in a new light. Original research and direct observation were only beginning to become part of a standard method, yet a major advance had been made.

During this important time, four individuals born within fifteen years of each other and each trained in medicine emerged as the great names of Renaissance zoology. First among them is Konrad Gesner. Gesner was a Zurich physician of wide-ranging interests whose work habits could only be described as compulsive. Among his many publications, the *Historiae animalium* stands as the most authoritative zoological study since Aristotle. This enormous work consists of five folio volumes containing some thirty-five hundred pages and one thousand woodcuts. The first four volumes, published in the period 1551–58, deal with quadrupeds, birds, and fish. The fifth volume, dealing with snakes and scorpions, appeared posthumously in 1587. The Library has the first edition of Gesner's first four volumes, published in Zurich.

Gesner used Aristotle's principles to arrange the animals and then treated them alphabetically within those groups. Each animal was discussed in eight different aspects—ranging from habitat and description to the animal's usefulness to man. Gesner also attempted to have a picture of each animal accompany the description, "so that students may more easily recognize objects that cannot be very clearly described in words," and some consider his emphasis on illustration as a zoological aid to be his most original contribution. These woodcuts were done by the eminent artists of his time and, although sometimes crude, they were sufficiently realistic to act as a valuable supplement to Gesner's dense and comprehensive text. Gesner was indefatigable and conscientious in whatever he did. His reputation for both work and accomplishment was widely recognized and admired, and his contemporaries called him "Plinius germanicus," the German Pliny.

In 1565, Gesner died of the plague, having contracted the disease from his patients. His *Historiae animalium* stands as the best purely zoological work of the Renaissance period, and it endured as the most authoritative

Konrad von Megenberg's popular *Buch der Natur* depended heavily on its illustrations. This illustration shows the "wonder-people" as well as other monstrous animals—illustrating legends that hark back to Pliny and before. Megenberg distinguished between human "monstruosi," wonder-people with souls or human beings born deformed or malformed, and monstruosi without souls, which he says are not truly human. He does not account for the origin of these soulless wonder-people nor does he explain how to differentiate, except to say that the latter are not born of women. *Buch der Natur,* 1481. Konrad von Megenberg.

138 De Aquatilibus.

BALAENA ERECTA GRANDEM NAVEM SVBMERGENS.

Videntur & alia quædam cete ex eodem Balænis adnumeranda, quæ ipse simpliciter cete nominat, cum præter magnitudinem balænis præcipuè conuenientem, nullam in se corporis partem raram aut monstrosam habeant. Eiusmodi sunt:

CETVS INGENS, QVEM INCOLAE FARAE INSVLAE ICHthyophagi tempestatibus appulsum, unco comprehensum ferreo, securibus dissecant & partiuntur inter se.

NAVTAE IN DORSA CETORVM, QVAE INSVLAS PVTANT, anchoras figentes sæpe periclitantur. Hos cetos Trolual sua lingua appellant, Germanicè Teüffelwal.

SIMILIS

Konrad Gesner's massive *Historiae animalium* stands as the best purely zoological work of the Renaissance period and remained the standard reference work for two centuries. It contained over one thousand woodcuts. These vivid images show three different types of whales—each a fearsome monster. The top image shows a ship being sunk by a whale. The middle depicts a captured whale being flensed. The bottom image curiously shows endangered sailors on the back of a whale, having anchored to the beast in the mistaken belief that it was an island. The coarseness of these woodcuts adds to the ferocity of the sea monsters. Gesner included scores of such fanciful creatures in his *Historiae animalium,* indicating that scientific zoology was still in its infancy. *Historiae animalium,* 1551–58. Konrad Gesner.

standard reference work for nearly another two centuries. The work saw many editions and translations, with *The Historie of Foure-footed Beastes* by Edward Topsell being the earliest and best English version. The Library has the first edition, printed in 1607 in London. Topsell also did a translation of Gesner's fifth book, *The Historie of Serpents,* which was published in London in 1608. The Library has this edition as well as a 1658 London publication of Topsell's whose two volumes contain not only the above two books of Gesner's but Thomas Moffett's *Insectorum sive minimorum animalium theatrum,* originally published in London in 1634 in Latin.

Gesner's counterpart in Italy was Ulisse Aldrovandi, an inveterate collector and student of nature. An impulsive youth, Aldrovandi traveled throughout western Europe and parts of the Middle East. At the age of twelve he set off for Rome with no money, leaving his native Bologna without a word to anyone. At sixteen he journeyed to Spain and later to Jerusalem. Aldrovandi later settled down and remained a professor at Bologna for forty years. His work on animals is as comprehensive and probably as worthwhile as Gesner's, although he has been described as far less critical than Gesner. On the other hand, the illustrations in his books are better than Gesner's and are grouped together on separate pages. Like Gesner, he offers as much information as possible about every aspect of an animal and arranges his beasts in somewhat logical groupings rather than simply alphabetically. Aldrovandi accumulated massive amounts of material over the years but it was not until he was seventy-seven years old that he published his first work, that being a three-volume work on birds, given the collective title *Ornithologiae.* The Library has the first edition, published between 1599 and 1603 in Bologna. Together the three volumes total over a thousand folio pages. With the exception of another volume on insects, the remainder of Aldrovandi's published works were completed from his notes by his pupils and successors. Of these, the Library has in first edition the *De piscibus* (1613) and *Serpentum* (1640) and the second edition of *De quadropedibus solidipedibus* (1639), all published in Bologna.

Two Frenchmen of the same period also made significant zoological contributions, but in a more specialized or limited area of investigation. Guillaume Rondelet was the younger of the two and made his mark with his study of aquatic animals. He lived and worked as a physician in Rome for much of his life. In 1554 he published in Lyons his first work, *De piscibus marinis,* which described and illustrated not only fish but seals, whales, and mollusks, as well as worms. The next year, Rondelet added a second part to include even more aquatic species and titled it *Universae aquatilium historiae.* The Library has both books bound together in one volume under the title *De piscibus marinis.*

Like Rondelet, Pierre Belon studied certain animal groups to the exclusion of others. He, too, published a book on aquatic animals, two in fact, in which the term "fishes" is applied not only to seals and hippopotamuses but to beavers and otters as well. (There is speculation that such a classification was a reflection of existing culinary practices among the

Catholic faithful, as an obviously broad and liberal interpretation increased the types of animals that might be eaten during a time of fasting.) His best work, however, was a later work on birds, *L'histoire de la nature des oyseaux,* published in Paris in 1555. The Library has this work in its first edition. Here Belon not only describes and illustrates various birds, but he makes a landmark attempt at a comparative anatomical investigation. Belon's famous comparison, in text and illustration, of the skeleton of a bird and that of a man pointed out their astonishingly homologous nature. Belon's book on birds was richly illustrated and led him to boast that "no one has yet shown them so true to life as we." Both Belon and Rondelet exhibited considerably more independence and originality than did the great compilers Gesner and Aldrovandi. Both did a substantial amount of field research—Belon traveling through Greece, Turkey, Syria, and Egypt. Interestingly, each of the four seems to have met at least one of his contemporaries. Aldrovandi went on collecting expeditions with Rondelet, who had met Belon while they were both in Rome. And it is fairly certain that Aldrovandi met Gesner at some time during their parallel careers.

These sixteenth-century zoologists were succeeded by individuals whose specialized interests focused their researches on a single aspect or area of

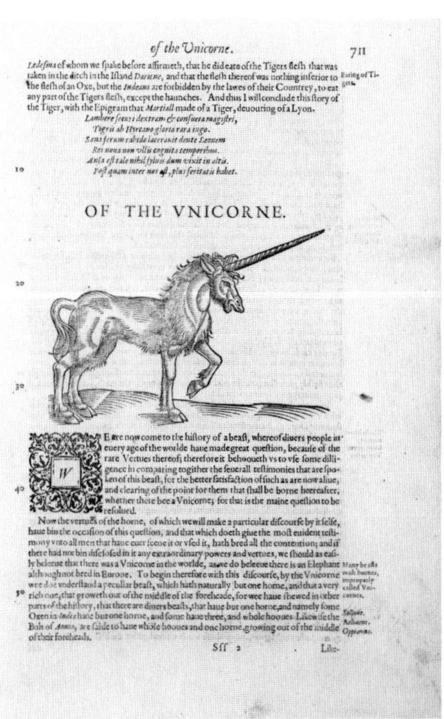

of the Vnicorne. 711

Ledeſma of whom we ſpake before affirmeth, that he did eate of the Tigers fleſh that was taken in the ditch in the Iſland *Dariene*, and that the fleſh thereof was nothing inferior to the fleſh of an Oxe, but the *Indeans* are forbidden by the lawes of their Countrey, to eat any part of the Tigers fleſh, except the haunches. And thus I will conclude this ſtory of the Tiger, with the Epigram that *Martiall* made of a Tiger, deuouring of a Lyon. — Eating of Tigers.

Lambere ſecuri dextram & conſueta magiſtri,
Tigris ab Hyrcano gloria rara iugo.
Saeua ferum rabido lacerauit dente Leonem
Res noua non vllis cognita temporibus.
Auſa eſt tale nihil ſyluis dum vixit in altis.
Poſt quam inter nos eſt, plus feritatis habet.

OF THE VNICORNE.

WEe are now come to the hiſtory of a beaſt, whereof diuers people in euery age of the worlde haue made great queſtion, becauſe of the rare Vertues thereof; therefore it behooueth vs to vſe ſome dilligence in comparing togither the ſeuerall teſtimonies that are ſpoken of this beaſt, for the better ſatisfaction of ſuch as are now aliue, and clearing of the point for them that ſhall be borne heereafter, whether there bee a Vnicorne; for that is the maine queſtion to be reſolued.

Now the vertues of the horne, of which we will make a particular diſcourſe by it ſelfe, haue bin the occaſion of this queſtion, and that which doeth giue the moſt euident teſtimony vnto all men that haue euer ſeene it or vſed it, hath bred all the contention; and if there had not bin diſcloſed in it any extraordinary powers and vertues, we ſhould as eaſily beleeue that there was a Vnicorne in the worlde, as we do beleeue there is an Elephant although not bred in Europe. To begin therefore with this diſcourſe, by the Vnicorne wee doe vnderſtand a peculiar beaſt, which hath naturally but one horne, and that a very rich one, that groweth out of the middle of the foreheade, for wee haue ſhewed in other parts of the hiſtory, that there are diuers beaſts, that haue but one horne, and namely ſome Oxen in *India* haue but one horne, and ſome haue three, and whole hooues. Likewiſe the Buls of *Aonia*, are ſaide to haue whole hooues and one horne, growing out of the middle of their foreheads. — Many beaſts with hornes, improperly called Vnicornes. *Solinus. Aelianus. Oppianus.*

Sſſ 2 Like-

The English translation of Gesner's *Historiae animalium* was Edward Topsell's *The Historie of Foure-footed Beastes* and *The Historie of Serpents.* Here Gesner and Topsell show the unicorn, a beast whose literary tradition goes as far back as the Bible. Although Gesner says that "There is nobody who has ever seen this animal in Europe," he defers to authority and tradition, saying, "But one has to trust the words of wanderers and far-going travelers, for the animal must be on earth, or else its horns would not exist." The prized and magical horns that were then available in many an apothecary shop are now believed to have been tusks of the narwhal, an arctic fish. *The Historie of Foure-footed Beastes*, 1607. Edward Topsell.

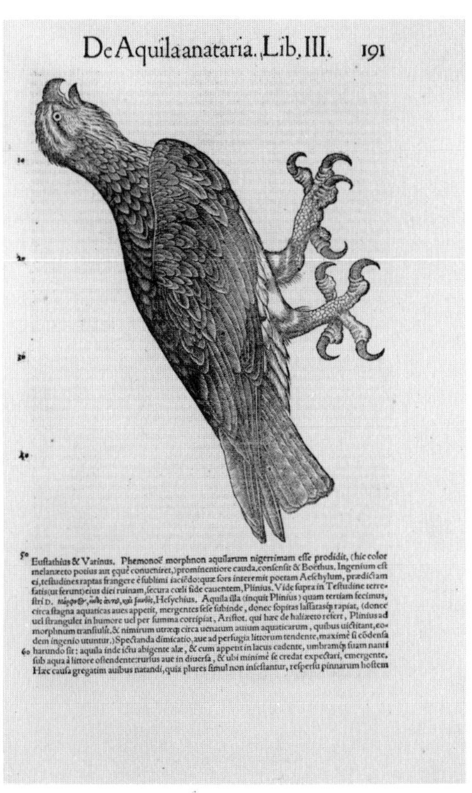

De Aquila anataria. Lib. III. 191

Euſtathius & Varinus. Phemonoë morphnon aquilarum nigerrimam eſſe prodidit, (hic color melanæeto potius aut ęquè conueniret,) prominentiore cauda, conſenſit & Boethus. Ingenium eſt ei, teſtudines raptas frangere è ſublimi iaciẽdo: quæ ſors interemit poetam Aeſchylum, prædictam fatis (ut ferunt) eius diei ruinam, ſecura cœli fide cauentem, Plinius. Vide ſupra in Teſtudine terreſtri D. [illegible], Heſychius. Aquila illa (inquit Plinius) quam tertiam fecimus, circa ſtagna aquaticas aues appetit, mergentes ſeſe ſubinde, donec ſopitas laſſatasq; rapiat, (donec uel ſtranguler in humore uel per ſumma corripiat, Ariſtot. qui hæc de haliæeto refert, Plinius ad morphnum tranſiulit. & nimirum utræq; circa uenatum auium aquaticarum, quibus uictitant, eodem ingenio utuntur.) Spectanda dimicatio, aue ad perfugia littorum tendente, maximè ſi cõdenſa harundo ſit: aquila inde ictu abigente alæ, & cum appetit in lacus cadente, umbramq; ſuam nanti ſub aqua à littore oſtendente: rurſus aue in diuerſa, & ubi minimè ſe credat expectari, emergente. Hæc cauſa gregatim auibus natandi, quia plures ſimul non infeſtantur, reſperſu pinnarum hoſtem

Historiae animalium, 1551–58. Konrad Gesner.

Ornithologiæ Lib. II. 215

Ulisse Aldrovandi's work is considered an improved Gesner, although it never compared in popularity. In this comparison of eagles, Aldrovandi's is sleeker and more true-to-life, reflecting the advantage of over three decades of zoological advances. His illustrations are also more thorough, showing skeletons and certain anatomical details. *Ornithologiae*, 1599–1603. Ulisse Aldrovandi.

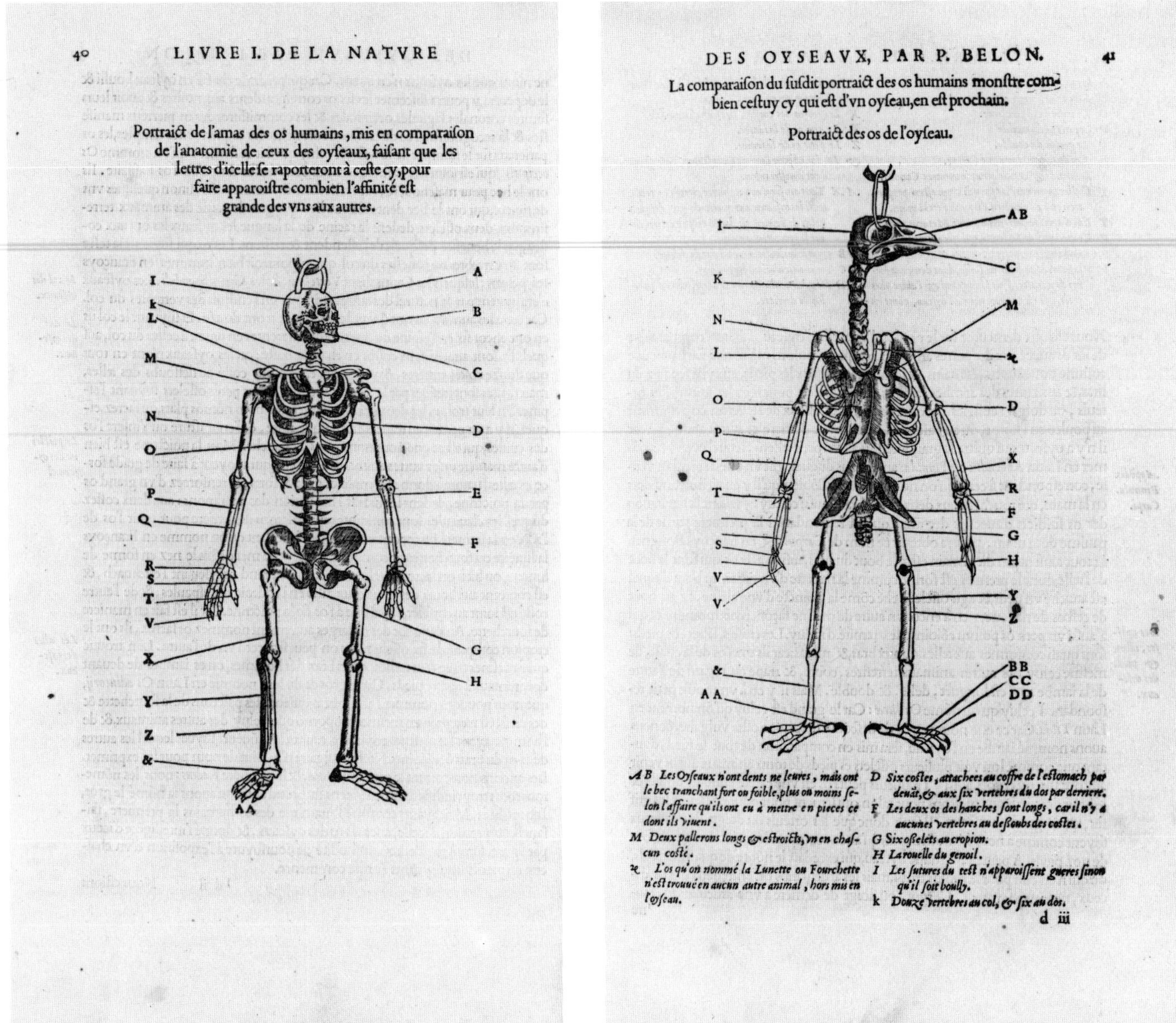
40 LIVRE I. DE LA NATVRE

Portraict de l'amas des os humains, mis en comparaiſon de l'anatomie de ceux des oyſeaux, faiſant que les lettres d'icelle ſe raporteront à ceſte cy, pour faire apparoiſtre combien l'affinité eſt grande des vns aux autres.

DES OYSEAVX, PAR P. BELON. 41

La comparaiſon du ſuſdit portraict des os humains monſtre combien ceſtuy cy qui eſt d'vn oyſeau, en eſt prochain.

Portraict des os de l'oyſeau.

AB Les Oyſeaux n'ont dents ne leures, mais ont le bec tranchant fort ou foible, plus ou moins ſelon l'affaire qu'ils ont eu à mettre en pieces ce dont ils viuent.

M Deux pallerons longs & eſtroicts, vn en chaſcun coſté.

ꝛ L'os qu'on nommé la Lunette ou Fourchette n'eſt trouué en aucun autre animal, hors mis en l'oyſeau.

D Six coſtes, attachees au coffre de l'eſtomach par deuāt, & aux ſix vertebres du dos par derriere.

F Les deux os des hanches ſont longs, car il n'y a aucunes vertebres au deſſoubs des coſtes.

G Six oſſelets au cropion.

H La rouelle du genoil.

I Les ſutures du teſt n'apparoiſſent gueres ſinon qu'il ſoit boully.

k Douze vertebres au col, & ſix au dos.

d iii

Pierre Belon displayed these drawings of the skeletons of a man and a bird on opposite pages of this book to show homologous structures. As the originator of comparative anatomy, Belon demonstrated the remarkable skeletal similarities among the various vertebrates—from man to fish—similarities which had heretofore been entirely unsuspected. Belon was also a vigorous traveler and was one of the first explorer-naturalists. He was murdered in the Bois du Boulogne under mysterious circumstances, perhaps killed by robbers. *L'histoire de la nature des oyseaux*, 1555. Pierre Belon.

zoology. Consequently, no major zoological work of any real scope appeared during the greater part of the seventeenth century. In the fields of comparative anatomy and microscopy, however, considerable progress was made.

During the previous century, Vesalius's realistic and factual anatomy had placed the study of human anatomy on the firmer basis of exact observation. Eighty-five years later, William Harvey's contributions did the same for human physiology. Both Vesalius and Harvey are treated at length in the chapter on medicine. Sandwiched temporally between these two giants is the aristocratic lawyer of Bologna, Carlo Ruini, whose detailed investigation of the anatomy of the horse is said to rival Vesalius's work in precision and beauty. Ruini's masterpiece, *Anatomia del cavallo*, is the first comprehensive work devoted exclusively to the structure of an animal other than man. The Library has the first edition, published in 1598 in Bologna. The beautiful engravings in this two-volume set are said to be modeled after those of Vesalius. Ruini's work became extremely popular and places him among the founders of comparative anatomy.

The centerpiece of seventeenth-century comparative anatomy is Giovanni Borelli's study *De motu animalium*. Published in the year of his death, this major work applied the mechanics of Galileo to the movements of animals, describing muscular action in terms of a system of levers. Every part of Borelli's work is highly original. His brilliant mechanical interpretation and explanation of animal movement left his imprint on many fields, medicine being a notable example. The Library has the two-volume work, published in Rome in 1680–81, in first edition.

Among the seventeenth-century proliferation of instruments for measurement and observation, the invention and development of the microscope had the greatest significance for the field of zoology. The ability to magnify the infinitely small and to peer into heretofore unseen worlds populated by what seemed to be living organisms offered science a virgin realm of discovery. The microscopists of the seventeenth century literally created new fields of study, such as cytology and histology, during this classic period in microscopy.

The first printed work to contain microscopic illustrations is the *Persio, tradatto in verso sciolto e dichiarato* of Francesco Stelluti. Stelluti was a friend of Galileo's and a founding member of the Accademia dei Lincei. He had turned his microscope to study the honeybee and produced a full-page illustration showing the many aspects and structural details of the bee. The appearance of a honeybee in Stelluti's translation of the satirical poems of Persius is explained by the politics of his time. Stelluti dedicated the book to Cardinal Francesco Barberini, whose uncle was Pope Urban VIII. The Barberini family emblem was the bee. The Library has Stelluti's work in first edition, published in Rome in 1630.

Thirty-five years later the multitalented Robert Hooke produced a work based almost totally on microscopic observation. Hooke's treatise was published in English with the imprint of the Royal Society of London in 1665 and was called the *Micrographia; or, Some Physiological Descriptions of Minute Bodies Made by Magnifying Glasses*. The Library has the first edition, which contains fifty-seven microscopic and three telescopic observations. Among them, those relevant to zoology are Hooke's first observations of the compound eye of the fly and the structure of the bee's sting, as well as that of the flea and the louse. In his Observation No. 18, he described the structure of cork and first used the word *cell*.

Hooke's *Micrographia* was an immediate success, and with it, the microscope became accepted as an indispensable aid to zoological research. The contributions of his contemporaries Leeuwenhoek, Malpighi, and Swammerdam made the second half of the seventeenth century an exciting and progressive period for zoology.

Anton van Leeuwenhoek, a Dutch linen draper born in 1632, became one of the most famous and successful amateur scientists of all time.

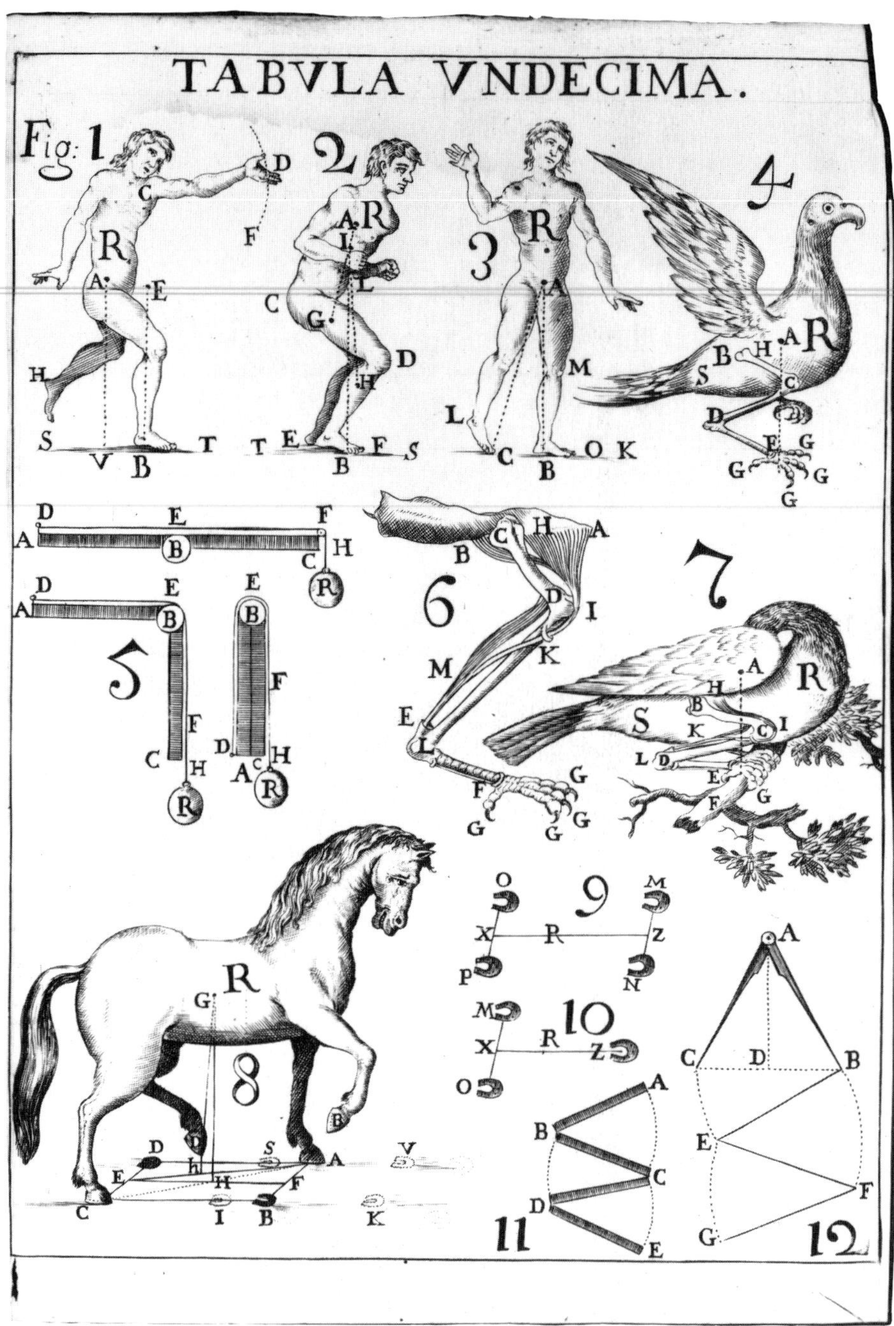

In a new way of looking at animals and man, Borelli sought to examine all movements of the body on a mechanical basis and investigated the actions of muscles according to the laws of statics and dynamics. This new approach to animal physiology had its origin with Borelli's contemporary, René Descartes. Descartes's "mechanistic" philosophy marked an early attempt at a zoological generalization or theory. Although Borelli's inclusion of man in this illustration seems to imply that the difference between him and animals is one of degree and not of kind, Descartes held that man has a reasoning soul and is therefore complete, separate, and different from animals. Borelli was not only a physiologist but a mathematician of note and an astronomer. He was also very active in the politics of Counter-Reformation Italy. *De motu animalium*, 1680–81. Giovanni Alfonso Borelli.

Leeuwenhoek had little schooling but sufficient income to allow him to pursue his hobby of grinding lenses. Leeuwenhoek used a simple microscope with a single lens, unlike Hooke's compound ones, and his skill was such that he could magnify up to nearly two hundred times. With his precise instruments, Leeuwenhoek spent his entire life observing and describing, and he lived to be ninety. In 1673, the Royal Society of London received some of his observations through an intermediary. Delighted with the work of this unknown Dutchman, the society asked him for more. Over the years, Leeuwenhoek sent 375 letters to the Royal Society, each one containing precise and careful descriptions of his observations. Leeuwenhoek applied his microscope to anything and everything—from an ordinary drop of water to tooth scrapings, ant eggs, and crystals. His discoveries are too numerous to detail, but two of them deserve mention. In 1674 he described observing "little animalcules" moving about in a specimen of water he had taken from a marshy lake. By his excellent description, the animal is now recognized to be a one-celled animal belonging to the phylum *Protozoa*. In 1677 he became the first to describe the little animals (spermatozoon) he observed in human and animal semen. This discovery was to support preformationist views concerning the origin of life.

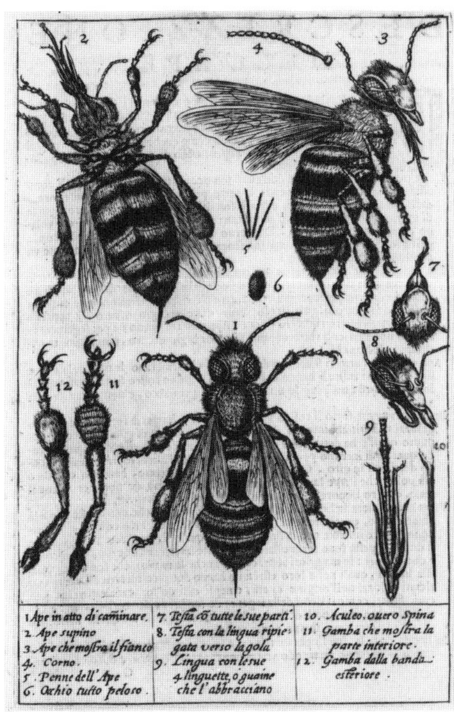

Zoological investigation was aided immensely by the invention of the microscope. The Academy of the Lynx conducted the first systematic investigation of living things with a microscope, and one of its members, Francesco Stelluti—a friend of Galileo and a founding member of the Academy—in 1625 produced a single printed sheet showing a bee in great detail. In 1630, Stelluti used this same engraving to dedicate a book of poems to Cardinal Francesco Barberini. The cardinal's uncle, Pope Urban VIII, was also a Barberini, and the family's crest showed three bees. The detail of Stelluti's external anatomy of a bee is extremely accurate. *Persio*, 1630. Francesco Stelluti.

Leeuwenhoek never wrote a scientific paper or a book, and his work was mainly published by the Royal Society in the form of his sometimes homely letters (which the society translated and edited). The Library has these as part of its collection of the *Philosophical Transactions* of the Royal Society. His letters also were published in Dutch and Latin collections, and the Library has the five-volume Dutch collection, variously composed of first through third editions, entitled *Ontledingen en ondekkingen . . . brieven,* published in Leiden and Delft between 1696 and 1718, as well as the 1695 Delft edition of selected letters, *Arcana naturae detecta.*

Leeuwenhoek's countryman Jan Swammerdam endured a sad and tortured personal life, but he bequeathed to science an unpublished storehouse of entomological information. As an anatomist of invertebrate life, Swammerdam was a genius. His mastery of even the most minute and complicated anatomical details of the smallest of insects was incomparable. Although his most famous microscopic discovery was that of the red blood corpuscle, he is recognized by most as the father of entomology. Swammerdam studied medicine at Leiden University but never practiced it, preferring to collect and study insects. His lifelong disputes with his father, who thought his son worthless, brought great distress to the young man. Throughout his life, Swammerdam was subject to what has been described as a sort of acute religious depression, and his scientifically productive years were very few. He died at forty-three, having given up his work seven years earlier to pursue a fanatical religious asceticism that eventually killed him. His short life's work lay unread and unrecognized until the great physician and teacher Hermann Boerhaave published it at his own expense. Boerhaave contributed a biography of Swammerdam and gave the entire work the title of *Bybel der natuure*. The Library has this two-

In 1665, Robert Hooke produced his monument to microscopy, *Micrographia*. Hooke was a genius of mechanics, and having produced an improved compound microscope, he used his new device to investigate more closely the world around him. Insects seemed to get most of his attention and his beautiful, detailed engravings are all the more impressive given the folio size of the book. Here he shows "a very beautiful creature," the blue fly. Hooke was an extremely disputatious man who suffered all his life from congenital infirmities. Despite this, he was a broadly talented individual—architect, artist, and rival of Newton in physics—who was the first "curator of experiments" of the Royal Society and later its secretary. His *Micrographia* is one of the most readable major works in the history of science and conveys the excitment of first discovery in a charming and informal manner. *Micrographia*, 1665. Robert Hooke.

volume work, published in Leiden in 1737–38, in first edition. The huge folio contains many elegantly thorough illustrations that demonstrated the extreme complexity of even the lower animals. It stands as the first major scientific study of insects, including their classification and transformations.

Swammerdam had great respect for an Italian contemporary he described as "the most indefatigable searcher into the miracles of nature." This was the Tuscan court physician and poet Francesco Redi. Redi was a thoroughgoing experimentalist who took nothing for granted. "I have taken the greatest care to convince myself of facts with my own eyes by means of accurate and continued experiments before submitting them to my mind as a matter for reflection," he said. This admirable scientific approach led him to attack directly the doctrine of spontaneous generation. The notion that certain lower forms of life arose spontaneously from decaying matter had been held and supported since Aristotle first endorsed it. Indeed, such eminent scientists as Jean Baptiste van Helmont had argued that spoiled wheat gave rise to mice.

Redi's originality and independence of thought are demonstrated in his treatment of what was considered the classic proof of spontaneous generation—the seemingly spontaneous emergence of maggots from rotting meat. Exercising what today are called "biological controls," Redi put various meats in both covered and uncovered flasks. The meat in the open vessels became both wormy and putrid, whereas that in the covered vessels became only putrid. Redi observed the experiment closely and realized that the maggots in the uncovered flasks were the developing larvae from fly eggs. He noted that in the flasks that were covered with gauze he "never saw any worms on the meat, though many were to be seen moving about on the net-covered frame."

In 1668 Redi offered his *Esperienze intorno alla generazione degl'insetti,* in which he refuted the notion of spontaneous generation. The Library has this work, published in Florence, in first edition. Compared to the exquisite and accurate drawings of Swammerdam, its illustrations of insects are somewhat crude. But in this work, Redi penetrated to the core of one of nature's secrets, stated earlier by William Harvey on the frontispiece of his own book: "Omne vivum ex ovo."

Despite Redi's evidence, the doctrine of spontaneous generation was not put to rest until the work of Louis Pasteur, some two hundred years later. As proposed in the seventeenth century and earlier, the idea of spontaneous generation was a relatively simple, straightforward notion with no real scientific basis. It should not be compared to theories arising from contemporary research on the origin of life which indicates that some sort of generation or chemical evolution from the "primordial organic soup" of distant times must have occurred.

Although seventeenth-century zoology became increasingly characterized by specialized investigation and vigorous collecting, it was nevertheless a science in disarray, held back by a chaos of nomenclature. No generally agreed upon and widely understood system of classification existed. Though mankind had been giving animals names since Adam and Eve, the conceptual difference between giving a particular beast a certain name and classifying it according to a set of definite principles represents an essential and fundamental distinction. This distinction is integral to the scientific method, for no science can progress without a body of rigorous technical terms. To zoology, classification became the orderly common language necessary to such an essentially descriptive science. Only after passing through the homely process of being systematized could zoology proceed to the higher level of the grand generalization.

The first real systematic arrangement of animals was produced by John Ray near the end of the seventeenth century. In his work entitled *Synopsis methodica animalium quadrupedum,* published in London in 1693, which

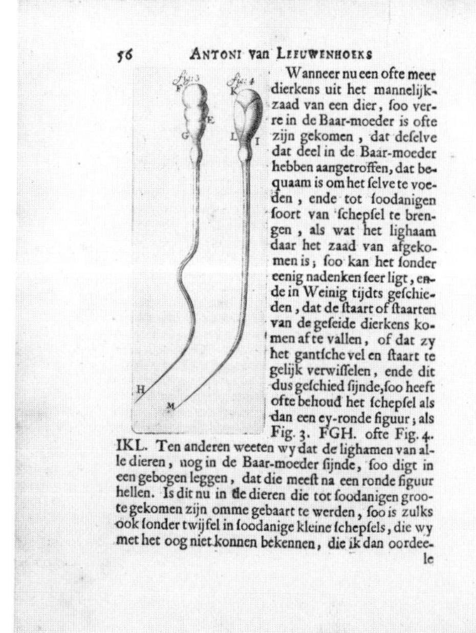
56 ANTONI van LEEUWENHOEKS

Wanneer nu een ofte meer dierkens uit het mannelijk-zaad van een dier, ſoo verre in de Baar-moeder is ofte zijn gekomen, dat deſelve dat deel in de Baar-moeder hebben aangetroffen, dat bequaam is om het ſelve te voeden, ende tot ſoodanigen ſoort van ſchepſel te brengen, als wat het lighaam daar het zaad van afgekomen is; ſoo kan het ſonder eenig nadenken ſeer ligt, ende in Weinig tijdts geſchieden, dat de ſtaart of ſtaarten van de geſeide dierkens komen af te vallen, of dat zy het gantſche vel en ſtaart te gelijk verwiſſelen, ende dit dus geſchied ſijnde, ſoo heeft ofte behoud het ſchepſel als dan een ey-ronde figuur; als Fig. 3. FGH. ofte Fig. 4. IKL. Ten anderen weeten wy dat de lighamen van alle dieren, nog in de Baar-moeder ſijnde, ſoo digt in een gebogen leggen, dat die meeſt na een ronde figuur hellen. Is dit nu in de dieren die tot ſoodanigen groote gekomen zijn omme gebaart te werden, ſoo is zulks ook ſonder twijfel in ſoodanige kleine ſchepſels, die wy met het oog niet konnen bekennen, die ik dan oordee-
le

Unlike Hooke, who investigated relatively large subjects with his microscope, Anthony van Leeuwenhoek was interested primarily in microscopic life. Although his investigations were somewhat desultory, one theme he did pursue throughout his life was that of sexual reproduction. During his fifty years of observation he described the spermatozoa of insects, fish, birds, and mammals. Here, in a Dutch version of his letters to the Royal Society of London published during his lifetime, he shows the spermatozoa of a dog. Figure 3 is a living spermatozoon, distinguished from figure 4, which is dead and has changed shape. Altogether, Leeuwenhoek's most important scientific contribution was his steadfast assertion that the moving objects he observed were a form of animal life. *Ontledingen en ontdekkingen*, 1696–1718. Anthony van Leeuwenhoek.

88 Although trained as a physician, Jan Swammerdam preferred studying insects to treating people. While his accomplishments in the field of medical research were known to his contemporaries, his work on insects remained mostly a private passion, and it is in this almost secret work that Swammerdam distinguished himself. His *Bible of Nature*, published one hundred years after his birth, contains some of the most remarkable details of insect anatomy ever produced. Using a single lens microscope and dissecting tools that he made himself, Swammerdam was able to use his special dissecting skills to great advantage. His considerable gifts are apparent when it is realized that the subject whose nervous system he details and describes was often an insect less than a quarter of an inch long. Swammerdam was anti-Aristotelian in the premises of his insect study, believing that insects were no less perfect or worthy of study than supposedly higher animals. Here he shows a male gnat. *Bybel der natuure*, 1737–38. Jan Swammerdam.

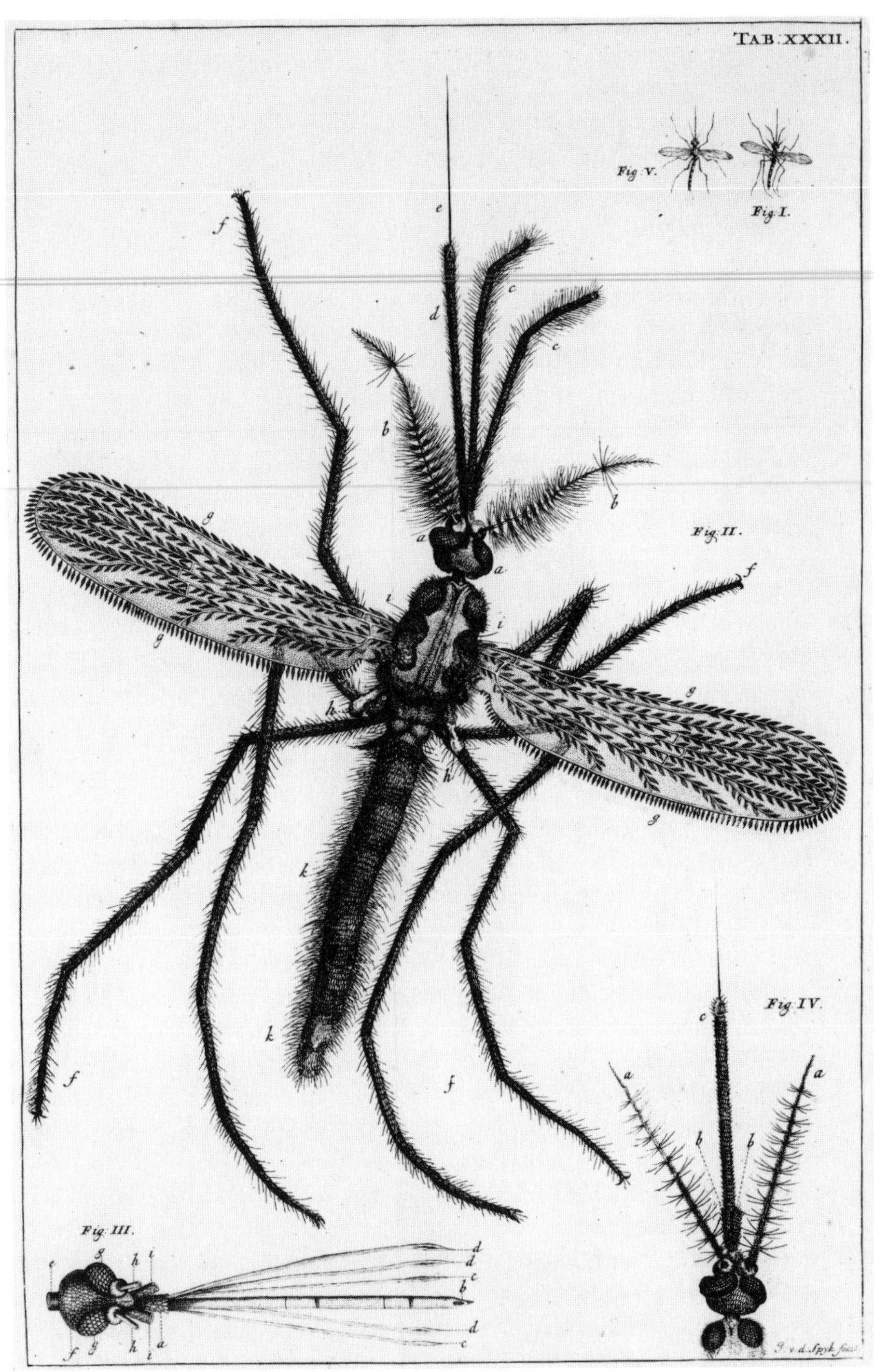

the Library has in first edition, Ray offered his personal view that the structure of an animal's body was what mattered most in classifying it. Ray's emphasis on physical structure focused particularly on an animal's teeth and feet as classifying determinants. Although he did not set up a system of classification with a concise and systematic nomenclature, his emphasis on animal structure and the distinction he made between species and genus went a long way toward producing a sound taxonomy. Until Ray, the term *species* had a decidedly indefinite usage. It was he who gave it its modern and concrete meaning, applying it only to groups of similar individuals that exhibit constant attributes from generation to generation.

Ray was an especially religious man, and his brand of "natural theology" found an essential harmony between God and nature. His deeply held religious convictions rarely dimmed his objectivity, however, and he came to believe that the stability of species was not absolute—an idea certainly ahead of its time. His interpretation of fossils as the petrified remains of extinct creatures was not accepted for a hundred years. Ray was an especially astute and wise naturalist whose ideas and work paved the way for Linnaeus, the founder of modern taxonomy.

Carl Linnaeus systematized both zoology and botany and gave all naturalists a common, useful language. For both disciplines, he established the classification principles of kingdom, class, order, genus, and species and thus gave science a basic scheme for organizing all of the natural world. Furthermore, he offered a new system of nomenclature, the binomial system still in use today, which gives each organism a double scientific name.

Linnaeus came from a peasant village in Sweden. His father was a clergyman and amateur botanist named Nils Ingemarson, who adopted the family name Linnaeus after a mighty linden tree near his home. This particular linden was especially large and old and was held in high regard by the locals as a sort of sacred tree. Given the essential role of plants and trees in the herbal medicine of the time, naming one's family after a respected tree seems not at all inappropriate. In fact, a famous contemporary of Linnaeus, Hermann Boerhaave, had the delightful habit of always raising his hat to an elder tree in deference to its renowned medicinal qualities.

The gifted young Carl was assisted by a series of patrons and obtained a degree in medicine. From the time he was very young, Linneaus possessed a natural and energetic interest in plants and herbs. His ability was such that he lectured in botany at Uppsala although not yet a graduate, and he received a grant in 1732 to visit Lapland as a scientific collector. This field trip took him through some of the more wild and unexplored regions of northern Sweden, and in five months he had traveled 4,600 miles. Later, after traveling through England and west Europe, he began to formulate the basic principles of what was to be his life's work.

Linnaeus was an autodidactic genius with an encyclopedic memory and a passion for classifying everything he knew. By the age of twenty-five he had dedicated himself to establishing an original classification system with which he would methodically and systematically organize the entire natu-

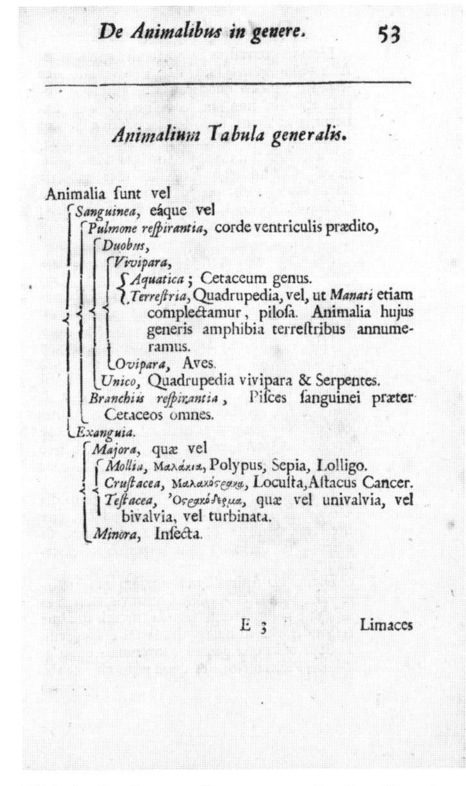

De Animalibus in genere. 53

Animalium Tabula generalis.

Animalia ſunt vel
Sanguinea, eáque vel
Pulmone reſpirantia, corde ventriculis prædito,
Duobus,
Vivipara,
Aquatica; Cetaceum genus.
Terreſtria, Quadrupedia, vel, ut *Manati* etiam complectamur, piloſa. Animalia hujus generis amphibia terreſtribus annumeramus.
Ovipara, Aves.
Unico, Quadrupedia vivipara & Serpentes.
Branchiis reſpirantia, Piſces ſanguinei præter Cetaceos omnes.
Exanguia.
Majora, quæ vel
Mollia, Μαλάκια, Polypus, Sepia, Lolligo.
Cruſtacea, Μαλακόστρακα, Locuſta, Aſtacus Cancer.
Teſtacea, Ὀστρακόδερμα, quæ vel univalvia, vel bivalvia, vel turbinata.
Minora, Inſecta.

E 3 Limaces

This is the first really systematic classification of animals. John Ray began this arrangement by returning to Aristotle's distinction between red-blooded animals (vertebrates) and those that are not red-blooded (invertebrates). From there, his system always proceeded on the basis of the structure of an animal's body. The vertebrates are therefore divided into those that breathe through lungs and those that have gills. The former are subdivided into viviparous and oviparous animals (birds and reptiles). Further subdivisions of viviparous animals are based on the number of toes, type of teeth, and other structural characteristics. John Ray's fundamental taxonomy had prepared the way for Linnaeus, who began his work where Ray left off. *Synopsis methodica animalium quadrupedum et serpentini generis*, 1693. John Ray.

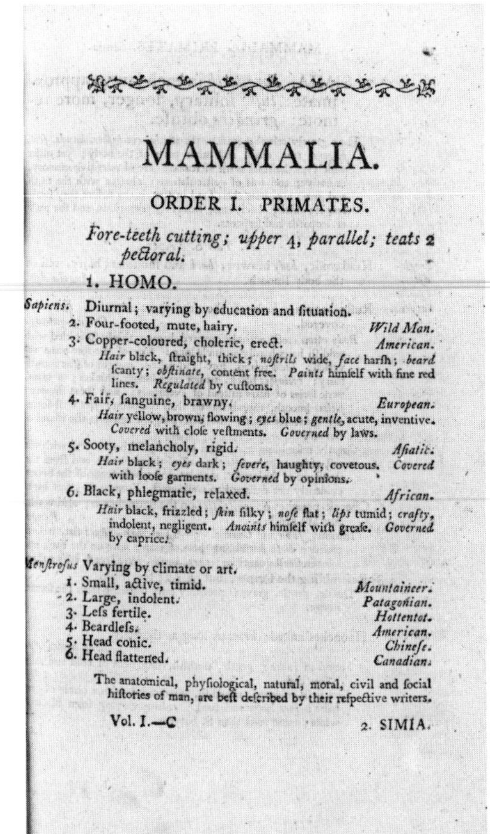

MAMMALIA.

ORDER I. PRIMATES.

Fore-teeth cutting; upper 4, parallel; teats 2 pectoral.

1. HOMO.

Sapiens. Diurnal; varying by education and ſituation.
2. Four-footed, mute, hairy. *Wild Man.*
3. Copper-coloured, choleric, erect. *American.*
Hair black, ſtraight, thick; *noſtrils* wide, *face* harſh; *beard* ſcanty; *obſtinate*, content free. *Paints* himſelf with fine red lines. *Regulated* by cuſtoms.
4. Fair, ſanguine, brawny. *European.*
Hair yellow, brown, flowing; *eyes* blue; *gentle*, acute, inventive. *Covered* with cloſe veſtments. *Governed* by laws.
5. Sooty, melancholy, rigid. *Aſiatic.*
Hair black; *eyes* dark; *ſevere*, haughty, covetous. *Covered* with looſe garments. *Governed* by opinions.
6. Black, phlegmatic, relaxed. *African.*
Hair black, frizzled; *ſkin* ſilky; *noſe* flat; *lips* tumid; *crafty*, indolent, negligent. *Anoints* himſelf with greaſe. *Governed* by caprice.

Monſtroſus Varying by climate or art.
1. Small, active, timid. *Mountaineer.*
2. Large, indolent. *Patagonian.*
3. Leſs fertile. *Hottentot.*
4. Beardleſs. *American.*
5. Head conic. *Chineſe.*
6. Head flattened. *Canadian.*

The anatomical, phyſiological, natural, moral, civil and ſocial hiſtories of man, are beſt deſcribed by their reſpective writers.

Vol. I.—C 2. SIMIA.

To the diverse, teeming subject matter of zoology, Linnaeus brought an ordering intellect. In 1735, his *Systema naturae* offered a classifying method whereby every animal (and plant) would be arranged and identified in sequence of class, order, genus, and species. Here, in an 1806 English edition, Linnaeus boldly classifies man as an animal of the genus *Homo* and species *sapiens*, the bionomial designation translating, "Man, the reasoner or the wise." His descriptions of the known races are interesting, the American Indian described as "regulated by customs," the Asiatics as "governed by opinions," the Africans as "governed by caprice," and his race, the Europeans, as "governed by laws." *A General System of Nature*, 1806. Carl von Linné.

ral world. At the age of twenty-eight he saw his epochal *Systema naturae* published and almost overnight became the internationally recognized arbiter of species.

The first edition of *Systema naturae* was printed in Leiden in 1735. Linnaeus had shown his manuscript to Hermann Boerhaave, who had just published Swammerdam's work, and it was through him that Linnaeus acquired J. F. Gronovius as a patron. Gronovius published *Systema naturae* at his own expense. It has become one of the rarest first editions in the history of science. A facsimile made from the copy in the collections of the Amsterdam Zoological Library in 1964 is in the collections of the Library of Congress. The enlarged second edition, published in Stockholm in 1740, is the earliest original imprint the Library holds. Over the years, Linnaeus revised and amplified his work so much that the tenth edition comprises 2,500 pages. The Library's collections include this tenth edition, published in 1758, which remains even today as the bible of taxonomy.

Compared to the rambling, imprecise picture books that preceded *Systema naturae,* Linnaeus's work may seem mechanical, artificial, rigid, and overly technical. Yet his system gave naturalists a method of accurately reducing the identification and classification of an animal or plant to a few details or sentences. For the first time zoologists and botanists could use a common language and method and through them communicate clearly with one another. Its principles formed the basis for all future progress in the natural sciences.

Although Linnaeus publicly held that all species are constant and immutable, he did modify his anti-evolutionary views late in his career. It was Linnaeus, after all, who classified man as an animal, *Homo sapiens,* characterized by his ability to think. He has been described as an arrogant, self-satisfied individual, whose naive optimism perceived nothing but harmony and order in the natural world. Always a religious person, he believed his system to be the result of divine guidance—and that it was his duty to discover and to elaborate God's natural order. Linnaeus was honored by everyone throughout his life and was made a noble in 1757, when he took the name Carl von Linné.

Georges Louis Leclerc Buffon was born the same year as Linnaeus but has little else in common with the great Swede. Buffon's family was well-to-do and offered him the advantages of a good education and extensive travel. Amid the love affairs, duels, and romantic adventures of his youth, Buffon had the time and talent to translate both Newton's *Fluxiones* and Stephen Hale's *Vegetable staticks.* He made his name in science in zoology not with new discoveries but rather because of the nature and scope of his original thinking. Unlike Linnaeus, he was unsystematic, theoretical, and imaginative. His main work, *Histoire naturelle,* was the labor of a lifetime. It eventually reached forty-four volumes and established him as one of the great synthesizers.

Buffon might have remained only a talented dilettante had he not been appointed keeper of the Jardin du Roi, or had he not taken his duties there so seriously. The catalog of the institution's collection that he drew up led him to take up his *Histoire naturelle*—a catalog of all of nature. The substance of his natural history reflected Buffon's personality—his writing was brilliant if somewhat oratorical, and his descriptions and anecdotes appealed to the popular as well as to the scholarly reader. The volumes were well illustrated with color engravings and their open discussion of animal sexuality and such natural phenomena as eunuchs made them best-sellers. To the special delight of French society of the Enlightenment, Buffon treated man as he would any animal.

Buffon did more than popularize zoology, however, for his influence was felt by the real scholars of his time. Perhaps most noteworthy and influential was the scope of his original ideas, the grand designs he saw in nature. From Buffon's comprehensive perspective, all of nature was a whole and any divisions were simply imposed by the mind of man. He drew no hard line between animal and vegetable life but united the organic world with the inorganic, stating that life "is not a metaphysical characteristic of living creatures, but a physical quality of matter." His treatment of man as part of all of animal creation and his adherence to the mutability of species foreshadowed later evolutionary ideas.

Buffon's technique was at times unscientific, as he was inclined to overlook detail and to opt for the easy generalization. Buffon's looseness with facts incensed Thomas Jefferson, for one. In his book *Notes on the State of Virginia,* written in 1781 and printed in Paris in 1784–85, Jefferson upbraided Buffon for his description of the American Indian as having "little sexual capacity" and being indifferent to his children. The Indian, Jefferson wrote, "is neither more defective in ardor, nor more impotent with the female, than the white reduced to the same diet and exercise." And he stated that he has seen "even the warriors weeping most bitterly on the loss of their children." To refute Buffon's contention that New World mammals had degenerated in size, Jefferson sent the skeleton and skin of a seven-foot moose as well as those of an American caribou, deer, and elk directly to Buffon in Paris.

Buffon's legacy was his inspiring and sometimes daring ventures into the theoretical realm, and this work no doubt stimulated both Lamarck and Cuvier. The Library has the first edition of Buffon's *Histoire naturelle* published in Paris during the period 1749–67 in its original first fifteen volumes, plus five volumes of supplements. The Library's complete, forty-four-volume set is from a later edition.

Nearly two centuries after Buffon first began his natural history series, a unique graphic work appeared based on his *Histoire naturelle.* Pablo Picasso and the publisher Ambroise Vollard planned to publish a series of animal etchings accompanied by texts from Buffon. Picasso began work selecting and interpreting various animals in etchings in 1936, long after he had promised the illustrations to Vollard. After Vollard's death in 1939, the book was taken up and published by Martin Fabiani in 1942. The resulting work, *Eaux-fortes originales [de] Picasso pour des textes de Buffon* was published in Paris in an edition of 226 numbered copies, one of which the Library has in its collections. The book has thirty-one etchings and aquatints by Picasso and, though it can be considered an illustrated book, it is really a suite of etchings to which text has been added.

In contrast to the rich, urbane Buffon, whose literary workday was punctuated by a visit from his wigmaker, who would arrange his coiffure, Jean Baptiste de Lamarck struggled all of his life with money problems, being the eleventh child of an aristocratic but impoverished family. Also in contrast to the early scientific accomplishments of Buffon was Lamarck's amazingly late entry into a field in which he was to become a pioneer. Born in 1744, Lamarck was fifty years old before he began any real work in zoology.

Having first served in the only two professions sufficiently honorable for the sons of faded nobility, the church and the army, Lamarck pursued his botanical interests in obscurity until the French Revolution. The same revolutionary government that took away the great Lavoisier's life gave Lamarck a new scientific life, appointing him professor of "insects and worms" at the reorganized Muséum National d'Histoire Naturelle (formerly the Jardin du Roi) in 1793. Lamarck had held various minor botanical positions at the Jardin du Roi for five years (1788–93), and he was given the new post almost by default since all the botanical positions were filled by others and there were no zoologists available for this post, which Lamarck renamed professor of "invertebrates."

Over the years Lamarck brought order to a field that had been essentially ignored: the field of invertebrate zoology. Even Linnaeus had avoided attempting any real invertebrate classification, calling all manner of invertebrate animals simply "worms." With the publication in Paris in 1801 of his *Systeme des animaux sans vertèbres,* Lamarck began to give order to this neglected life form. Having separated animals into vertebrate and invertebrate, he then classified the latter according to their respiratory, circulatory, and nervous systems. The Library has this book in first edition. Lamarck continued this work and between 1815 and 1822 his huge seven-volume *Histoire naturelle des animaux sans vertèbres* was published in Paris. A first edition of this work, which brought him recognition as a systematist from his contemporaries and established him as the founder of modern invertebrate zoology, is in the Library's collections.

But it was a work generally ignored by his peers, the *Philosophie zoologique,* that was to earn Lamarck the respect of posterity. In this two-volume theoretical work, more than half of which concerns itself with speculations on the nature of life itself, Lamarck demonstrates his intellectual indebtedness to Buffon, and in doing so he takes the first definitive step toward a theory of biological evolution. With the same broad, sweeping view of all of life that characterized Buffon, Lamarck offers in his *Philosophie zoologique* a bold rationale for the successive, evolutionary development of life. All living forms were constantly changing and developing, always to a level of higher complexity, he said. To a static science still enthralled with method, Lamarck's ideas were revolutionary. He did his cause no good, however, when he offered to explain the mechanism of natural change. He argued that an organism is modified structurally simply by the use or disuse of a certain part of its body. Those parts used would

Where Linnaeus was systematic and specific, his contemporary, Buffon, was synthetic and sensational. Buffon's forty-four volume *Histoire naturelle* took all of nature as its subject and offered more original ideas and generalizations then zoological facts. He viewed all life and matter as related and argued that species are mutable. It is significant that nearly two centuries later Pablo Picasso chose Buffon's natural history to inspire him, perhaps because both were interpreters of nature. Here, Piscasso's deer effortlessly conveys the essence of the animal and pays little regard to anatomical correctness. *Eaux-fortes originales [de] Picasso pour des textes de Buffon,* 1942. Pablo Picasso.

develop accordingly and those unused would eventually wither. These traits were then passed on to its descendants, he explained. Despite his weak gropings to explicate this thesis, at its core was the firm notion that the interaction of natural forces was responsible for the variation of species.

Lamarck's evolutionary ideas had no influence on his time, as he lived in an age dominated by the theological notion of the fixity of species, and unfortunately Lamarck seems best remembered today for his incorrect idea that acquired characteristics are inherited. His life was a long one but full of sadness and struggle. He was widowed four times, saw most of his children die, and was poor most of his life. He was buried in a pauper's grave. His famous *Philosophie zoologique* was first published in 1809 in Paris. The Library of Congress has the 1830 edition in its collections.

Lamarck was at times his own worst enemy. His vociferous attack on the new chemistry of Lavoisier won him no friends, and in fact the majestic and dictatorial Georges Cuvier became his lifelong opponent. Cuvier was a child prodigy who had entered the Academy of Stuttgart at the age of fourteen not knowing a word of German and had won a prize for German studies four months later. He was brilliant, disciplined, and energetic, with a forceful but appealing personality that won him the favor of monarchist and revolutionary alike. The imperious Cuvier became the most famous European scientist of his time, literally dictating biological doctrine by the weight of his authority.

Cuvier began his career at the Muséum National d'Histoire Naturelle, where his genius for anatomical detail and his reverence for facts made him the ultimate anatomical arbiter. His most comprehensive work was *Le règne animal,* published in Paris in 1817. The Library has this four-volume work in first edition as well as a much larger version (eleven volumes of text and ten volumes of atlases) published in Paris after his death. In it, Cuvier offers the results of a lifetime of research on both living and fossil animals and divides the animal kingdom into four main groups: vertebrate, mollusca, articulate, and radiate. This breakdown transformed the animal taxonomy of Linnaeus into a truly natural system of classification. Along with his earlier *Leçons d'anatomie comparée,* published in Paris between 1800 and 1805, *Le règne animal* established comparative anatomy, at the core of which was Cuvier's correlation theory, which stated that there was a necessary relationship between one part of the body and another. A famous story illustrates the cool rationalism of his theory. When one of his students dressed in a devil's costume, and woke him with the pronouncement, "Cuvier, I have come to eat you," the other students are said to have heard the great man calmly reply before going back to sleep, "All creatures with horns and hooves are herbivores. You cannot eat me."

It was through Cuvier's expert use of the correlation theory that he founded modern paleontology as well. In his *Recherches sur les ossemens fossiles,* published in Paris in 1812, Cuvier was able to reconstruct entire animals from a few fossil bones and demonstrate that these animals were indeed extinct. Despite his pioneering work with fossils, however, Cuvier remained a steadfast anti-evolutionist. It was on this point that he dis-

In his comprehensive work on the animal kingdom, Cuvier extended the Linnaean system by grouping related classes into broader groups called phyla, using the internal structure of animals as a guide. Here Cuvier shows different species of bats, classified correctly as mammals and carnivores, and displays their teeth and wing structure. His anatomical knowledge was so vast that given a piece of well-preserved bone, he could reconstruct an entire animal. He viewed an animal as a closed system of mutual and reciprocally related parts, none of which could change without affecting the others. Cuvier steadfastly opposed Lamarck's ideas of the evolutionary development of life and never waivered from his belief in the fixity of species. *Le règne animal*, 1836–49. Georges Cuvier.

agreed with Lamarck so strongly, attributing the extinction of species to a series of catastrophic events rather than to a process of evolutionary change. Cuvier was a man who seldom went beyond observed facts, and this particular foray into the realm of theory indicates his speculative abilities were flawed or at least limited. It has been said that Cuvier talked evolution while denying it, and it is the irony of his life that he provided later theorists with some of the essential foundation material for the construction of the modern edifice of biological evolution. The Library has Cuvier's 1812 work on fossils in facsimile as well as the third edition (1825) and the first edition of his work on comparative anatomy (1800–1805).

The first half of the nineteenth century was distinguished not only by major advances in comparative anatomy and paleontology and the beginnings of some progress toward a theory of natural evolution, but also by a major breakthrough in embryology. During the eighteenth century, embryological research was nearly at a standstill, owing to the dominance of the preformation theory. According to this theory, the embryo contained the complete organism in miniature, and since it was much easier to study the grown or mature organism than to study it at its beginnings no attempt was made to study the embryo itself. The startling breakthrough in embryology was the work of one man, Karl Ernst von Baer. Baer was born in Estonia, educated in Germany, and accepted a chair at St. Petersburg. His early work at Königsberg not only founded modern embryology as an independent field of research but was an essential factor in the larger and later evolution equation. It was his work on comparative embryology that offered the means of demonstrating the affinity of different animal forms.

Baer began his scientific career as a physician but turned to research when confronted with the realities of practicing medicine. His comparative studies in 1826 of the ovaries of dogs and other animals led him almost accidentally to the discovery of the mammalian egg—a minute yellowish speck that he found in both the follicle fluid and the oviduct. Most had expected the eggs of mammals to be relative to the known egg size of birds, reptiles, and fish. Baer made his epochal discovery known in a brochure entitled *De ovi mammalium et hominis genesi,* published in Leipzig in 1827. The Library has this work in facsimile only. In his larger work, *Über Entwickelungsgeschichte der Thiere,* he first presented an exhaustive summary of all the known facts on embryology and then surveyed the embryonic development of all the vertebrates. The first volume was published in 1828, and the Library has a copy from the first edition. Volume two was published in an unfinished form in 1837, and the Library has a facsimile copy of it. Volume three was published posthumously in 1888 and contained the material missing from the second volume, edited by L. Stieda, but the Library has no copy of it. All three volumes were published in Königsberg.

Baer was a complex man of wide scientific interests who began more writing projects than he completed. A man of great wit and vigor, he continued his researches until his death at eighty-four and yet is said to have once suffered from such severe depression that he did not leave his

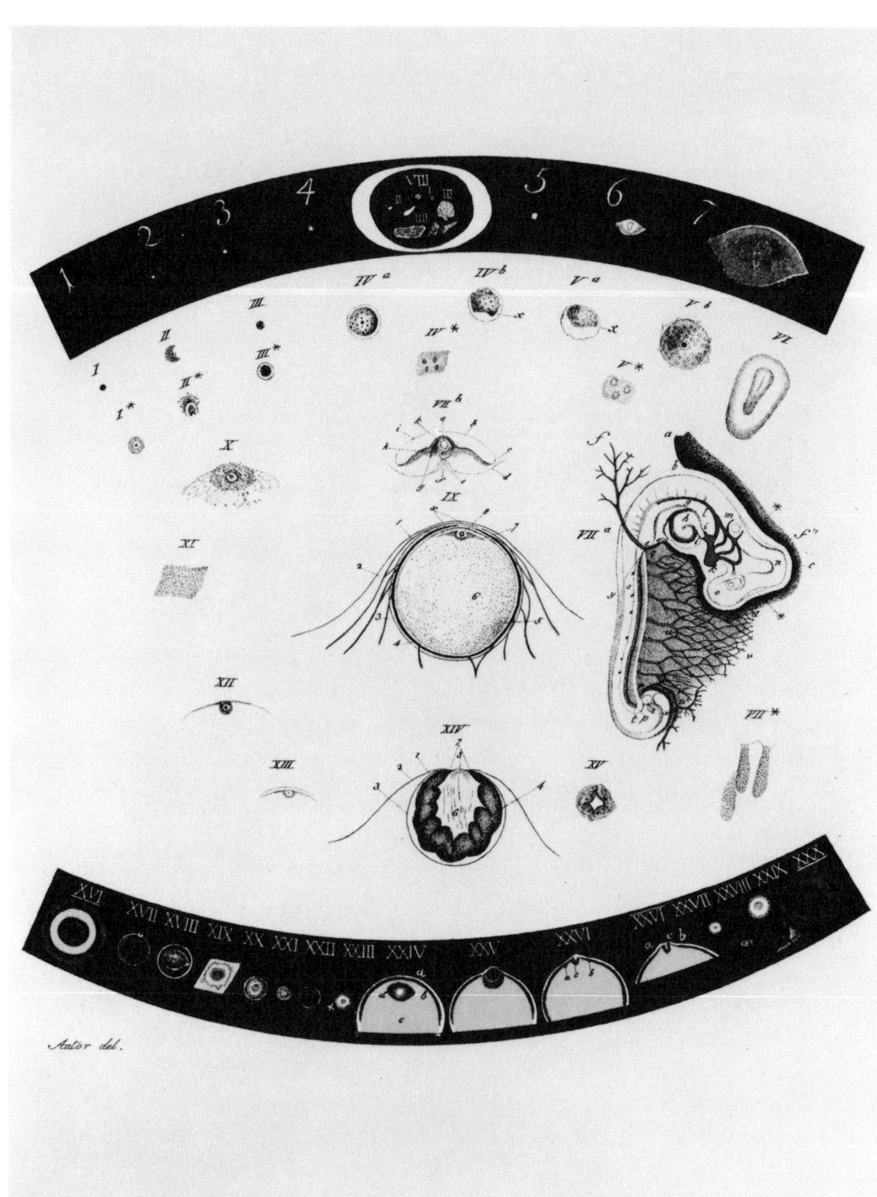

In 1826 Baer discovered the egg of the mammal in the ovary, indicating once and for all that the reproductive processes for mammals (and man) were not fundamentally different from other animals. Baer's later work was on comparative embryology, and his studies revealed that in the early stages of development most vertebrates resembled one another—only later to become more differentiated. Here Baer compares the mammalian egg with eggs of such lower animals as a lizard, a frog, and a crab. *De ovi mammalium et hominis genesi*, 1827 [1966]. Karl Ernst von Baer.

home for a year. The significance of his embryological discoveries and of the generalizations like the germ theory that he formulated cannot be overemphasized. His *De ovi mammalium* gave zoology an important unifying doctrine stated best by Baer himself: "Every animal which springs from the coition of male and female is developed from an ovum, and none from a simple formative fluid." His later synthetic work on the development of animals provided the basis for their systematic study.

One of the major intellectual elements essential to the formulation of zoology's "big idea" of evolution was the formal elaboration of the cell theory. In 1842, the eccentric Matthias Schleiden published the idea that a plant is a community of cells and that the cell is the essential unit of the individual plant (as discussed in chapter 2). During his many travels,

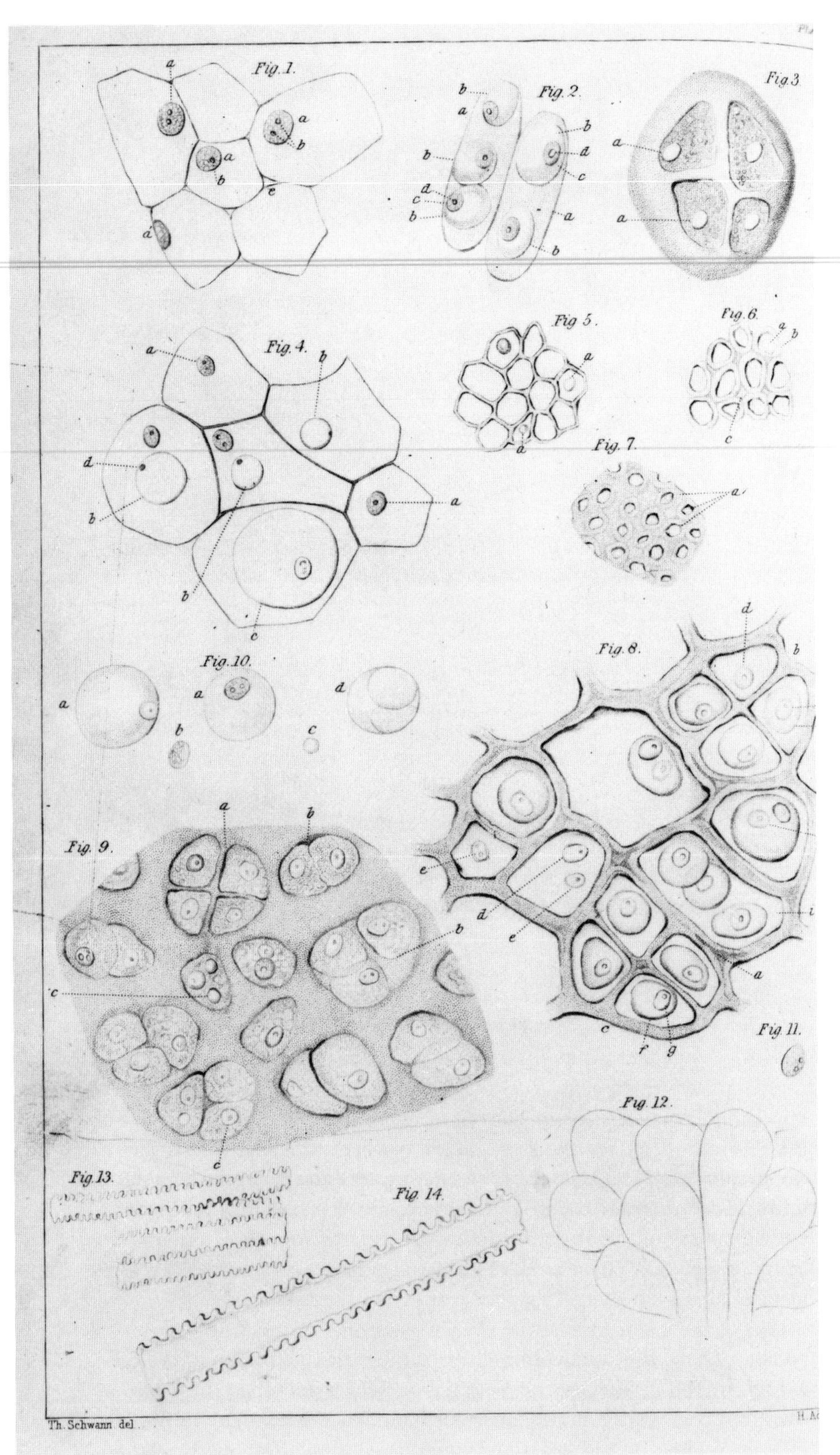

Fig. 1.
Fig. 2.
Fig. 3.
Fig. 4.
Fig. 5.
Fig. 6.
Fig. 7.
Fig. 8.
Fig. 9.
Fig. 10.
Fig. 11.
Fig. 12.
Fig. 13.
Fig. 14.
Th. Schwann del.

Schleiden met a timid, pious scientist in Berlin with whom he somehow struck up a friendship. This quiet introvert was Theodor Schwann, a young student who had been a favorite pupil of the great German physiologist Johannes Peter Müller. In October 1837 Schleiden communicated the unpublished results of his research on plant cells to Schwann, who accepted Schleiden's cell-formation theory in its entirety, applied it to animals, and expanded it into a general theory of the basis of life. Schwann's famous treatise was first published in Berlin in 1839 and was titled *Mikroskopische Untersuchungen über die Übereinstimmung in der Struktur und dem Wachstum der Thiere und Pflanzen*. In it, Schwann points out the fundamental differences between plants and animals and then proceeds to demonstrate their similarities. In the third and final section of this work, he expounds his general cell theory, stating that the cell is the general or universal unit of all life. "One common principle of evolution is laid down for the most highly differentiated elementary parts of the organisms, and this principle of evolution is the cell-formation." Schwann's theory was accepted almost immediately and universally. The Library has the English translation of Schwann's work published in London in 1847—*Microscopial Researches into the Accordance in the Structure and Growth of Animals and Plants*. Schwann's cell theory was a zoological milestone, for it offered a major unifying idea—that the varied forms and phenomena of life are a connected whole—an idea that led to a greater understanding of all living things.

Charles Darwin was a contemporary of Schleiden and Schwann and died within a year of both of them. He owes an intellectual debt to them and to another contemporary, Charles Lyell, as well as to such varied scientific predecessors as Thomas Malthus, Georges Cuvier, Jean Lamarck, Georges Buffon, and even his own grandfather, Erasmus Darwin. Despite these debts, Darwin's theory of evolution was virtually the work of one man. His greatness resides in his theory, which is saying a great deal, for evolution is to biology what Newton's theory of gravitation was to physics. Within their fields, such theories have an immense—sometimes total—unifying and expositional power. They also have a substantial spillover effect in both physical and metaphysical domains. When a grand theory alters how we think or how we regard ourselves and our environment, it is surely a deep, rare, and essentially revolutionary idea.

This is what the Darwinian revolution accomplished, as did its historical predecessors, the Copernican revolution and the Newtonian revolution. The personal story of Darwin's early advantages and unpromising youth, of his chance presence on the *Beagle's* voyage, of his flash of insight and long hesitancy to publish, of his honorable actions with Wallace, and of his chronic ill health and reclusive old age, are all worth noting, for their retelling places a theoretical doctrine in its proper human context.

Charles Darwin was born the fifth of eight children, with two very famous and well-to-do grandfathers, Erasmus Darwin, the physician-poet, and Josiah Wedgewood, of porcelain fame, both of whom died before he was born. During his formal education, he appeared ordinary at best, and his father pushed him first toward medicine and then the church, fearful of

Opposite page:
The cell theory offered zoology a major unifying idea, the idea that animals and plants alike are formed out of cells, that the egg itself is made up of cells, and that the organism develops by division of the egg cell. Here Schwann illustrated the nature and origin of animal and plant cells. *Microscopical Researches into the Accordance in the Structure and Growth of Animals and Plants*, 1847. Theodor Schwann.

having a scandalous wastrel on his hands. The indifferent Charles proved inept in both fields and was intrigued only by his natural science hobbies. These he pursued on his own, although he did receive a good background in geology from Adam Sedgwick. On the recommendation of a friend and through the good offices of an uncle, who helped gain his father's permission, Darwin obtained in 1831 the position of unpaid naturalist aboard the small cruiser, H.M.S. *Beagle,* which was to circumnavigate the world on a mainly cartographic mission.

The five-year voyage was eventful both to Darwin and to posterity, for it gave him his real training as a naturalist and shaped all of his future work. The Library has the *Journal of Researches in the Geology and Natural History of the Various Countries Visited by H.M.S. Beagle* in first edition. Published in London in 1839, the *Journal* contained experiences and observations Darwin took from his diaries and notebooks, and it became a popular success. Among Darwin's many findings and impressions of the long trip, one particularly intriguing observation that he first made on the Galapagos Islands kept presenting itself to him. What could account, he asked, for the slight but obvious differentiation of species that he found on each island? He had, for example, noted fourteen different species of finch, now known as "Darwin's finches," on the Galapagos and observed that these birds existed nowhere else. According to the creation dogma of his time, each separate species would have had to have been created for each island. To Darwin, this seemed unreasonable. More likely, he reasoned, these finches were descendants of birds that had originally strayed from the mainland and that then developed their different beaks to suit better the different feeding conditions of each island.

Darwin was both a perfectionist and a very patient man, and he was not about to publicize an incomplete idea. After returning from his famous voyage in 1836, he came to his idea of "natural selection" as the mechanism of evolution in 1838 while reading Thomas Malthus's *Essay on the Principle of Population* (which the Library has in its 1798 London first edition). He then proceeded to gather data to test his hypothesis—keeping his idea mainly to himself and endlessly collecting and classifying more and more information.

Darwin served as secretary to the Geological Society of London from 1838 to 1844 and became close to Charles Lyell, whose *Principles of Geology* so influenced Darwin, and Lyell was one of the few with whom Darwin discussed his evolutionary ideas during this time. In 1842 Darwin wrote a preliminary sketch of his ideas and by 1844 he actually began to write a book on the subject. By 1858 he was still at work on his book when Lyell's warnings that he would be preempted if he did not publish soon came true. In June 1858 Darwin received a letter from Alfred Russel Wallace that shocked him. Wallace, whose career paralleled Darwin's—he also was largely self-taught, had explored an island region and was struck by the same species differentiation as was Darwin, and had read and been influenced by Malthus—was asking in his letter for Darwin's opinion of a theory of evolution remarkably similar to his own. Although Darwin had spent twenty years in deliberation and preparation of his theory, it was

One of the most popular German works on natural history, medicine, and science during the fifteenth century was *Buch der Natur*. Its depiction of the universe is most unusual in its representation of the planets as contained within straight horizontal bands separating the earth below from the heavens above. *Buch der Natur,* 1481. Konrad von Megenberg.

From *Der scaepherders kalengier,* 1516.

Few books can rival the striking title page of Apianus's *Caesar's Astronomy*. This work is thought to have taken eight years to prepare and contains thirty-seven colorful and detailed volvelles (movable, rotating circles) useful as planetary tables. Although his astronomy was geocentric, Apianus did observe correctly that a comet's tail always points away from the sun. Astronomy aside, the beauty and extravagance of this volume make it a unique prize. *Astronomicum caesareum*, 1540. Petrus Apianus.

A study of trees from Konrad von Megenberg's *Buch der Natur* shows some trees growing naturally and others cultivated in pots. This is one of the first printed books in which a woodcut shows plants mainly to illustrate and not only to decorate. *Buch der Natur*, 1481. Konrad von Megenberg.

This difficult-to-recognize plant is the *Plantago* or Plantain, one of three genera in the family Plantaginaceae (order Scrophulariales). The snake at the bottom and the scorpion at the top indicate that this plant is an antidote for their bites or stings. Such devices were used frequently in medieval times, since few could read, and its inclusion here indicates that the *Herbarium* was taken from a very old source. *Herbarium*, 1484. Apuleius Barbarus.

Alii porno. Alii Polypleron.
Naſcitur in paludibus plurimum & pratis.

AD CAPITIS DOLOREM.

Herbae Plãtagis radix collo ſuſpẽſa dolorẽ mire tollit. AD VENTRIS DOLOREM.

Herbae Plantaginis ſucus tepefactus fometãdo uentris dolorẽ tollit mire: & ſi tumor fuerit: tu/ſa & impoſita tollet tumorem.

Jean Jacques Rousseau's popular botanical work, *Essais élémentaries sur la botanique* (Paris, 1771), contributed to the late eighteenth-century movement toward a more scientific look at plants and flowers. In 1805, that work was reissued with sixty-five plates printed in color after paintings by Pierre Joseph Redouté. The dandelion shown here, like all of the plates in this work, was produced as a stippled engraving printed in color and finished by hand. Rousseau was not all science when he considered a striking flower and likened it to a beautiful woman: "The sweet fragrances, the lively colors, the most elegant shapes seem to vie with one another for the right to hold our attention. One need only to love pleasure to abandon oneself to such sweet sensations." *La botanique de J. J. Rousseau,* 1805. Jean Jacques Rousseau.

Pierre Joseph Redouté has been called the Raphael of flowers, the greatest flower painter of all time. This spectacular rose from *Les roses* demonstrates his genius. The harmonious beauty and elegance of his artistry in no way detracts from the forthright depiction of a botanical specimen. The Library's copy of *Les roses* is one of only five large folios ever made and was Redouté's own copy. Redouté's success and popularity seemed independent of the political turbulence of his times. During his long career, he served as Master of Drawing to a succession of queens, empresses, and claimants, from Marie-Antoinette in the 1780s, through Napolean's Josephine, to Marie-Amelie in the 1830s. At his death, a critic paid him a unique tribute, saying, "He composed a bouquet with the intelligence and the happiness of a young girl at her first ball; and yet he brought about those delicate masterpieces with the thick hands which resembled the feet of some antediluvian animal." *Les roses,* 1817–24. Pierre Joseph Redouté.

Printed and published in London, Mark Catesby's *Natural History of Carolina, Florida, and the Bahama Islands* was the earliest book with colored plates of American birds. This vibrantly colored flamingo was drawn, engraved, and hand-colored by Catesby, who could not afford to hire an engraver. Nearly half of the volume's 220 hand-colored engravings are of birds. *The Natural History of Carolina*, 1731–43. Mark Catesby.

Alexander Wilson's *American Ornithology* initiated the serious study of birds in America. This founder of American ornithology was born in Scotland and wrote poetry while working as a weaver and peddler. Although much less well-known, Wilson's work preceded Audubon's by nearly twenty years and certainly inspired Audubon. *American Ornithology*, 1808–14. Alexander Wilson.

Aside from depicting birds in motion and rendering the subtle implication of vitality and movement to those he painted motionless, perhaps the most striking thing about Audubon's bird drawings is their scale. Each drawing is over three feet high. This combination of artistry and scale enabled him to convey the grand dimensions of the American continent and its indigenous birds, here the Golden Eagle and the American White Pelican. *The Birds of America*, 1827–38. John James Audubon.

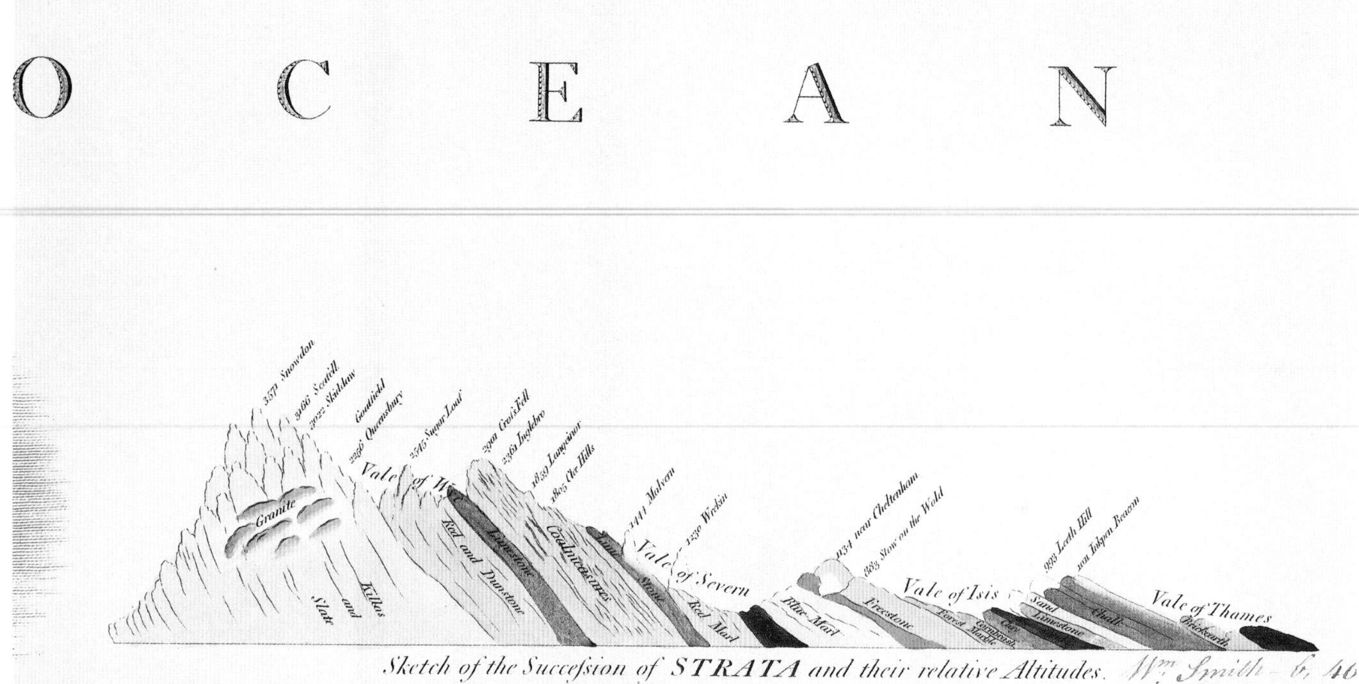

This detail and key from William Smith's large-scale geological map of England show his contribution to geology. It was the first such map of any country and was based on scientific principles discovered by Smith himself. The conditions of his native land were particularly favorable to Smith's work since it was possible to find sedimentary rocks of every age—from Precambrian to Tertiary—that were relatively undisturbed. Smith not only discovered this regular succession in the strata of England but he determined that many of the individual beds contained a particular fossil content that could be used to distinguish them from other beds. Smith made these discoveries known in a unique and very understandable way, namely by using different colors in his map to indicate the succession of sedimentary beds or groups of beds. The depth of his understanding is revealed by his novel use of a deeper shade of color along the base of a formation that indicated how the beds were superimposed. Smith thus added dimension to his map by indicating a useful structural factor. *A Delineation of the Strata of England and Wales with Part of Scotland*, 1815. William Smith.

MOUTH of the SEVERN
NEL
BRIDGWATER BAY
BRENDON HILL
MARLBRO DOWNS
LAMBOURN DOWNS
LISLEY DOWN
STUDLAND BAY
ISLE of WIGHT
SANDOWN BAY
SPITHEAD
FOREST of BERE
WINDSOR

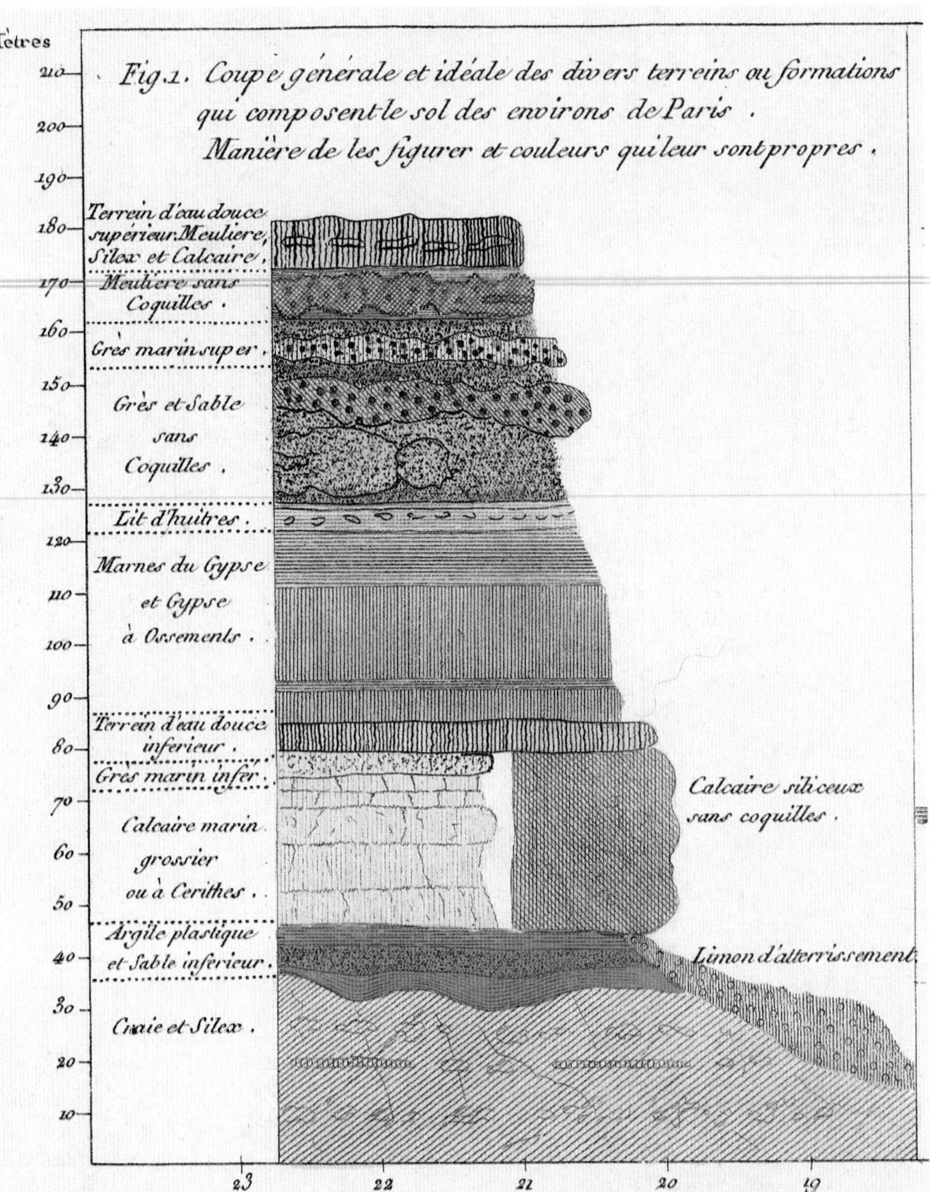

112 This geological map depicts the stratigraphy of the country around Paris, which came to be known as the Paris basin. It was the result of the collaborative work of Georges Cuvier and Alexander Brongniart, although Brongniart did most of the work. Their geological studies of that area revealed alternating layers—one rich in marine shells, one with remains of land animals only, and another with no fossils at all—which indicated that the sea had twice drained from the land. The larger meaning of this succession of distinctly different strata became clear when they noticed that the stratigraphically lower (and thus older) animals were also lower on the taxonomic scale and often quite unlike those in the higher beds. The two scientists were then able to demonstrate how fossils might be used to determine accurately the geological chronology of a particular area. *Essai sur la géographie minéralogique des environs de Paris*, 1811. Georges Cuvier and Alexander Brongniart.

obvious that Wallace had arrived at the same conclusion, albeit independently and almost overnight.

Taking the good counsel of his friends, Darwin suggested joint publication of a paper to Wallace, with the result that in 1858 the *Journal of the Proceedings of the Linnean Society* published their paper, "On the Tendency of Species to Form Varieties; and On the Perpetuation of Varieties and Species by Natural Selection," which can be found today in the Library's complete set of the society's *Journal.* The article contained both Wallace's paper and passages from Darwin's unpublished book, as well as an abstract of a letter Darwin wrote to Asa Gray (Harvard professor of natural history) in 1857 that capsulized his theory.

The Wallace experience apparently provided Darwin with sufficient motivation to publish, and near the end of 1859 his book appeared. Only one-fifth as long as he would have preferred, his 502-page book was titled *On the Origin of Species by Means of Natural Selection; or, The Preservation of Favoured Races in the Struggle for Life.* Only 1,250 copies were printed in London that year and the Library of Congress collections contain one of them. With this work, Darwin demonstrated the breadth of his genius and left Wallace far behind. Where Wallace had taken two days of insight to produce his short paper, Darwin had amassed twenty years of example and evidence. Darwin remained ever the scientist, always critical of unsubstantiated fact, whereas Wallace was by nature more imaginative and uncritical. Wallace's later absorption with seances, his refusal to apply his own theory to mankind, and his odd crusades against such things as vaccination did little to maintain his scientific reputation on a par with Darwin's.

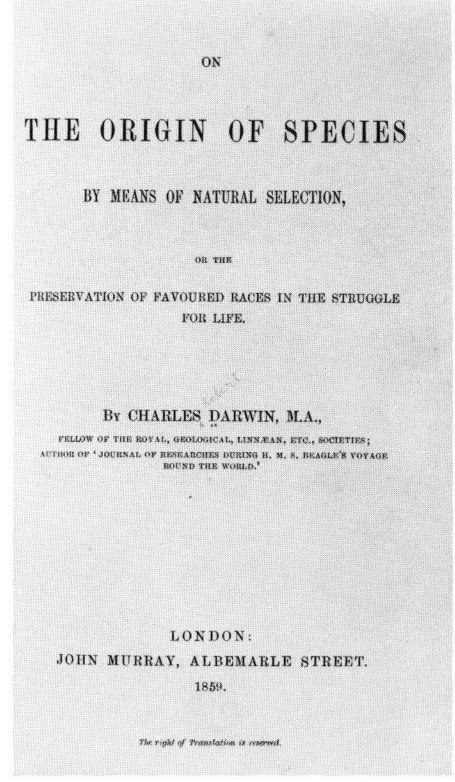
ON

THE ORIGIN OF SPECIES

BY MEANS OF NATURAL SELECTION,

OR THE

PRESERVATION OF FAVOURED RACES IN THE STRUGGLE FOR LIFE.

BY CHARLES DARWIN, M.A.,

FELLOW OF THE ROYAL, GEOLOGICAL, LINNÆAN, ETC., SOCIETIES;
AUTHOR OF 'JOURNAL OF RESEARCHES DURING H. M. S. BEAGLE'S VOYAGE ROUND THE WORLD.'

LONDON:
JOHN MURRAY, ALBEMARLE STREET.
1859.

The right of Translation is reserved.

Darwin's *Origin of Species* gave zoology a unifying, encompassing idea based on natural causes and processes. Overall, it stated that change is the natural order of things and that the present is the product of the past. *On the Origin of Species*, 1859. Charles Robert Darwin.

Darwin continued to grapple boldly with the most difficult of problems, producing in 1871 *The Descent of Man, and Selection in Relation to Sex.* In his *Origin of Species,* Darwin was intentionally oblique when dealing with the issue of human beginnings, but in this later work, published in London, Darwin masses his evidence that mankind too is the product of evolution from lower forms of life. The Library of Congress collections contain this work in its first edition.

Darwin's theory of evolution is both simple and ingenious. It states basically that change is the natural order of things and that the present is but the product of the past. First, Darwin applied Malthus's thesis to the entire animal and plant world, saying, "As many more of each species are born than can possibly survive, and as, consequently, there is a frequently recurring struggle for existence, it follows that any being, if it vary however slightly in a manner profitable to itself under the complex and sometimes varying conditions of life, will have a better chance of surviving, and thus be naturally selected. From the strong principle of inheritance, any selected variety will tend to propagate its new and modified form." The first assumption upon which this statement was based, namely that organisms in fact continually vary, was demonstrated by Darwin himself. The second assumption, that surviving organisms will genetically transmit their advantageous variations, was later proven by the science of genetics, which Gregor Mendel was even then developing.

The deep and pervasive influence of Darwinism was felt not only in every field of zoology and botany but in nearly every discipline of the arts and sciences alike. At times, it has seemed that the nonscientific fields of politics and religion have been even more affected by it than the sciences themselves. Darwin gave mankind a great new concept of nature and of life itself, one based on natural causes and processes, and one that consequently was truly comprehensible. The legacy of his genius is by no means a gloomy, determined future but rather a vision that celebrates the variety of life. Darwin's view is positive and optimistic about our ability to understand more about ourselves and our world. The dynamism essential to evolutionary theory assures that zoology, like all other sciences, will never be a closed book. The study of animals is an intrinsically changing and changeable discipline.

Apart from the classifiers, experimentalists, and theorists who made zoology a real science, two other types of zoological investigators made it an exciting and inspiring endeavor. They are the explorers and the illustrators. In the eighteenth century, it became the practice to take naturalists and artists on ocean expeditions to observe, collect, and document the plant and animal life of more exotic parts of the world. The earliest of these major ocean voyages was the Pacific trip of Capt. James Cook, begun in 1768 under the auspices of the Royal Society. The young Joseph Banks accompanied Cook on the first of his three trips to the South Seas, and Banks, who later became president of the Royal Society, took along the naturalist Daniel Solander and two artists. From this first major expedition, which was motivated primarily by sincere astronomical and geographical goals in addition to its ambitions for empire and trade, there arose the tradition of taking the naturalist and artist to his subject rather than the other way around. Later exploratory voyages famous for their natural history were almost certainly inspired by Cook's circumnavigation and the wealth of exciting, new information provided by his trips. The five-year expedition of Humboldt to the Americas in 1799 was a triumph for that natural historian and inspired the young Charles Darwin. Cook's great voyages during the eighteenth century were the forerunners of all subsequent ocean expeditions. His trips caused a sensation in Europe and resulted in numerous publications, many of which contain valuable zoological information. The Library has in first edition Cook's own accounts of his trips to the South Pole and the Pacific, published in 1777 and 1784, respectively. Also in the Library's collections in first edition are the three volumes John Hawkesworth compiled from Cook's journals and Banks's papers, entitled *An Account of the Voyages . . .*, published in London in 1773. But the first complete version of Cook's journals was not published until 1955. The Library has this work, *The Journals of Captain James Cook on His Voyages of Discovery,* which was edited by John Cawte Beaglehole and published in four volumes in Cambridge. Other publications that resulted from this voyage contributed to the science of botany.

In addition to the literature of the Cook voyages, the Library has the report of the Wilkes's expedition, *Narrative of the United States Exploring Expedition,* published first in five volumes in Philadelphia in 1844. This

The people encountered on Captain Cook's first voyage around the world were often more strange and exotic to the voyagers than the flora and fauna. Cook described this highly ornamented chief as "punctured, or curiously tattowed, from head to foot." In the center of his feathered headdress is a mother-of-pearl shell decorated with tortoise-shell. Santa Christina is one of the five islands which make up the Marquesas Islands north of Tahiti and Pitcairn in the central Pacific Ocean. *A Voyage towards the South Pole*, 1777. James Cook.

Pacific trip was the first of its kind authorized by the U.S. Congress. Also among the published results of the more famous expeditions is *The Zoology of the Voyage of H.M.S. Beagle,* admirably edited by Charles Darwin and published in London in five volumes from 1839 to 1843. The Library has a complete set. It also has the massive, forty-volume *Challenger Report,* issued between 1880 and 1895 and written mostly under the direction of the famous biologist John Murray. The H.M.S. *Challenger* left Portsmouth, England, on December 21, 1872, for a three-and-a-half-year expedition around the world. This highly organized voyage accumulated a huge amount of oceanographic, botanical, and zoological information, which was published over a fifteen-year period. Thirty-seven of its volumes deal solely with zoology.

By the time of the *Challenger* expedition, the wildlife photographer had replaced the natural history artist. Until then, however, it was the zoologi-

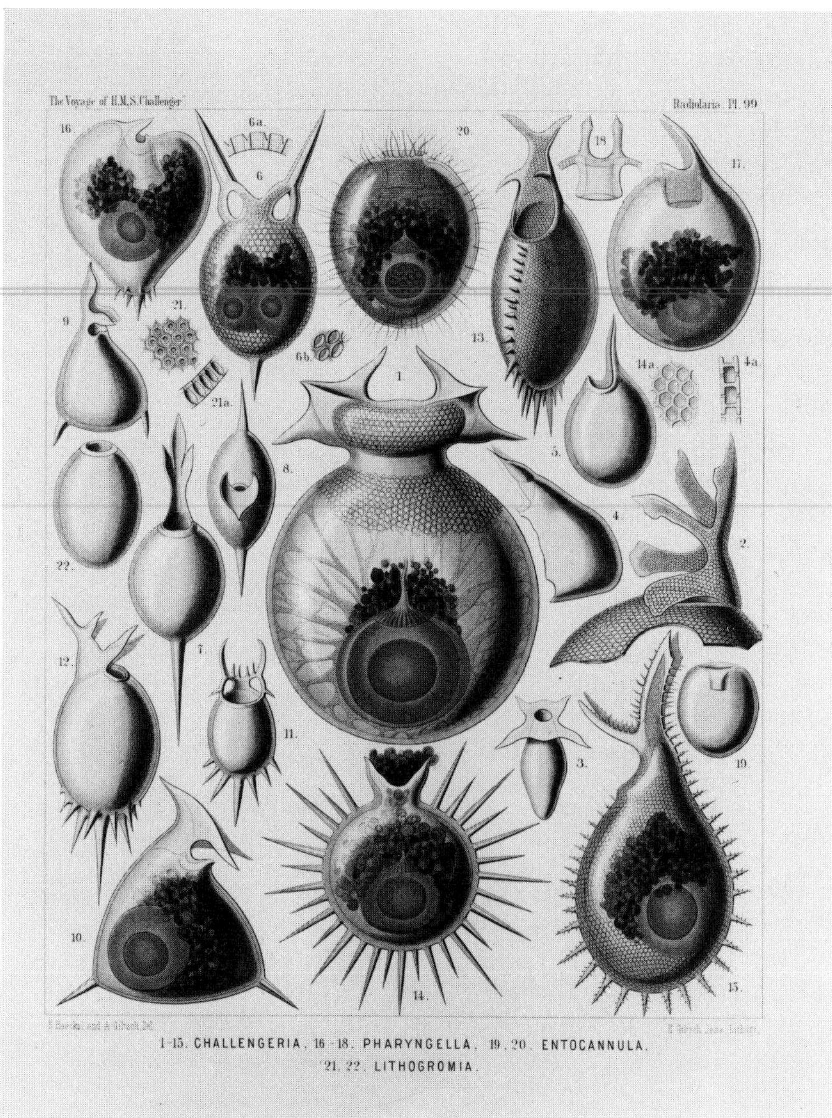

Many of the zoological discoveries made during the *Challenger*'s world voyage were of organisms too small to be seen by the naked eye. Here, Ernst Haeckel depicts fifteen species of radiolaria, all named after the famous vessel, as well as seven other kinds. Altogether, Haeckel described 144 new kinds of these one-celled marine animals characterized by their spikes and spines. *Report on the Scientific Results of the Voyage of H.M.S. Challenger*, 1880–95.

cal illustrator who filled the important and primary role of depicting what various animals actually looked like. By the beginning of the eighteenth century, zoological illustrators were becoming increasingly intrigued by the wealth of new animal subjects available in the expanding colonies of English and French North America.

Colonial America was a zoological wonderland, where new species were constantly being discovered. The first major work of this colonial period was produced by Mark Catesby, who was born in England. He died well before the American Revolution broke out, but his work is genuinely American. The Library has the two-volume *Natural History of Carolina,* published in London in 1731–43, in first edition. This large work contains 220 hand-colored etchings of birds and other animals done by the largely self-trained Catesby. Acknowledged as the first real naturalist in America, Catesby contributed tremendously to Europe's knowledge of the New

World and provided later illustrators with a solid body of work upon which they could build.

The first significant American-born naturalist illustrator was William Bartram, whose *Travels through North & South Carolina, Georgia, East & West Florida* had a remarkably strong effect on English romantic thought and writing. As the son of John Bartram, who had been appointed American botanist to King George III, William pursued his father's interest in natural history and became a very competent ornithological artist. The Library has his *Travels,* a small book published in Philadelphia in 1791, in first edition. He was perhaps the first American ecologist—one whose balanced sense of nature saw every plant and animal as a living entity intimately related to its environment.

From *The Natural History of Carolina,* 1731–43. Mark Catesby. See p. 106.

Catesby's book provided valuable natural history information as well as exuberant illustrations. The green snake in the picture, he says, catches flies and insects, is harmless, and is easily tamed. The plant, *Caffena vera Floridanorum,* around which the snake is coiled, is a favorite of the Indians who make a drink from its leaves, "which they drink in large quantities as well as for their health as with great gust and pleasure." *The Natural History of Carolina,* 1731–43. Mark Catesby.

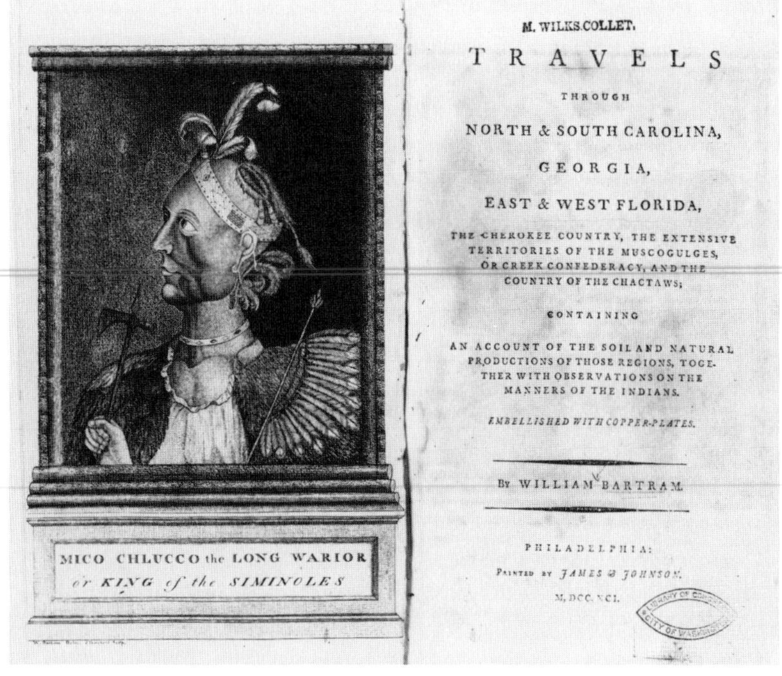

MICO CHLUCCO the LONG WARIOR
or KING of the SIMINOLES

M. WILKS COLLET.

TRAVELS

THROUGH

NORTH & SOUTH CAROLINA,

GEORGIA,

EAST & WEST FLORIDA,

THE CHEROKEE COUNTRY, THE EXTENSIVE TERRITORIES OF THE MUSCOGULGES, OR CREEK CONFEDERACY, AND THE COUNTRY OF THE CHACTAWS;

CONTAINING

AN ACCOUNT OF THE SOIL AND NATURAL PRODUCTIONS OF THOSE REGIONS, TOGETHER WITH OBSERVATIONS ON THE MANNERS OF THE INDIANS.

EMBELLISHED WITH COPPER-PLATES.

By WILLIAM BARTRAM.

PHILADELPHIA:

PRINTED BY *JAMES & JOHNSON.*

M, DCC, XCI.

Bartram's *Travels* is a loosely organized book filled with generalized botanical, zoological, and anthropological descriptions. In a somewhat rambling and personal manner, Bartram would offer a description of a water rat or a wild turkey, then one of a rhododendron, and then conclude the passage by recounting an exciting adventure with a crocodile. Throughout, Bartram's directness, enthusiasm, and innocence enliven his writing. He stated his view of the world in the first page of his introduction: "This world, as a glorious apartment of the boundless palace of the sovereign Creator, is furnished with an infinite variety of animated scenes, inexpressibly beautiful and pleasing, equally free to the inspection and enjoyment of all his creatures." *Travels through North & South Carolina*, 1791. William Bartram.

From *American Ornithology*, 1808–14. Alexander Wilson. See p. 107.

Born in Scotland, Alexander Wilson is the greatest of Audubon's precursors. After his arrival in America in 1794, Wilson met William Bartram, who encouraged him to undertake a serious study of birds. Wilson was never well off financially and was tubercular besides, but he nevertheless produced nine folio volumes that rendered obsolete all previous books on ornithology. *American Ornithology* initiated the serious study of birds in America.

Like Catesby, Wilson taught himself engraving, but he found his engravings to be much weaker than his drawing. He therefore commissioned Alexander Lawson, a fellow Scot, to prepare plates from Wilson's drawings. Wilson's work on American birds has both decorative and scientific significance. *American Ornithology* was published in Philadelphia between 1808 and 1814 and was the product of ten years of feverish traveling, drawing, and writing.

The Library of Congress collections include Wilson's nine-volume work in first edition along with a similarly titled book that has become inseparable from it. Between 1825 and 1833, Charles Lucien Bonaparte, who was the nephew of the Emperor Napoleon, issued his own "supplement" to Wilson, entitled *American Ornithology; or, The Natural History of Birds Inhabiting the United States, Not Given by Wilson.* The Library has all four volumes, published in Philadelphia, in first edition. Bonaparte had settled in Philadelphia and was a respected naturalist, although no artist. He therefore commissioned Alexander Lawson to engrave the plates of Ramsay Titian Peale and Alexander Rider, who contributed the majority of the watercolors in the four volumes.

John James Audubon was their contemporary and indeed had met both Alexander Wilson and Charles Lucien Bonaparte. Wilson eventually became his feuding rival, while Bonaparte was to offer encouragement and publishing advice. Despite an inauspicious early career and a string of artistic hardships and difficulties, Audubon became the most famous of all bird artists, and probably the best known natural history illustrator that ever was. As an artist of birds, however, he is simply without peer. His magnificent four-volume elephant folio *The Birds of America* is breathtaking, showing many birds as large as life. The production of this unique folio is a story of perseverance and hard work, qualities which color the story of Audubon's entire life and career.

Everything about Audubon's beginnings, not at all ordinary, has a romantic aspect. He was born in 1785 in Les Cayes, Santo Domingo (now Haiti), the illegitimate son of a French naval officer who had fought with George Washington at Yorktown. He was raised in France and took over his father's Philadelphia plantation at the age of eighteen to escape conscription into Napoleon's army. His repeated failures in business are attributed to his preference for the artist's pen and brush over the ledger. Audubon had been drawing birds since he was fifteen, and he found himself drawn increasingly to a life of excursions, observations, and drawings. Gradually, he formulated what he later called his "Great Idea"—to travel throughout America and to document, in a life-size format, every species of continental bird. At the age of thirty-five he finally gave himself over to his ornithological passion and set out his travels. Four years later he had compiled an extensive portfolio of bird drawings and was in search of a publisher. During this time, his wife helped support the family with her earnings as a governess, and Audubon drew portraits and painted street signs. The prospects of finding an American publisher were not good. Not only were Wilson's nine volumes available, but Bonaparte's supplement was being published, and it was Bonaparte who advised Audubon to seek a publisher in Europe.

Left:
Golden Eagle. From *The Birds of America,* 1827–38. John James Audubon. See p. 108.

Right:
American White Pelican. From *The Birds of America,* 1827–38. John James Audubon. See p. 109.

In 1826 Audubon sailed to Liverpool and then on to Edinburgh and London, all the while obtaining subscribers for a future book of engravings. His spectacular *Birds of America* was produced and published in England. The first volume appeared in 1827 and the entire work was completed by 1838. The four volumes contained 435 hand-colored aquatint plates. Audubon was extremely fortunate in his selection of an engraver, the sympathetic and talented Robert Havell. William Lizars had begun the project in Edinburgh and actually did the first ten plates, after which Havell took over completely in London. The great size of the book, an elephant folio, certainly is a major factor in its stunning effect. Over three feet high, the plates give a real sense of the openness and large scale of America.

The exact number of complete sets of *The Birds of America* elephant folio is not known, but evidence suggests there were fewer than 200 and more than 175. The Library of Congress is fortunate in having in its collections two complete sets from this original printing. In fact, the Library appears in the subscription list Audubon appended to the fifth and final volume of his *Ornithological Biography*. The list is divided into American and European subscribers, and first among the American subscribers is "Library of Congress of the United States." In an exhaustive history of the elephant folio, Waldemar H. Fries says that Edward Everett rather than Librarian John S. Meehan signed for this subscription. Everett was a congressman from Massachusetts and chairman of the Joint Committee on the Library. Fries also notes that the Library's second set was given to it in 1929 by the Army War College, which had somehow inherited the copy from the State Department, another subscriber.

The Library also has in first edition Audubon's companion to the elephant folio, the *Ornithological Biography,* published in Edinburgh between 1831 and 1839. This five-volume work contained the bird descriptions not found in *The Birds of America,* in which it had been deliberately decided to forego any text in order to circumvent the British Copyright Act of 1709, which mandated that free copies of books with text be furnished to nine libraries in the United Kingdom. The production of each copy of the elephant folio was extremely expensive, and a subscriber paid approximately a thousand dollars for a complete set.

Many have accused Audubon of being a successful self-advertiser and a poor zoologist. To this could be added criticism of his quaint spoken English or his naive prose style. Nonetheless, the strikingly beautiful plates in *The Birds of America* capture the lifelike qualities of their subjects, so essential to successful zoological illustrations. Audubon knew it was the suggestion of movement that brought his drawings to life, and he was able to capture the liveliness of the birds sensitively and subtly. His work is an American zoological treasure.

As with Darwin's theory or the classification schemes of Linnaeus, Audubon's illustrations are in their way transcendent. Each of these achievements is the result of an individual's partial but tenacious intellectual grasp of a portion of nature's secrets.

Zoology was the latest of all the major disciplines to bloom scientifically. Not until well past the middle of the last century did zoology embrace the oneness of life. Before then, its story was a confused tale of false starts, dead-ends, and stultifying dogmas. Confronted by the lushness and rich variety that characterized the phenomenon of animal life, men were for a very long time defeated by their subject. With Darwin's great mental leap, however, zoology was given a single encompassing idea through which we might perceive and interpret the mysteries of animal life. Darwin's recognition of the oneness of nature gave zoology its tradition of unity and enabled us to see and to accept the primal bond of kinship between ourselves and the animal world.

ANDREAE VESALII
BRVXELLENSIS, SCHOLAE
medicorum Patauinæ professoris, de
Humani corporis fabrica
Libri septem.
CVM CAESAREAE
Maiest. Galliarum Regis, ac Senatus Veneti gratia &
priuilegio, ut in diplomatis eorundem continetur.
-IO-

BASILEAE.

4. *Medicine: The Healing Science*

The story of medicine is a never-ending tale of mankind's journey of self-discovery. On the walls of the Temple of Delphi is the Greek precept, "Know thyself"—and no more exciting or worthwhile pursuit is available to man. From its beginnings, medicine sought not only to ameliorate the hurt or to cure the disease, but, more significantly, to attain a real understanding of its most mysterious and complex subject—the human animal.

No other scientific discipline has as its object of study such a dynamic, delicate, and self-aware living mechanism as the individual human being. And no other scientific pursuit can in any way approximate the essential identity of the observer and the observed that characterizes medicine. The astronomer is not a planet; the botanist shares little with his sedentary but colorful subjects; the geologist treads upon the earth yet is separate from it. But the physician who gazes upon and studies the subject of his science looks at and touches himself. No other science has this intrinsic empathy, this actual species-sharing experience that allows the observer almost to participate in the physical or even the emotional existence of the one being observed. After all, what doctor has never felt ill or has always been free of pain or of guilt? At bottom, both doctor and patient are essentially living laboratories, separated only by the knowledge possessed and the social role performed by the former.

The irony of medicine, however, is that despite this intimate familiarity and identification of observer and observed, medical history is perhaps best characterized by a misunderstanding and estrangement of man from himself. Until relatively recently, man knew more of the motion of the planets than he did about the coursing of blood through his veins. Man's eyes turned first to the heavens and only later did he begin to gaze with reflection upon himself. All of this is to say that medicine as a scientific activity is both young and unique.

Its relative youth is attributed to its long-term dominance by magic and religion and even more to the esteem accorded traditional medical beliefs and to their remarkable tenacity. Few sciences have deferred so totally to the doctrinaire and the dogmatic. Save for the early Greek emphasis on reason over dogma, medicine was dominated by the torpid and the retrograde until the Renaissance. Notably, it was not until the Renaissance with its greater emphasis on experience over authority that individuals began to regard the human body with requisite detachment and curiosity. The regular dissection of cadavers demanded a certain sangfroid that could be obtained through the rationalizing imperative of the pursuit of truth. It

Opposite page:
De humani corporis fabrica, 1543. Andreas Vesalius.

was in this Renaissance context of respect for nature and a return to observing the natural world that medicine began to wrench itself free from the stifling inertia of its past. By far the most concrete example of a traditional religious proscription inhibiting medical advance is the Koran's prohibition of any dissection of the body or even of its representative depiction. This led of course to the great voids in anatomy and physiology in Arab medicine. The western medieval mind restricted medicine's progress in a similar way. To the Scholastics, the body was relatively unimportant in the overall scheme of things, it being the temporal, corruptible counterpart to the immortal soul. The human body thus connoted all that was imperfect in the material world. With medieval overemphasis on the hereafter, any elaborate concern with the corporeal needs of mankind was considered decidedly superfluous.

Some might argue that medicine and modern society in general have swung too far in the direction of the material present, but there is no doubt that medicine today is more connected with the essential needs of life than any other scientific discipline. This necessary concern, even preoccupation, with the fundamental physical needs of mankind makes medicine at once familiar and special. Of all the sciences, none makes such good use of the irrational or the intuitive, using methods sometimes questionable, even quixotic. No forthright physician would disagree with the statement, "If it works to the benefit of the patient, use it," and few can claim never to have used a placebo in some manner. Nor has any other science so incorporated what is in reality an art as such an integral part of its practical methodology. The modern physician still conducts what is called the "art of diagnosis" based on what might be described as the intuition of the skilled craftsman.

Since the story of medicine is so essentially a human story, it has—more than most disciplines—all of the texture and variety characteristic of human history. Its literature therefore reflects all the drama and richness endemic to human interaction and the erratic and colorful course of mankind. For unlike any other science, it is at home both in the laboratory and in everyday life, on the battlefield or on the birthing-chair, in life and in death. It concerns itself with the unseen microbe that plagues its human host as well as the irksome thorn festering in the thumb. It seeks to understand the mystery of human behavior by studying the mind as well as the body. In sum, medicine is as complicated and as simple as its human subject.

In the following treatment of some of the more significant items in the medical collections of the Library of Congress, certain historical themes recur that are peculiar to the study of medicine. Most obvious is the tenacity of basically false traditional conceptions as to how our bodies work. The corollary to this is certainly how ignorant man was for how long about himself. As an adjunct to this theme, and offering some explanation, is the reality of how difficult the job of piecing together the puzzle of man is. Certain things are simply too complex to easily sort out and others have demanded the attainment of a certain level of technology to explore. Discovering the basic nature of infectious diseases had to await the microscope, for example. As for the function and purpose of our organs and the nature of their relationships, imagine the difficulty confronting the Renaissance anatomist attempting to find and then to determine the purpose of one of the smaller glands like the pituitary or the pineal gland, whose functions even today are unclear. One can almost imagine the frustrated investigator giving up and referring once again to the ancient texts of Galen which purported to explain all.

Another theme that emerges concerns medicine as a relatively young science that is nevertheless practically as old as man himself. Medicine was one of the earliest true professions, the medieval universities having had but three faculties—law, divinity, and medicine. Significant too is the fact that so many great men of science began as physicians—Roger Bacon, Copernicus, Agricola, and William Gilbert, to name a few.

Although medicine would have made few major advances had it not embraced the scientific method, it might be said to owe it the smallest debt of all the sciences. For unlike most scientific disciplines, medicine sometimes admits of what might be considered the unscientific, albeit grudgingly and under cover. A recognition of the mind-body nexus and all of its implications and ramifications should remind us that as long as medicine deals with such a dynamic, evolving, and self-conscious subject as the individual human being, no prescribed boundary or definitive system could contain it. Repeatedly in the history of medicine, yesterday's dogma has become today's error—with supposed error often emerging as new truth. In medicine, change must be endemic.

The history of medicine traditionally begins with Hippocrates. The heir of a medical tradition whose origins were in Babylon, Assyria, Italy, and ancient Egypt, Hippocrates was a Greek physician from the island of Cos. Few particulars are known of his life, but the founding of a school of medicine on that island is recognized as his most significant achievement. Based on reason and observation, the school's medical philosophy and practice regarded disease as a physical phenomenon with natural, not supernatural, causes. Called the Hippocratic tradition, this rational system became the cornerstone of medicine after its Renaissance revival. The Greek medical writings known as the Hippocratic collection, numbering between fifty-nine and one hundred works, are now considered to be the collected work of several generations of his school. Although it is not known for certain whether Hippocrates wrote any of the works, they are nonetheless written in the best Hippocratic tradition of reason, observation, and moral conduct.

The first complete Latin edition of Hippocrates was published in 1525 in Rome under the auspices of Pope Clement VII. The Greek "editio princeps" followed the next year in Venice. The earliest of these collected works at the Library of Congress is the 1546 *Opera quae ad nos extant omnia,* printed in Basel in Latin. The Library also has the 1624 *Opera*

This title page is from a 1546 Latin edition of the works of Hippocrates of Cos published by Hieronymus Froben. Froben's father, Johann, was the Basel printer famous for his close friendship with Erasmus. After their meeting in 1514, Froben printed nearly all of Erasmus's work. The family device, two serpents winding up a hand-held staff on which a bird is perched, is said to have been taken from Matthew 10:16, "Be ye therefore wise as serpents, and harmless as doves." Interestingly, the family's device is most appropriate to this collection, resembling somewhat the staff of Asclepius, the only true symbol of the medical profession. *Opera quae ad nos extant omnia*, 1546. Hippocrates.

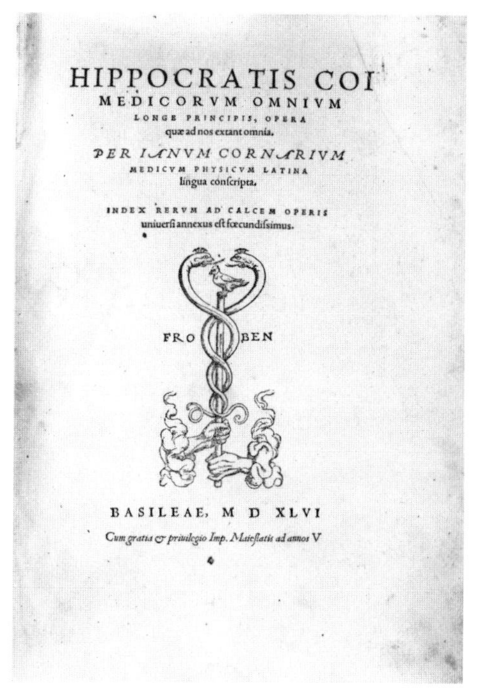

HIPPOCRATIS COI
MEDICORVM OMNIVM
LONGE PRINCIPIS, OPERA
quæ ad nos extant omnia.
PER IANVM CORNARIVM
MEDICVM PHYSICVM LATINA
lingua conſcripta.
INDEX RERVM AD CALCEM OPERIS
uniuerſi annexus eſt fœcundiſsimus.

BASILEAE, M D XLVI
Cum gratia & priuilegio Imp. Maieſtatis ad annos V

omnia quae extant published in Frankfurt, which presents the text in Greek and Latin in parallel columns. Finally, among the collected works in the Library is the magnificent *Oeuvres complètes,* edited by Émile Littré and published in Paris between 1839 and 1861. The Library has volumes 1–6 of the ten-volume set. This triumph of modern scholarship contains both the Greek text and the French translation, critical notes, and special introductions to each separate treatise. It was Littré's lifework.

Hippocrates and his school became as famous for their commonsense medical sayings as for the philosophy and method they exemplified. Up to the Middle Ages these adages were regarded as the quintessence of Hippocratic medicine. Once rediscovered and printed in the early sixteenth century, the *Aphorisms,* collected in seven books, became the most famous of all Hippocratic writings. The earliest edition of this work in the Library of Congress is a 1530 Latin edition, *Hippocratis aphorismi,* published in Paris. The first and most famous of these aphorisms states, "Life is short, and the art long; the occasion fleeting; experience fallacious, and judgment difficult."

Hippocrates lived during the classic period of Greek history and was the contemporary of Sophocles, Euripedes, Aristophanes, Pindar, Socrates, Plato, Herodotus, and Thucydides—a remarkable group of peers. His renown has equaled that of any of these giants and he remains our exemplar of the scientific tradition. To medicine, he offered freedom from superstition, a systematized body of knowledge, and a tradition of the highest standards of conduct. The Hippocratic oath that defines the duties of a physician is still taken by medical students upon completion of their training.

The post-Hippocratic decline of Greece and the rise of Rome saw medicine go from the formal to the doctrinaire to the moribund. One of the few real contributions to medicine from this time came from a very learned Roman, Aulus Cornelius Celsus. Although he may not have been a practicing physician, he was a scholar of the first rank and one of the first great encyclopedists. His *De medicina,* the work for which he is best known, was originally part of a much larger, encyclopedic work on many subjects, including medicine. Written during the reign of Tiberius, the work received little praise in its time and was actually lost for many centuries. With the revival of learning and the rediscovery of *De medicina* in the early fifteenth century, the book became one of the first medical texts to be printed. It was published in Florence in 1478. The Library's copy is a 1497 edition published in Venice. Written in elegant Latin, *De medicina* consists of eight books and systematically offers an exposition of prevailing medical knowledge. In most things, Celsus held strictly to Hippocrates and was later called "the Roman Hippocrates." His book also qualifies as the first real medical history text, Celsus having impartially included descriptions of developments from the post-Hellenistic period to his own time. During the Renaissance, its Latin translations of Greek medical works made it a popular and valuable source book of medical terminology. Celsus garnered the medical knowledge of Greece, Egypt, and Rome and set it forth in one place both for practical use and for future generations.

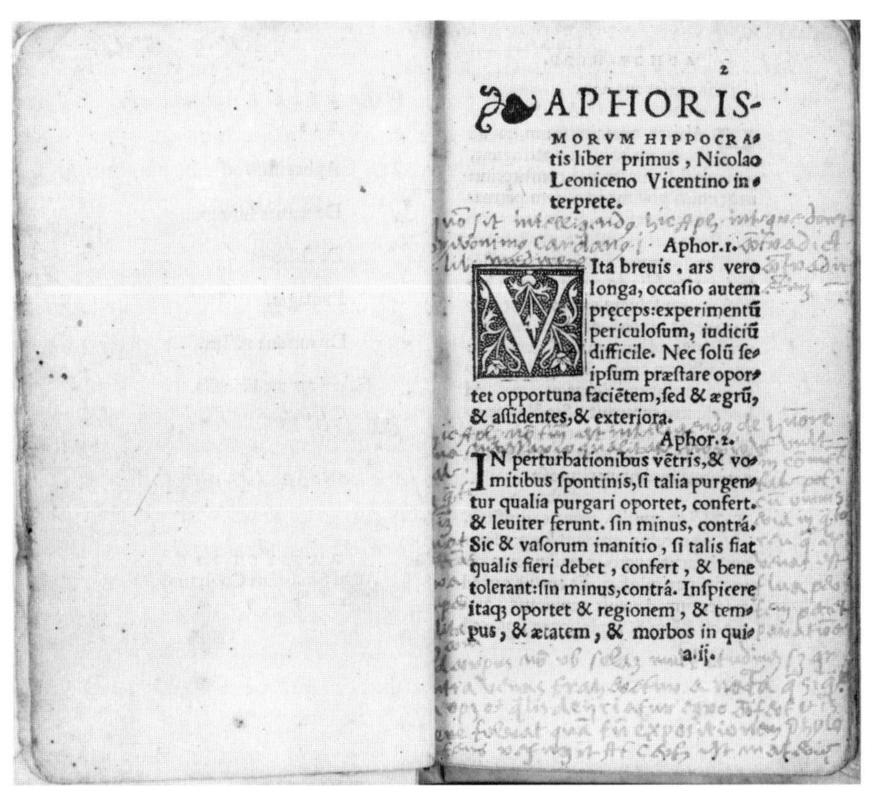

2

APHORISMORVM HIPPOCRATIS liber primus, Nicolao Leoniceno Vicentino interprete.

Aphor. I.

Vita breuis, ars vero longa, occaſio autem pręceps: experimentū periculoſum, iudiciū difficile. Nec ſolū ſe ipſum præſtare oportet opportuna faciētem, ſed & ægrū, & aſſidentes, & exteriora.

Aphor. 2.

IN perturbationibus vētris, & vomitibus ſpontinis, ſi talia purgentur qualia purgari oportet, confert. & leuiter ferunt. ſin minus, contrà. Sic & vaſorum inanitio, ſi talis fiat qualis fieri debet, confert, & bene tolerant: ſin minus, contrà. Inſpicere itaq; oportet & regionem, & tempus, & ætatem, & morbos in qui-

a.ij.

The earliest edition of the *Aphorisms* of Hippocrates in the Library of Congress is this tiny 1530 edition printed in Paris. The heavy annotations on the first page of this pocket-sized edition show it was well-used. The aphorisms are mostly commonsense enjoinders to follow the rule of moderation in all things. They begin: "Life is short, and the art long, the occasion fleeting; experience fallacious; judgment difficult." *Aphorismi*, 1530. Hippocrates.

Up to now, nothing has been said of any specific medical treatments or diseases. This is so primarily because the Hippocratic school was not distinguished by its ability to right a particular medical wrong—although it offered much that was helpful. Rather, the Hippocratic school was concerned more with the learning process itself—that of observing and describing the medical situation in its totality—than with a particular treatment of a particular condition. The Hippocratic physician would focus on diagnosis and prognosis—searching all the while for real understanding and knowledge. As for treatment, the same physician was a great believer in what has loosely been translated from the Greek as "life force"—that natural, recuperative process or power that the body contains. Although the Hippocratic physician was not a total do-nothing, who could name a disease but do little to cure it, his powers were certainly limited, especially from the patient's point of view. By the time of Pliny and his contemporary Dioscorides, who compiled the first materia medica (described in the chapter on botany), medicine had undergone a substantial change. Rather than practicing the careful, laissez-faire method of Hippocrates, the Greek physician employed by the rich Roman offered a combination of herbalism and magic to help his ailing patient, looking always for external cures. By the second century A.D., medicine had lost its Hippocratic impetus and had begun a steadily retrogressive movement away from the paradigms of reason and observation, a tendency so symptomatic of a culture in decline.

It was during this period of Roman decadence that a Greek physician, Claudius Galenus, known as Galen, acquired his extraordinary reputation.

Galen was educated in philosophy and medicine and gained considerable medical experience traveling throughout the eastern provinces of the Roman Empire. At one time, he was physician to the gladiators at his native Pergamon. Soon after his move to Rome he became the most celebrated and well-connected physician in the city, eventually serving as court physician under Emperor Marcus Aurelius. His fame and reputation as a medical authority were to reach such proportions that his writings would become the bible of medicine for nearly fifteen hundred years. It could be said that his personality—his style was one of bland self-confidence—made his reputation and that his prolific writing guaranteed its permanence. Galen never stopped writing or dictating and produced about four hundred individual works. Much of this was destroyed in a fire in A.D. 192. What remained, however, was sufficient to ensure his glory, for Galen had cast his medical work in a rigid all-encompassing (and all-answering) mold. For well over a millennium, his work was regarded as medical dogma by an unquestioning horde of physicians who used it as one would a dictionary. Up to the time of Vesalius, Galen was regarded as the final authority from whom there could be no appeal. In tone and substance his writings were omniscient and authoritative, and were naturally much admired in an age when people preferred to accept rather than to discuss.

Despite their popularity, Galen's works were not among the first generation of printed books. The first Latin translation of his complete works was published in Venice in 1490, and the first edition in the original Greek was printed in Venice in 1525 in five volumes. The Giunta Press of Venice published several Latin editions, and the Library has volume 3 of the 1541 *Opera omnia,* titled *Extra ordinem classium libri.* Two books form this large volume, one of which is considered to be a spurious work of Galen. The volume has a woodcut border on the title page, initials, and headpieces, all of which depict scenes from the life of Galen. Among his individual works in the Library's collections are eleven selected editions published during the sixteenth and seventeenth centuries on such topics as anatomy, the pulse, food, and orthopedics.

Galen's writings were an extension of his system of pathology, which combined the humoral notions of Hippocrates with the four-element theory of Pythagoras. These, in turn, were overlaid by a strictly teleological system that said that everything in nature was designed by God beforehand. Teleological "proof" replaced empirical evidence in Galen's system—he perceived the cosmic design and therefore had an answer for everything. Despite Galen's arrogant and seemingly omniscient stance, it was largely his followers who were to blame for the sterile state of medicine that descended upon Europe after his death. Galen himself was an acute observer with wide experience of native remedies, and his work contained much that was good and even progressive. But his authoritarian philosophy was reinforced by that of the coming age, and medical thought simply crystallized.

With the fall of Rome, the centers of culture and science shifted to the East—first to Constantinople and then further east to Persia. Medicine too

Opposite page:
This bordered title page shows medical scenes from the life of Galen, physician to Marcus Aurelius. Galen's anatomy was based on his dissecting knowledge of pigs rather than humans and, appropriately, the bottom panel shows him in sixteenth-century dress dissecting a pig before various sages. *Extra ordinem classium libri*, 1541. Galenus.

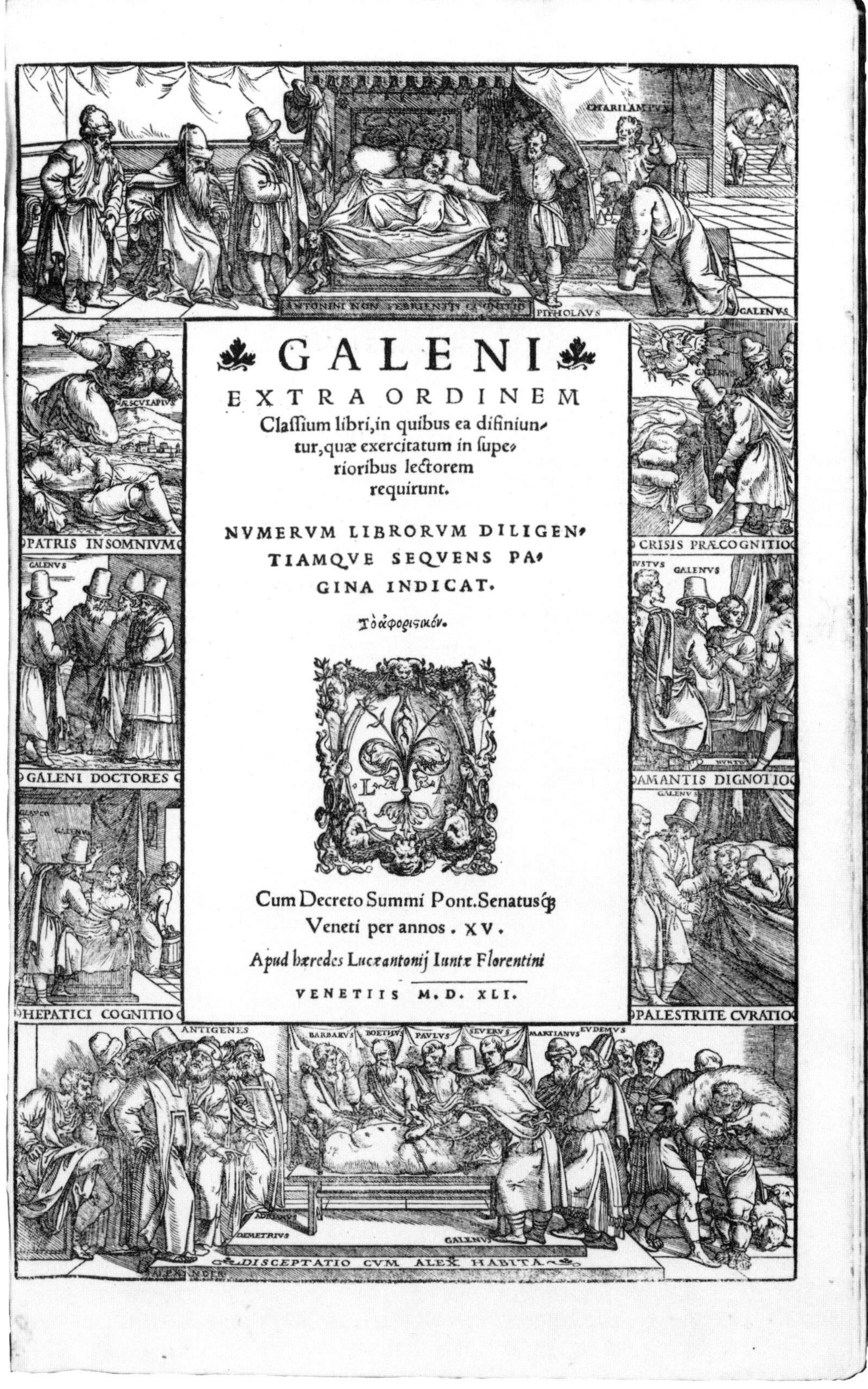

GALENI
EXTRA ORDINEM
Classium libri, in quibus ea disiniuntur, quæ exercitatum in superioribus lectorem requirunt.

NVMERVM LIBRORVM DILIGENTIAMQVE SEQVENS PAGINA INDICAT.

Τὸ ἀφοριστικόν.

Cum Decreto Summi Pont. Senatusq;
Veneti per annos. XV.
Apud hæredes Lucæantonij Iuntæ Florentini
VENETIIS M. D. XLI.

In this late fifteenth-century collection of medical texts, the author pays homage to the Arab influence on medicine by showing Pietro da Montagnana, a Paduan teacher of Arabic medicine, surrounded by Arabic as well as Greek texts. In addition to the works of Aristotle, Hippocrates, and Galen on the top shelf, are those of Avicenna, Haly Abbas, Rhazes, Mesue, and Averroes. On the teacher's left is an open copy of Pliny's *Natural History* and at his feet lie Peter of Abano's *Conciliator* and the works of Isaac Judaeus and Avenzoar. Below are three patients waiting to have their urine examined. *Fasciculus medicinae*, 1495. Joannes de Ketham.

made this eastward journey, returning to the part of the world where it was born. As a threatening stillness settled over the West, medicine found a safe and secure port with the ascendant Arabs. This historical period of Arab medicine, which coincided with the most flourishing period of Islam, lasted more than seven centuries. During that time, medical knowledge was carefully maintained by the Arabs and underwent a period of coalescence. Arab scholars collected, compiled, and translated Greek works into Arabic—all the while using and testing what they read. They were not mere depositories of knowledge, simply guarding the flame of ancient wisdom and then passing it on, but made many original contributions. These scholars and physicians are called Arabic since that was the language in which they wrote, but many were in fact Persian and Jewish.

They produced their great compilatory medical texts in manuscript form during this period of Arabian medicine but remained largely unknown to the West until the twelfth-century translators of Spain retranslated the Arabic into Spanish. These medical translators were generally Jewish physicians, men who were subsequently called the great intermediaries of the Mediterranean. By the mid-fifteenth-century invention of printing, medical knowledge had come full circle, and the wisdom of the Greeks saw print in Latin, Greek, and various vernacular languages under the name of several Arab authors. Particulars of medicine aside, what was most significant about this recapture of Greek knowledge was its philosophical foundation, as assimilated and passed on by Islam. This foundation enabled pre-Renaissance Europe to regain the spirit of inquiry so essential to the scientific method and to the progress of any body of knowledge—in science or the arts.

During the tenth century when Arab medicine was flourishing, its most famous physician was a Persian who was called Rhazes in the West. (Rhazes is the Latinized version of Abú-Bakr Muhammad ibn Zakarīyā' Al-Rāzī). A true follower of Hippocrates, Rhazes was a first-rate clinician as well as teacher, author, and court physician. During his long career, he produced over two hundred works, few of which survive. One of his most famous, an enormous encyclopedic compendium of all known medical knowledge, was first printed at Brescia in 1486 and was called *Liber continens*. The Library does not have a copy of this huge incunabulum in its collections, but it does have a 1501 compendium of medicine in the form of a commentary of Rhazes's famous ninth book of the *Almansor*. The ninth book, printed separately as *Liber nonus almansoris,* was a textbook of pathology and treatment, and was often commented upon in Western medical schools. The Library's Rhazes, published in Lyons in 1501, is called *Clarificatorium . . . super nono Almansoris* and was probably the course of lectures given by its commentator, Jean de Tournemire, as an introduction to the study of Rhazes's ninth book. The manner in which Rhazes is said to have lost his sight illustrates the perils of being the physician of a powerful ruler in those days. Legend has it that Al-Mansûr, the ruler of Bukhara, was displeased by some of Rhazes's failed chemistry experiments and ordered him beaten with his own book until either the head or the book was broken. Such mistreatment of court physicians was not uncommon in the West, either.

Another Persian who at times feared for his life was the illustrious physician called Avicenna (who was also known in the West as Ibn Sina). A child prodigy who was able to recite the entire Koran at ten years of age, Avicenna lived two centuries after Rhazes and became the physician of several Muslim rulers. He wrote on many subjects, and his medical work is based largely on that of Hippocrates and Galen. Like Rhazes, he also contributed many original medical observations and descriptions. His most famous work, *Canon medicinae,* went through many editions and influenced medical thought and practice for centuries. The first complete edition was published in Milan in 1473. The edition in the Library of Congress collections was published in Padua in 1479. The *Canon of*

Medicine is composed of five large books and has been described as the final codification of all Greco-Arabic medicine. This massive work was written with a tone of absolute authority and its title, *Canon* (a regulation or dogma decreed by the Church), indicated the manner in which Avicenna wished it to be regarded. The Library also has his *Cantica de medicina* with commentary by Averröes. It was translated from Arabic into Latin and printed in Venice in 1484.

A century after Avicenna's death, Arab medicine began its decline as the Muslim empire came apart. Muslim disunity, pressure from the Christians in Spain, and assaults by the Turks and Mongols to the east were among the forces that worked against Arab hegemony. It was during this period that the greatest of all Spanish Muhammadan physicians, Avenzoar, and his pupil Averroës prospered. Avenzoar was born in Seville and practiced medicine at Cordova, which had become a seat of learning and commerce and was by far the cultural center of Europe. Cordova in the twelfth century was a city full of doctors and could claim, at one time, to have fifty-two hospitals for its one million inhabitants. Avenzoar (or, Ibn Zuhr) was known to the city as the "Famous Wise Man," who had the courage to challenge the teachings of both Galen and the revered Avicenna. He was an essentially practical physician and took great exception to all of the philosophy he found in Avicenna. He is represented in the Library's collections by his *Liber Teisir; sive, Rectificatio medicationis et regiminis* first published in Venice, 1491. Originally written in Arabic, it was translated first into Hebrew and then into Latin and served as a practical handbook for the physician. Avenzoar reinforced the Arabic antipathy toward actual contact with the human body by proscribing surgery as unworthy of a physician. This distance between medicine and surgery was to increase during the Middle Ages and the Renaissance, much to the detriment of the discipline, and Avenzoar must take some negative credit for this.

His friend and pupil Averroës (or, Ibn Rushid) was both philosopher and physician, and he stressed the theoretical aspects of medicine. His major medical work, called the *Colliget,* is a very general medical encyclopedia. First published in Ferrara in 1482, the *Colliget,* or *Book of Universals,* is found in the Library's collections bound with the previously described *Rectificatio medicationis* by Avenzoar. Since Averröes wrote during the decadent phase of Arab medicine, his work had a greater impact on the Christian culture of Europe than upon his own. In fact, he is best known as the commentator par excellence and the introducer of Aristotle to the Christian Scholastics. His name has become linked to one of the more interesting intellectual phenomena in the history of medieval philosophy. During the thirteenth century, Averroism became a platform for anticlerical opposition, the anticlericalists espousing its founder's penchant for uncompromising rationalism. Averröes the man was indeed a free spirit for his time, and his denial of the immortality of the soul (it rejoined universal nature after death he believed), gained him the condemnation of both Moslems and Christians.

Thirty-eight years after the death of Averröes, Cordova was reconquered by Ferdinand III of Castille, in 1236, and the Arabian presence in Spain was finally removed. In the East, the destruction of Baghdad by the Mongols sealed the fate of the Arab Empire and extinguished five centuries worth of Arab power and dominance. Assessing the Arab influence on medicine, most now agree that the Arabs were both preservers and contributors. Their primary role was undoubtedly that of keeper of the ancient flame of Hippocratic thought. By respecting these ancient texts, translating them, and preserving them, they performed an invaluable service to themselves, to the West, and to science. As a vigorous and pragmatic people, quite urbane and sophisticated when compared to the feudal, ascetic, medieval West, they were receptive to new, non-Arab ideas and tolerant of intellectual diversity. Their enthusiasm for the Greek texts was thus characteristic. Once the old knowledge was theirs, it did not lie fallow. Although Arabian tenancy in the house of the Greeks did not alter its basic structure, their stewardship was by no means passive. Their creative work in chemistry gave a scientific basis to pharmacology and had a lasting influence on the medical use of drugs. Arab physicians contributed many original medical observations and therapeutic techniques. More importantly, they gave particular emphasis to reason over dogma and encouraged the lay practice of medicine, in contrast to the practice in the West where cleric-physicians dominated the field of medicine. The major void in Arab medicine was a scanty knowledge of anatomy and physiology—attributable to the Koran's prohibition both of dissection and representative depiction of the human body.

During these fertile centuries in the East, medieval European medicine was in a state of relative dormancy. For the most part, the practice of medicine in the West in these early centuries of the Middle Ages was carried out by the clergy. The establishment of several monastic infirmaries, whose monks roamed far and wide dispensing medical aid, led people to look to the Church as much for physical as for spiritual balm. Because of this very success, various Church councils eventually forbade monks from practicing outside the monastery. The decline of active monastic medicine in the tenth century coincided with the beginnings of lay medicine, and foremost in its rise was the famous school of Salerno.

The seaside town of Salerno, south of Naples, was known even to the Romans as an ideal health resort, and with a nearby monastic infirmary, Montecassino, exercising its influence, the name Salerno became identified with the healing art. It was literally a school of medicine—the first since Alexandria—and became the focus of all secular medical activity. Legend has it that its origins lie with four enlightened founders, each of whom taught medicine in his native language. Not surprisingly, each also represents a major medical—and ethnic and religious—sect. So, history tells of the school being founded by the rabbi Helinus, the Greek master Pontus, the Saracen Adela, and the Latin master Salernus. In fact, the port of Salerno was a crossroads of races and civilizations, a city of decidedly international flavor. Cultivating a taste for tolerance and openness, the school reflected the city in its receptivity to new and different medical ideas and methods, and as such became a beacon of hope in a very dark age.

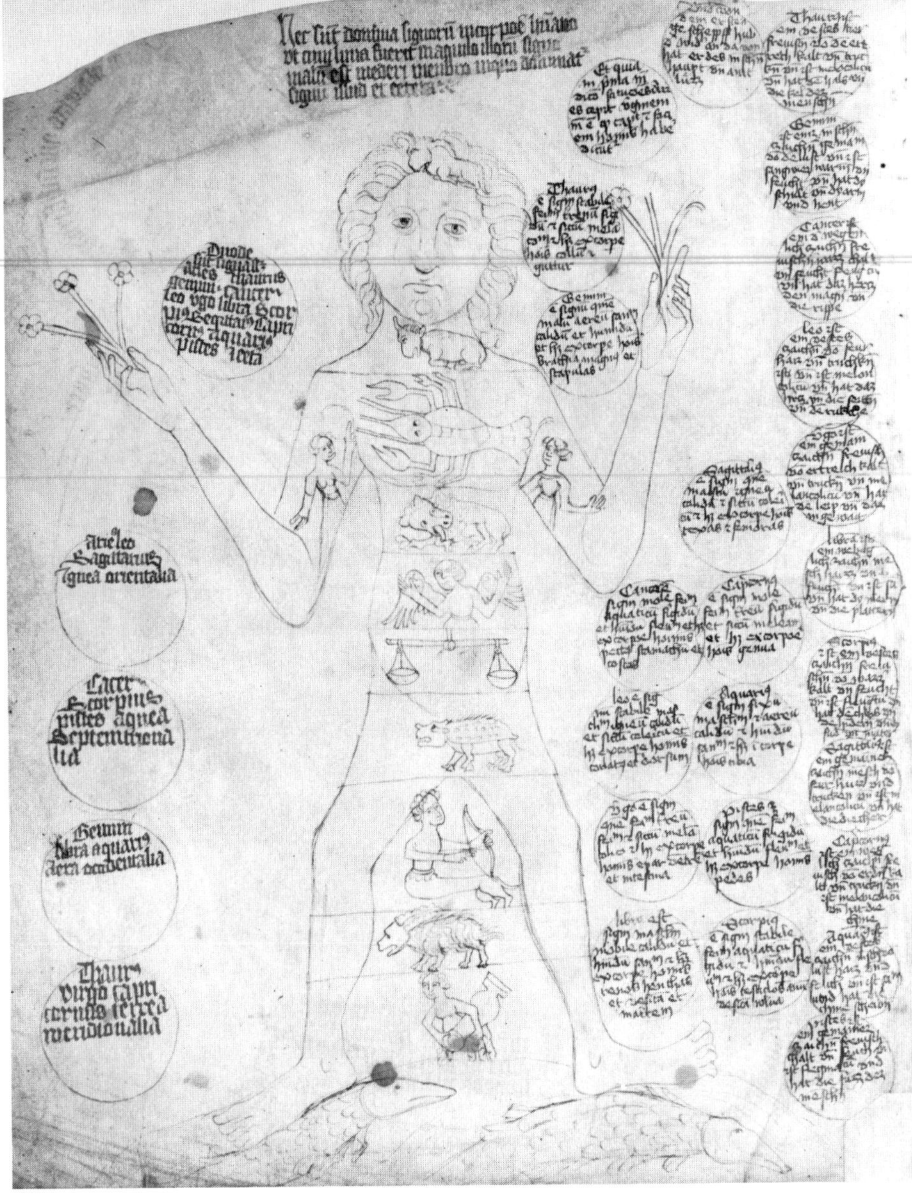

This late medieval vellum manuscript page was done about 1410 in South Germany. It is a crude representation of Zodiac-man, a typical illustration that accompanied other venesection plates, showing a naked man with the twelve signs of the zodiac, each of which relates to a specific part of the body. Encyclopedic manuscript containing allegorical and medical drawings. South Germany, ca. 1410.

By the eleventh century, the Arab (and therefore, Greek) influence had settled into the school's teachings and an organized corpus of practical medical knowledge became available to all. The medium for this information was truly inspired, however, for its compilers used verse form to convey their message. The famous poem *Regimen sanitatis Salernitanum,* with its seductive rhyme and sometimes humorous maxims, became the backbone of all practical medical literature up to the Renaissance. Among the Library's collection of seven incunabula of this long didactic poem, the earliest is a Louvain edition published sometime between 1483 and 1496. By the time of its first edition (Cologne, 1480), the book was publicizing a school in decline, Salerno having seen the end of its golden period well before the invention of printing. But *The Salerno Regimen of Health,* its catchy verses of diet and hygiene having been memorized by generations of physicians, became the town's chief medical and literary legacy.

The book's popularity was such that it saw almost three hundred editions and was translated into many languages. *Regimen* was for some time attributed solely to Arnaldus de Villanova, who lived from 1235 to 1315 and produced the first manuscript version, but its real origins were collaborative and are therefore anonymous. The decline of Salerno did not cause the field of medicine to suffer, its preeminence having been assumed by other new universities that were springing up north of the Mediterranean. Salerno's medical legacy was perpetuated by its literature—the *Regimen sanitatis Salernitanum* having ensured this by its popularity—and was carried on by others. Most importantly, Salerno had reintroduced the West to its Greek medical heritage, a legacy whose essential common sense and rationality were to blossom slowly into a real science, despite astrological and alchemical detours.

Many medical concepts and methods regarded now as detours or dead-ends were a part of the mainstream of medieval medicine and were no more questioned than are any of our modern nostrums. No medical procedure so typifies all that was bad about medieval medicine than does that of the essentially ignorant physician deliberately withdrawing a prescribed amount of blood from a particular part of the body of his sick patient. The medical tradition of bleeding was perhaps the most persistent and tenacious of all the therapeutics, lasting well into the nineteenth century. Despite its ancient origins—it has been traced as far back as the Egyptians—bleeding might not have taken such a firm hold nor been as easily accepted had it not been joined by the equally unscientific and specious ally astrology.

The astrological vein in the history of medicine is both wide and deep and it is perhaps best exemplified in the many popular "bleeding calendars" of the Middle Ages. Such calendars were found in manuscript, printed broadside, and book form and indicated the optimum days on which to draw blood and the body points from which to draw it. In a way, these calendars were a type of institutionalized irrationality in that they provided a degree of legitimacy to an essentially worthless and possibly detrimental act. Their very complexity—instructions varied for different phases of the moon, for different diseases, and at different body points—lent them an aura of sophistication and credibility. "The art is to know what vein to empty for what disease." Thus it is not too surprising that the first known piece of medical printing is the famous Mainz *Calendarium* for the year 1457, printed with the Gutenberg types of the Thirty-Six Line Bible. A unique copy was discovered in 1803 and consists only of its upper half. It now rests in the Bibliothèque Nationale in Paris. The earliest calendar in the collections of the Library of Congress is a vellum manuscript produced in South Germany around 1410. It contains five leaves of pen-and-ink drawings of the human body with indication lines showing the points to bleed. Another leaf shows the earth and seven planets. Two other leaves containing allegorical and religious illustrations are also included in this manuscript.

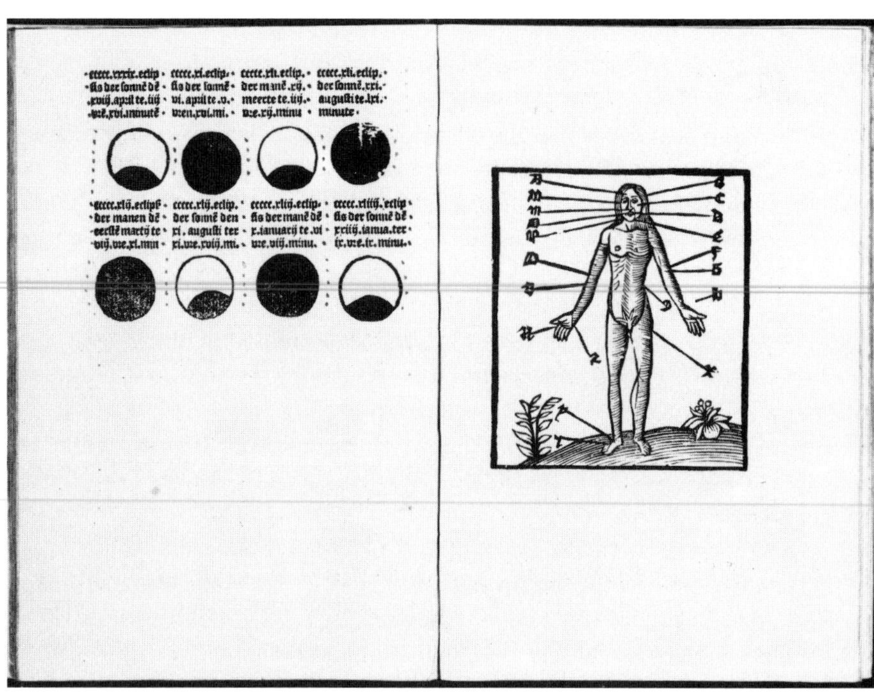

This bloodletting guide is from an early sixteenth-century reference manual called *The Shepherd's Calendar*. As a compendium of all sorts of practical information, it contained an obligatory section on health which offered zodiacal and physiological charts and drawings on bloodletting. The lines leading to the figure's body indicate points from which blood is to be let. *Der scaepherders Kalengier,* 1516.

The earliest printed calendar in the Library is the *Kalendrier des bergers,* first published in Paris in 1491 by Guy Marchant (printer of the famous *Danse macabre*). The Library's copy of this popular manual of medicinal and astrological lore is an amplified edition, published in the same city in 1497. This work has been described as the first book printed for the common people, and its title alone, *The Shepherd's Calendar,* tells us that it was not intended for scholars or the rich. As a compendium of practical information, it served the wider reading public as a secular bible for daily reference. One of its five sections is devoted to "physic and the governance of health" and contains zodiacal and physiological charts and drawings on bloodletting. The Library also has a 1516, Antwerp edition in Dutch, *Der scaepherders kalengier,* whose illustrations are identical to Marchant's work but are done in more lively color.

Another type of pre-Renaissance medical literature that became even more popular—and was more useful—than the bleeding calendar was the herbal. Aimed at the laity rather than the clergy, these books not only described the medicinal properties of plants but, in a broader way, functioned as encyclopedias of popular medicine. Such great herbals as Dioscorides's *De materia medica* (1478), the famous *Hortus sanitatis* of Mainz (1491), and the well-illustrated *Herbarius Latinus* (1499) are all part of the Library's collections and are discussed in the chapter on botany. One major work not mentioned in chapter 2 is attributed to Albertus Magnus, the great Dominican bishop and teacher of Thomas Aquinas. Albert's interest in science was wide-ranging and he seems to have dabbled in almost everything. Nearly two hundred years after his death, the *Liber aggregationis* or *De virtutibus herbarum,* which contains a group of medical writings attributed to Albert, was first published, in 1477. The Library has the 1481 edition published in Rome as well as the 1482 Bologna edition of this work, which exemplifies the extent to which astrology had crept into herbal recipes by the fifteenth century. An interesting sample of Albert's advice suggests that if marigolds "be gathered, the Sunne beynge in the sygne Leo, in August, and be wrapped in the leafe of a Laurell, or baye tree, and a wolves tothe be added thereto, no man shalbe able to have a word to speake agaynst the bearer thereof, but woordes of peace." Although in many cases the nostrum of the herbals were more magical than medicinal, there were a significant number of herbal recommendations both sound and efficacious. A good example is the juice of the willow leaves taken for fevers—its salicylic acid later became the base for aspirin.

Although the herbal pharmacopoeia and encyclopedias were very popular, they were not based on any sound theories of medicine or upon any real, scientific body of knowledge. In fact, the most elementary details of anatomy and physiology were unknown to the best physicians of the fourteenth century. Soon, however, the creative burst of Renaissance energy and imagination was to alter irreparably this tranquil but ignorant medical scene. It is one of the achievements of the Middle Ages that the universities were by then fully organized and entrenched in European intellectual life. The emergence of the universities, or "studium generale," as they were called, rivals on an intellectual plane the architectural magnificance of the medieval cathedrals. In this intellectual advance of Western civilization, medicine led the way with the school of Salerno. With Salerno's decline, other, broader institutions arose which taught law, philosophy, and theology in addition to medicine. The greatest and perhaps the oldest of these universities was Bologna, with rivals in Oxford, Salamanca, and Paris, to name a few. The significance to medicine of the university system is apparent when one notes that virtually every individual to be discussed in this chapter from this point on—from the fifteenth century on—is linked in some way with a university. To sum up the medieval legacy to medicine, then, is to describe the framework and the foundation within which and upon which the coming explosion of medical knowledge was to be organized and constructed.

As it seems so often in human history, things get worse before they get better, and so it was that the renaissance in medicine was preceded by a succession of terrifying plagues. The pestilence is said to have begun in the interior of Asia and later India during the first third of the fourteenth century, crisscrossed Europe several times during the following two decades, and even reached Russia by mid-century. Although the cities' ravaged populations were totally ignorant of its etiology, they were well aware of the contagious nature of the disease. (It was really several diseases, the worst being bubonic plague.) Many hygienic ordinances that reasonably regulated everyday life were attempted, but the hopelessness and medical impotence encountered daily led to the desperate embrace of the irrational as well. Prohibitions against cursing and bell-tolling were innocuous when compared to the fate of those Jews accused of causing the Black Death by poisoning wells. Supernatural explanations also were popular—the 1345

Ein vorred

Hie anfahen ist das büch genāt liber pestilen
tialis de venenis epidimie/von der kunst der vergifft des gemeynen sterbens
genant/von Jeronymo Brunschwig/so dan er von den aller bewertesten Ar/
tzten der fryen künst erfaren vnd collegiert hatt/Vnd ouch durcch sin practi
ca selber geübt vnd gesehen hatt.

Als die altē wisen
schribē dz vndāckbarkeit ein la
ster ist/dz alle andere laster vertribē
vñ vertilgē ist. So dan ich ingeden
cē bin vil güter enpfangener frünt

schafft/so mir geschehen vñ noch teg/
lichs geschicht von mynen genedigen
liebē herrē der keiserlichē statt Straß/
burg. Bezwingt mich Jeronymū zü
lob vnd ere vñ iren burgern zü offen/
baren vñ zemachen diß clein büchlin

A ii

The plague or Black Death that ripped through Europe in the fourteenth century is estimated to have killed one-quarter of the population. In this illustration from the *Pestbuch*, a plague victim, looking wary but comfortable, points out to his three attending physicians the characteristic swelling or boil under his armpit. Medicine was powerless to do anything to help, and ordinary life in the afflicted cities and towns became a trial of despair, dislocation, and disruption. *Pestbuch*, 1500. Hieronymus Brunschwig.

conjunction of Saturn, Jupiter, and Mars being interpreted as especially catastrophic.

As would be expected, plague literature is rich and varied, the most vivid and literary being Boccaccio's description of the plague of Florence. His introduction of the stories of the *Decameron* contains what has become a classic description of the plague's physical ravages and social destructiveness. Despite this vividness of detail, it is his pathos that is most arresting:

> How many valiant men, how many fair ladies, how many sprightly youths, whom, not only others, but Galen, Hippocrates or Aesculapius themselves would have judged most hale, breakfasted in the morning with their kinsfolk, comrades and friends and that same night supped with their ancestors in the other world!

The Library has the 1492 Venice edition of *Decamerone* as well as several other early editions of this masterpiece of world literature. A book that relates more closely to the medical aspects of the plague is the *Pestbuch,* written in German by the famous surgeon Hieronymus Brunschwig and published in Strasbourg in 1500. Although the text is in German, the title page uses the Latin *Liber pestilentialis de venenis epidimie.* Three centuries after these medieval scourges, a plague was to rip through seventeenth-century England with similarly catastrophic results. This time its chronicler was Daniel Defoe, whose 1722 *Journal of the Plague Year* presents a striking picture of those hard times. Defoe was only five years old when the 1655 plague that he chronicled took place, but his portrayal of those times was so vivid that many regarded it as the work of an eyewitness. The Library has a 1763 edition, retitled *The Dreadful Visitation,* published in Germantown, Pennsylvania. Significantly, it was during the plague year of 1665 that the young Isaac Newton retreated to the countryside for study and contemplation—and discovery—the universities having closed down because of the contagion.

The horrors of these times can only be imagined, but no doubt they were decades of grotesque extremes—of social chaos and apathy, of demoralization and hysteria, of hopelessness and violence. Little wonder, then, that such bizarre collective aberrations as the dancing mania and the Flagellants sprang up. During plague times, wrote Boccaccio, "Neither the advice of any physician, nor the virtue of any medicine prevailed." Said another contemporary, "The father did not visit the son nor the son the father. Charity was dead and hope abandoned." The passing of these times heralded also the end of the Middle Ages and brought a revival of learning. The invention of printing allowed for the rapid spread of new medical knowledge, and medicine was about to begin a remarkable new era.

The renaissance in medicine paralleled that of art and literature and recovered the traditions and concepts of classical Greece. Disease again came to be regarded as an imbalance of our physical natures that could be studied and treated. No more was ill health simply the consequence of sin, as the Scholastics had taught. The human body came to be regarded not only as a beautiful object and a worthy subject of art but also as a proper

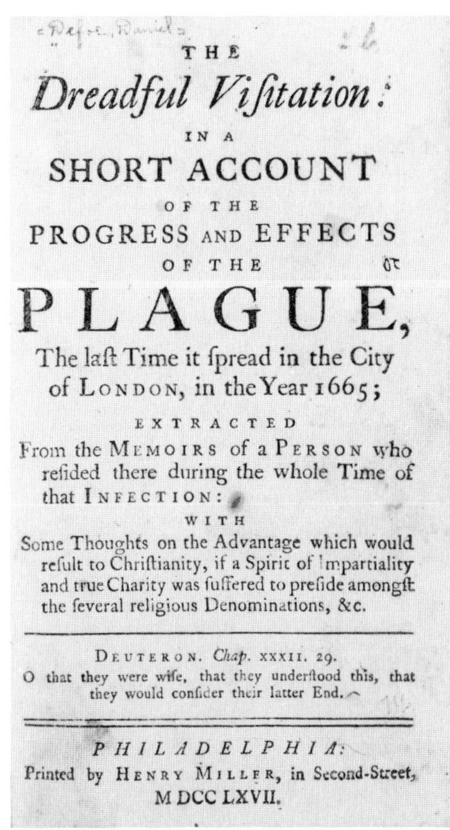

THE

Dreadful Vifitation:

IN A

SHORT ACCOUNT

OF THE

PROGRESS AND EFFECTS

OF THE

PLAGUE,

The laft Time it fpread in the City of LONDON, in the Year 1665;

EXTRACTED

From the MEMOIRS of a PERSON who refided there during the whole Time of that INFECTION:

WITH

Some Thoughts on the Advantage which would refult to Chriftianity, if a Spirit of Impartiality and true Charity was fuffered to prefide amongft the feveral religious Denominations, &c.

DEUTERON. *Chap.* xxxii. 29.

O that they were wife, that they underftood this, that they would confider their latter End.

PHILADELPHIA:

Printed by HENRY MILLER, in Second-Street, MDCCLXVII.

In 1722, Daniel Defoe, author of *Robinson Crusoe* and *Moll Flanders*, published this account of the infamous London plague of 1665. His portrayal of those times was so vivid that most thought it was the work of a real eyewitness, but Defoe was only five years old in 1665. Defoe tells his historical tale of horror in a journalistic manner, using the fictional "H.F." as his narrator. Nearly two and a half centuries later, Anthony Burgess wrote, "Its truth is twofold: it has the truth of the conscientious and scrupulous historian, but its deeper truth belongs to the creative imagination." *The Dreadful Visitation in a Short Account of the Progress and Effects of the Plague*, 1763. Daniel Defoe.

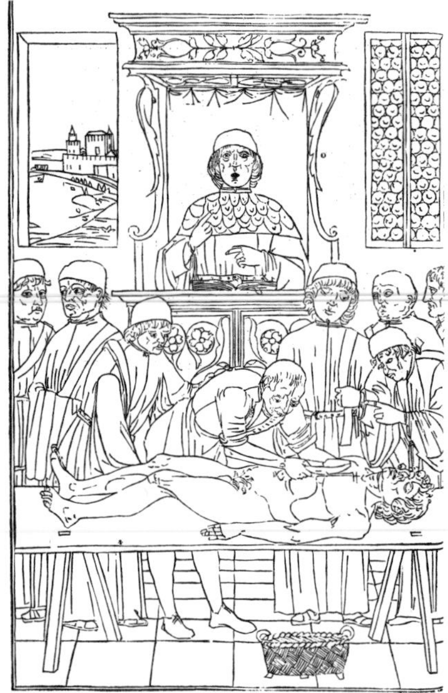

Without dissection of corpses, knowledge of human anatomy could not advance. In the early fourteenth century, Remondino de Luzzi, known as Mondino, practiced systematic dissections at Bologna and used the human cadaver to teach anatomy. In this fifteenth-century engraving, Mondino is shown in the lecture chair overseeing an assistant who is about to begin a dissection as students watch. The lesson begins with the traditional vertical incision. *Fasciculus medicinae*, 1495. Joannes de Ketham.

object for intense physical study. This change in attitude made dissection of cadavers a fit and proper scientific activity. Altogether, the humanist tone of the times allowed some of medicine's greatest minds to do their greatest work. The richness and depth of Renaissance medical literature is such that only major representative works in each general area can be discussed here.

No part of medicine experienced such a quantitative and qualitative leap during this time as did the study of human anatomy. During the preceding centuries, dissections were performed at universities only occasionally and were not a regular part of medical instruction. The first outstanding anatomist was the northern Italian Mondino, who taught at Bologna during the early years of the fourteenth century. It is to his work that the next two major texts of anatomy are related. *Fasciculus medicinae,* compiled by the German physician Johannes de Ketham, is recognized as the first medical text with realistic illustrations. The work is not original with Ketham and contains a series of separate treatises by others, most notably Mondino's *Anatomia* (which was added in Ketham's later editions). The Library's earliest copy is the second Latin edition (Venice, 1495), which differs markedly at times from the first edition (Venice, 1491). In 1493, an Italian edition was made whose smaller format mandated that new woodcuts be made, and these were used in the 1495 Latin edition also. It is the 1493 edition and its successors that have been hailed as marking the transition from medieval to modern medical illustration. The Library has a later Dutch edition, published in 1529 in Antwerp, as well.

As the first printed illustrated medical text, *Fasciculus medicinae* exhibits woodcuts of high quality. Among the didactic illustrations were conventional pictures of Zodiac Man, Blood-Letting Man, and Planet Man—drawings of the body that linked treatment and astrology. The illustration in the 1493 edition that differs most because of its new realism is that of the female anatomy. The work is significant not for its medical content but for its landmark illustrations and its inclusion of Mondino's dissecting manual.

Another illustrated anatomical work that harks back to Mondino is the *Commentario . . . super Anatomia Mūndini,* by Jacopo Berengario da Carpi. The Library has the first edition (Bologna, 1521) of this small, well-illustrated work. As a professor at Bologna, Berengario wrote this anatomical compendium to supersede Mondino's work, and amplify Mondino it does, for its fine engravings reveal many anatomical discoveries. Berengario can be considered the precursor of Vesalius, in that his accurate descriptions differed considerably from the traditional doctrine of Galen. His anatomical illustrations seem to incorporate techniques learned from Leonardo da Vinci.

Leonardo began the first of his anatomical drawings in 1489 (the date on the oldest sheet of his sketchbook), and he died just two years before Berengario's text was published. The magnificent Florentine was not only an anatomist but, among other things, a graphic artist, an architect, and an engineer. His dedication to reproducing accurately what he saw led him to disregard the anatomical dogma of Galen or of any other traditional

LEO PX

Carpi Commen
taria cū amplissi
mis Additioni
bus super anato
mia mūdini vna
cum textu eiusdē
in pristinū ⁊ verū
nitorē redacto.

This elaborate architectural title page introduces the work of Jacopo Berengario da Carpi, a precursor of Vesalius. Berengario's realistic work amplifies and goes beyond Mondino but is still part of the medieval tradition. The picture below—the sitting teacher lecturing at the foot of the corpse while an assistant does the actual dissection—is a typically medieval scene. *Commentaria . . . super Anatomia Mundini*, 1521. Jacopo Berengario.

authority. He performed scores of careful dissections and reproduced what he observed with an almost inhuman fidelity. He never produced his intended textbook on anatomy, however, and only his original sketchbook remains. Although Leonardo's sketchbooks were not published in their entirety until the end of the nineteenth century (the Library has the 1898 Paris folio A of *I manoscritti di Leonardo da Vinci della Reale biblioteca di Windsor: Dell'anatomia,* but it does not have the 1901 Torino folio B), it is known that copies were in circulation after his death.

While Leonardo's work certainly influenced those who saw his studies, there is no established connection between his anatomical work and that

ANDREAE VESALII 605
BRVXELLENSIS, DE HVMANI CORPORIS FABRICA LIBER SEPTIMVS, CEREBRO ANImalis facultatis ſedi & ſenſuum organis dedicatus, & mox in initio omnes propemodum ipſius figuras, uti & duo proximè præcedentes libri, commonſtrans.

PRIMA SEPTIMI LIBRI FIGVRA.

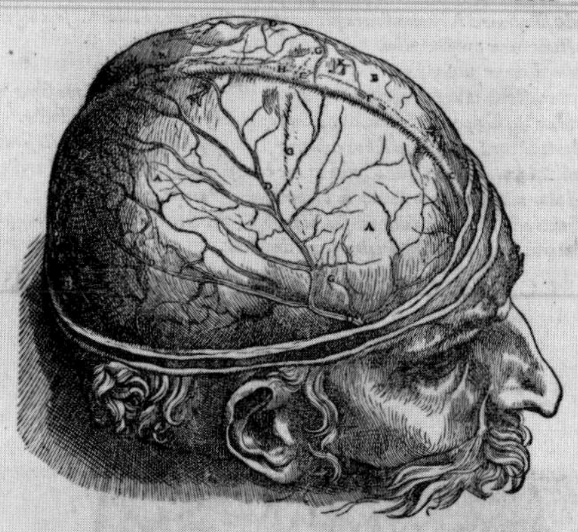

PRIMAE FIGVRAE, EIVSDEMQVE CHARACTERVM INDEX.

PRIMA ſeptimi libri figura humanũ caput ita adaptatũ exprimit, quemadmodum id cerebro oſtendẽdo opportunè à collo & inferiori maxilla diſſecãtibus liberatur. Prætereà tantam caluariæ partẽ orbiculatim ſerra abſtulimus, quanta quoq; omniũ quæ in caluariæ cõtinentur amplitudine uidendorũ gratia, auferri ſolet. quanta uerò illa ſit, liquidò dijudicabis, ſi ſeptimam figurã ſexti capitis libri primi examinaueris, quæ hinc ablatã caluariæ partem interna ſuperficie exprimit. Quẽadmodũ itaq; præſens figura ſectionis ſerie cæteras omnes inuicẽ ordine ſuccedentes præcedit, ita quoq; illã ſeptimi libri figurarũ primã non inopportunè inſcribimus, quæ durã cerebri cõmonſtrat membranam adhuc illæſam, neq; aliqua ex parte pertuſam, uulneratamúe. quamuis interim ipſius membranæ uincula diuulſimus, quæ per capitis ſuturas ad membranam efformandã porriguntur. quæ quod caluariã ſuccingit, περικράνιος nuncupabitur. atque cum his fibris pariter uaſcula ſunt effracta, quæ per caluariæ foraminula & ſuturas deducta, ipſi duræ membranæ, ac illi qua caluaria ſuccingitur, communes cenſentur. Cæterùm ex duobus qui figurã ambire conſpiciuntur orbibus, humiliorẽ, cutis & membranæ ipſi ſubditæ cõſtituunt, elatior autem ipſa eſt caluaria. Vniuerſum uerò hoc orbe complexũ, durã cerebri membranam refert, omnibus characteribus in figura cõſpicuis uniuerſim ſemelq; indicatam. at ſinguli characteres in hunc modum priuatim habent.

A, A Dextrũ duræ cerebri membranæ latus, ſeu eius mẽbranæ pars dextrã cerebri regionẽ ambiẽs.
B, B Siniſtrum duræ cerebri membranæ latus.
C, C, C Tertius duræ membranæ ſinus ſecundũ capitis longitudinem exporrectus, & hic nulla ex

Ee 3 parte

Andreas Vesalius was prompted to write his own anatomical textbook when he discovered that the prevailing works of the medical establishment, all based on Galen, were incorrect about many aspects of human anatomy. In true Renaissance spirit, Vesalius then decided that his book would adhere strictly to what his dissections actually revealed. The result is a work of both great medicine and great art. The realism and fineness of detail in this engraving as well as the startlingly unique manner of presentation is characteristic of the entire work. *De humani corporis fabrica*, 1543. Andreas Vesalius.

of Andreas Vesalius, the famous anatomist who was born five years before Leonardo's death. The name Vesalius has revolutionary connotations to the world of medicine. Although his work caused no real revolution in medicine during his lifetime, it was his epochal *De humani corporis fabrica* that put the fatal crack in the wall of Galenism. Others surely had doubted and even questioned the teachings of the revered Greek. Berengario made innovative corrections to Mondino (and therefore Galen), and Leonardo went his own way, discounting all traditional wisdom. But it was not until Andreas Vesalius, a Fleming of German origin who taught anatomy at the

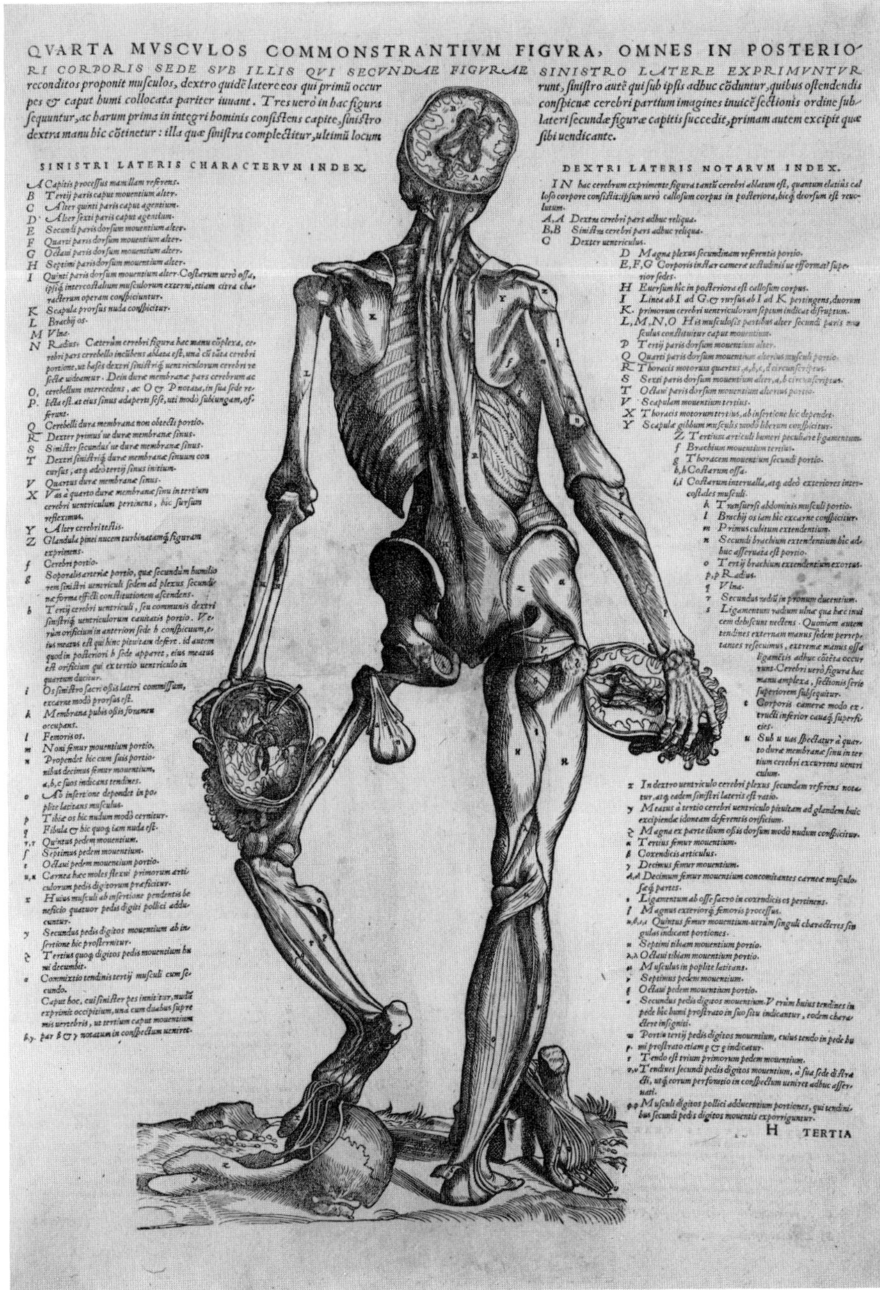

Vesalius's *Fabrica* is replete with these eye-catching "living" cadavers in everyday poses. This illustration is from his *Epitome*, an illustrated anatomical atlas that accompanied the *Fabrica*. In its realistic detail of the body's skeleton and musculature, this standing cadaver is both scientifically accurate and artistically brilliant. *De humani corporis fabrica librorum epitome*, 1543. Andreas Vesalius.

University of Padua, that a book was published that openly contradicted the essence of Galen's teachings. The *Fabrica* appealed to nature and not to any textbook for final authority. In it, Vesalius literally started over, shaping the science of anatomy anew.

The Library's copy of *De humani corporis fabrica* is the 1543 first edition published in Basel by Johannes Oporinus. As a classic in both the science of medicine and the art of printing, it is one of the prizes of the Library's collections. Considered by many the greatest medical book ever written, the *Fabrica* presented a completely radical view of medicine. "The human body," said Vesalius, was his "true Bible." To that bible Vesalius devoted five years of intense study, so that by the time he turned twenty-five he had finished a complete anatomical and physiological study of every part of the human body. His *Fabrica* is also one of the most beautiful books ever published. Its series of magnificent plates are from woodcuts made by a student of Titian, Jan Stephen van Calcar. Vesalius supervised their making and they are both lavishly beautiful and astoundingly accurate. Also illustrated by Calcar and published in the same year in Basel is Vesalius's anatomical atlas *Epitome,* also in the Library's collections. Intended for use by surgeons and students, this thin volume complements the *Fabrica.*

Many ironies surround the *Fabrica* and its author. Most obvious perhaps is the date of the book's publication—Vesalius's work appeared less than a week after Copernicus's *De revolutionibus.* Both, of course, were essentially revolutionary texts and became landmarks of the scientific revolution. A less well-known fact is that after publication, Vesalius quit anatomical research forever and became a court physician, leaving his beloved Italy. Although he achieved considerable fame during his lifetime, much was of the negative variety, and he was denounced by most of the conservative medical establishment. He was at times accused of heresy and even of vivisection. He died at sea at the age of fifty, returning from a pilgrimage to the Holy Land.

The resistance encountered by Vesalius—even his former teacher denounced him—underscores again the tenacity of traditional beliefs. For although Galen's teaching was based to a large degree on the anatomy of swine (he had done little dissection, except on animals), it held sway over that of Vesalius for some time. One who attacked the teachings of Galen more openly and boldly than even Vesalius was his pupil Gabriello Fallopio. Known as Fallopius, he became the most famous Italian anatomist of his time. He succeeded Vesalius to the chair of anatomy at Padua. As a student of the human body, he seems to have excelled in his study of the more intricate and tiny body parts—the ovarian tubes that now bear his name and the trigeminal nerves, to name only two. The Library does not have his most important work, *Observationes anatomicae,* which was published in Venice in 1561 at his own expense. It contained no illustrations and was the only work published in his lifetime. Following his early death, his lectures on anatomy were compiled by his pupil Volcher Coiter. Published in Nuremberg in 1575, *Lectiones de partibus similaribus humani corporis* is in the Library's collections. Fallopius criticized any authority he could prove wrong and did not spare his former teacher, pointing out mistakes Vesalius had made in his *Fabrica.* Like Vesalius and many other anatomists of his time, he too was accused of having performed human vivisection. Much of this criticism came from outside Italy, however. It is not surprising that the enlightened attitude of Renaissance Italy coincided with the amazing progress of its anatomy schools. At a time when anatomists in other parts of Europe rarely obtained cadavers, Vesalius could describe how Italian judges sometimes adapted the capital sentences given to criminals so as not to impair the anatomist's work.

During the Renaissance the field of surgery underwent a similar advance, no doubt in part because of its close alliance to anatomical research. In previous centuries, surgery had been regarded as a type of manual labor—an unskilled act with little dignity. By the sixteenth century, surgery was gaining stature in the medical profession as its practitioners were more and more men of higher training and not mere barbers. Ironically, though, one of the greatest figures of this time began his surgical studies in a barber shop. Ambroise Paré, recognized by many as the father of modern surgery, was a poorly educated barber's apprentice whose real training came while tending the wounded French soldiers during the Italian campaigns of 1536–45. It was this practical experience gained as an army surgeon that made him realize how wrong and actually detrimental many of the traditional surgical procedures were. Paré was a pioneer who had the courage to denounce false therapeutics. It was he who first demonstrated that simple soothing dressings and not searing with boiling oil or hot iron helped to heal gunshot wounds. Paré possessed a humble respect for the natural healing force, in the best of the Hippocratic tradition. When praised for his work, he is said to have noted, "I treated him, but God cured him."

Paré had little schooling and knew no Latin. His writings are therefore all in French, and they are very rare. The Library has his complete *Oeuvres,* first published in Paris in 1579. This work became extremely popular and displayed well Paré's honesty and originality. It reflected the virtues of the new age and was in no way tainted by stultifying scholastic medical traditions. Paré wisely saw that his lack of an education was a virtue in enabling him to view surgical problems with a fresh, unbiased, commonsense viewpoint. He often said with some sarcasm that he was denied the benefit of studying the medical masters. Unlike Vesalius, Paré saw his work accepted in his lifetime and witnessed the almost overnight abolition of wound cauterization. Although he rose to become surgeon to four French kings, Paré was most beloved by the common soldiers whose plight he eased by his courage and honesty.

Surgery during Paré's time experienced a period of great change and progress. New procedures and new instruments were being developed one after another, and such advances as the ligation of arteries made surgery more manageable. Among the more surprising surgical procedures performed during the sixteenth century were plastic surgery (notably the rhinoplasties performed by Gaspare Tagliacozzi until they were banned by the Church) and eye surgery. In the Library of Congress collections is the first book on eye surgery, the *Ophthalmodouleia* of George Bartisch. Published in Dresden in 1583, this large book contains many striking illustrations. Besides its intended purpose as an illustrated textbook on cataract operations, the volume offers a complete summary of Renaissance eye surgery, giving us a good idea of the state of the art in the sixteenth century. The woodcuts were done after watercolors Bartisch had made himself.

As anatomical and surgical knowledge increased, so did knowledge of obstetrics, although certainly not proportionally. Despite the obstetrical

146 The traumas depicted here had a special relevance to Ambroise Paré, who, as an army surgeon, was probably called upon to treat these and many more battlefield injuries. This illustration, taken from an English version of his *Oeuvres*, is a variation of the medieval "wound-man," a popular didactic figure in many medical texts throughout the Middle Ages. As a sort of graphic catalog of frequent traumas, the wound-man has been described by Karl Sudhoff as a "surgical caricature of St. Sebastian," his body pierced not only by arrows but by cudgels, maces, and knives. *The Workes*, 1649. Ambroise Paré.

328 *Of Wounds made by Gunſhot,* LIB. II.

have the ſhaft put into the head, others the head into the ſhaft; ſome have their heads nailed to the ſhaft, others not, but have their heads ſo looſly ſet on, that by gentle plucking the ſhaft, they leave their heads behind them, whence dangerous wounds proceed. But they differ in force, for that ſome hurt by their Iron only, others beſides that, by poyſon, wherewith they are infected. You may ſee the other various ſhapes repreſented to you in the preceding Figure.

In force.

CHAP. XVII. *Of the difference of the wounded parts.*

THe wounded parts are either fleſhy or bony; ſome are neer the joints, others ſeated upon the very joints; ſome are principall, others ſerve them; ſome are externall, others internall. Now in wounds where deadly ſigns appear, it's fit you give an abſolute judgment to that effect; leſt you make the Art to be ſcandalled by the ignorant. But it is an inhumane part, and much digreſſing from Art, to leave the Iron in the wound; it is ſometimes difficult to take it out, yet a charitable and artificiall work. For it is much better to try a doubtfull remedy, than none at all.

You muſt not leave the weapon in the wound.

CHAP. XVIII. *Of drawing forth Arrows.*

YOu muſt in drawing forth Arrows ſhun inciſions and dilacerations of Veins and Arteries, Nerves and Tendons. For it is a ſhamefull and bungling part to do more harm with your hand, than the Iron hath done. Now Arrows are drawn forth two wayes, that is, either by extraction, or impulſion. Now you muſt preſently at the firſt dreſſing pull forth all ſtrange bodies, which that you may more eaſily and happily perform, you ſhall ſet the Patient in the ſame poſture, as he ſtood when he received his wound; and he muſt have alſo his Inſtruments in a readineſſe, chiefly that which hath a ſlit pipe and toothed without, into which there is put a ſharpe Iron ſtyle, like the Gimblets we formerly mentioned for the taking forth of Bullets; but that it hath no ſcrue at the end, but is larger and thicker, ſo to widen the pipe, that ſo widened it may fill up the hole of the Arrows head whereinto the ſhaft was put, and ſo bring it forth with it, both out of the fleſhy as alſo out of the bony parts, if ſo be that the end of the ſhaft be not broken, and left in the hole of the head. That alſo is a fit Inſtrument for this purpoſe, which opens the other end toothed on the outſide, by preſſing together of the handle. You ſhall find the Iron or head that lies hid by theſe ſigns, there will be a certain roughneſſe and inequality obſervable on that part if you feel it up and down with your hand; the fleſh there will be bruiſed, livid, or black, and there is heavineſſe and pain felt by the Patient both there and in the wound.

The manner of drawing forth Arrows, and ſuch weapons.

A delineation of Inſtruments fit to draw forth the heads of arrows, & darts, which are left in the wound without their ſhafts.

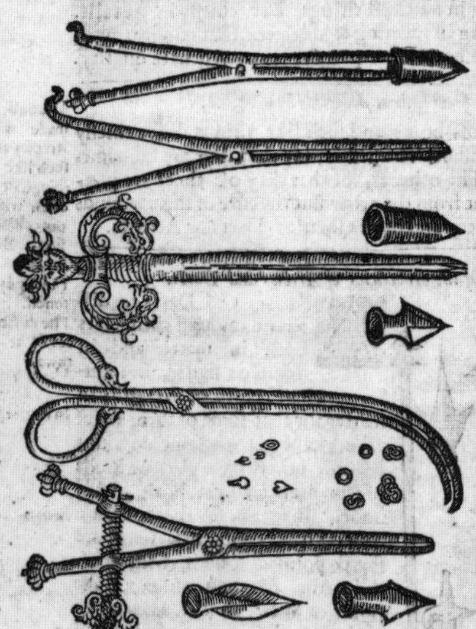

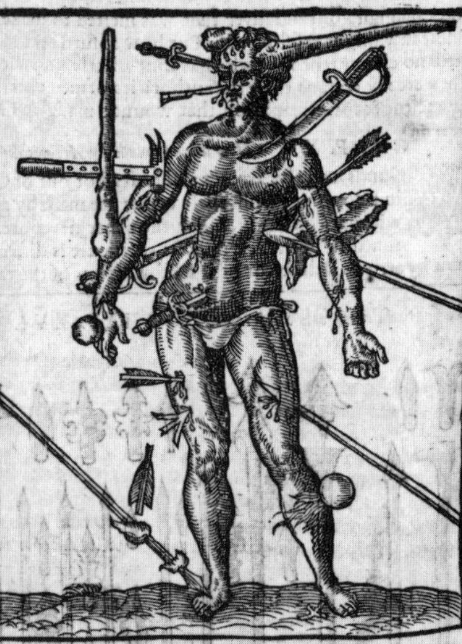

A hooked Inſtrument fit for to draw forth ſtrange bodies, as pieces of Male, and ſuch other things as it can catch hold of, which may alſo be uſed in wounds made by Gunſhot.

But if by chance either Arrows, Darts or Lances, or any winged head of any other weapon, be run through and left ſticking in any part of the body, as the thigh, with a portion of the ſhaft or ſtaffe ſlivered in pieces, or broken off; then it is fit the Chirurgeon with his cutting mullets ſhould cut off the end of the ſtaffe or ſhaft, and then with his other mullets pluck forth the head, as you may ſee by this Figure.

CHAP.

Das Ander Teil. 13

In dem Andern Teil wird angezeiget vnd beschrieben / vpn den mängeln vnd gebrechen der Augen / so angeerbet vnd von Mutter leibe angeboren werden / Als schielen / vbersichtigkeit / vnnatürliche kleine enge Augen / auch vnnatürliche grosse weite Augen / Vnd so eine vnnatürliche vbel vnd mißstendige farbe am weissen der Augen sey.

C Obwol

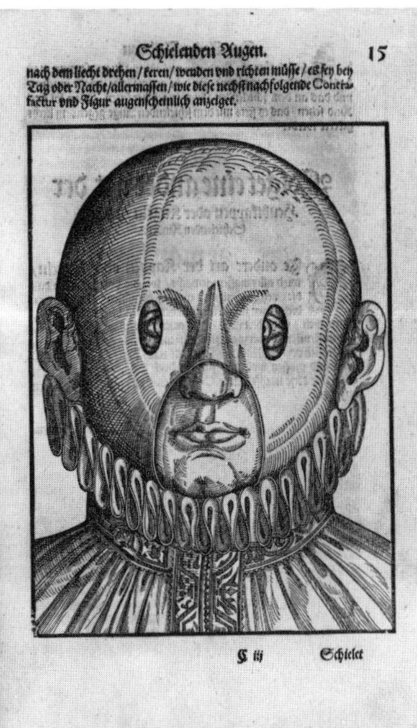
Schielenden Augen. 15

nach dem liecht drehen / keren / wenden vnd richten müsse / es sey bey Tag oder Nacht / allermassen / wie diese nechst nachfolgende Contrafactur vnd Figur augenscheinlich anzeiget.

C iij Schielet

These curious woodcuts are from the first Renaissance book on eye surgery and suggest a mechanical strategem for correcting crossed eyes. The *Ophthalmodouleia* not only contained anatomical illustrations of the eye and information on eye diseases but distinguished five types of cataracts and instructed how to remove them. *Ophthalmodouleia*, 1583. George Bartisch.

writings of Vesalius, Paré, and others, the most widely used book during the Renaissance was in effect a holdover from medieval times. In 1513, the *Rosengarten* of Eucharius Roeslin was published in Strasbourg. Written in the vernacular, its full title was *Der Swangern Frawen und Heb Ammē Roszgartē* or *The Rose Garden of Pregnant Women and Midwives*. Roeslin was a municipal physician at Worms and wrote his book, which is little more than a survey of Greek and Roman literature, primarily for midwives. The book's significance lies in the twenty woodcuts done by Conrad Merkel, a friend of Albrecht Dürer's. These quaint and somewhat primitive illustrations show such things as the birthing-chair and the in utero position of the fetus. The Library has Roeslin's first edition only in facsimile, but it does have the later Latin version, a small book called *De partu hominis*, in the 1551 Frankfurt edition. The collections also include an Italian version, published in Venice in 1538.

The book was widely circulated and was translated into English by Richard Jonas, eventually to appear as *The Byrth of Mankind*. In this extremely popular English text, Thomas Raynalde supplemented Jonas's translation, borrowing from many authors, Vesalius among them, and adding new plates as well as new text to Roeslin's original work. *The Byrth of Mankind* was first published in London in 1545, and the Library has this first edition. Various editions of the *Rose Garden* and its many versions were published well into the eighteenth century.

Renaissance obstetrics may be said to have benefited more from an increase in accurate anatomical knowledge than from any particular obstetrical breakthrough. It is safe to say that if the above texts are truly representative, childbearing during the Renaissance was as risky and as dangerous as it was during medieval times.

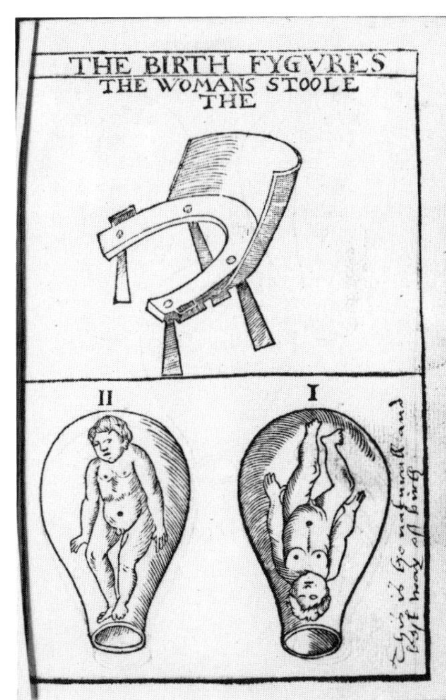

Although sixteenth-century medicine saw eye surgery and even plastic surgery being done, no real advances were made in obstetrics. The most widely used book, the *Rosengarten*, was a holdover from medieval times. In this English version, the obstetric stool or birthing chair is shown along with a normal (I) and an abnormal (II) presentation of the fetus in the uterus. *The Byrth of Mankind*, 1545. Thomas Raynalde.

No individual so personified the Renaissance spirit of revolt as did Philippus Aureolus Theophrastus Bombastus von Hohenheim, better known as Paracelsus. Few characters in the history of science are so colorful or so variously regarded. Posterity has at times judged him both charlatan and saint. Born in a quiet, conservative part of Switzerland, Paracelsus took his medical degree at Ferrara and never looked back. As with many great men, a formal education served only to make him keenly aware of both his own inadequacies and the limitations of his field. Of his medical peers, he said, "When I saw that nothing resulted from their practice but killing and laming, that they deemed most complaints incurable, and that they administered scarcely anything but syrup laxatives, purgatives and oatmeal gruel, with everlasting clysters [enemas], I determined to abandon such a miserable art and seek truth elsewhere." So Paracelsus took to traveling throughout Europe for ten years, gathering information from any willing source—gypsy, midwife, executioner, barber, alchemist. The knowledge of folk medicine he gained served only to reinforce his contempt for the orthodox medicine of his day—Galenism.

He arrived at Basel in 1526 with both a mystique and a reputation, numbering among his successfully treated patients the famous printer Froben and the illustrious Erasmus of Rotterdam. Soon after, he accepted a university chair and an appointment as municipal doctor, and it was from these traditional seats of establishment authority that he worked to undermine orthodox medical theory and practice. Such eye-catching tactics as publicly burning the revered works of Galen and Avicenna, lecturing his students in German rather than the customary Latin, and refusing to participate in the university's solemn ceremonies served as open symbols of his total assault on the medicine of his day.

The contributions of Paracelsus to medicine are not so obvious or demonstrable as those of Vesalius or Paré in anatomy and surgery. Though he wrote a great deal, Paracelsus saw little of his work published—mainly because of his infamous reputation and activities. And, in extreme contrast to Vesalius, much of his work contains an admixture of the mystical or the unscientific and the scientific—often to its detriment. More significant than any specific cure or medical discovery, however, was his constant and usually strident advocacy of a certain medical approach and method. He approached medicine in a Hippocratic manner, arguing that its practice should be based on nature and physical laws. He stressed practical, clinical experience over the teachings of authority. "I have not been ashamed to learn from tramps, butchers and barbers things which seemed of use to me," he said. He discarded the prevalent systems of "humours" and viewed disease as both a disturbance of normal functions and an invasion of the body from without. He is best known as the founder of iatrochemistry, following upon his belief that the body was in some way linked with the laws of chemistry and was therefore responsive to chemical cures. Particulars aside, the shearing off of science from religious and philosophical dogmatism is what is at the core of the work of Paracelsus. Central to this irascible iconoclast, then, despite his many forays into mysticism and the occult, was the liberation of the intellect from artificial and self-

This portrait of Paracelsus was done by his contemporary, the Mannerist master Tintoretto. Unlike flattering portraits found in many an author's collected works, this is a striking picture of a very real person. It especially contrasts with the well-known painting of a fat, jolly Paracelsus done by Quentin Metsys. Here Paracelsus appears gaunt and tired, although his eyes seem very much alive. He died at age forty-eight. *Opera omnia*, 1658. Paracelsus.

imposed restraints. Although Paracelsus was an extremist in action and temperament, his rationalism was modified by a healthy, probably intuitive respect for the limitations of the human intellect as well as an equal, usually mystical, regard for forces still undiscovered.

> We contain within ourselves as many natural powers as heaven and earth possess. Can the magnet draw the iron to itself even though it appears to be a dead thing? . . . Can the climbing vine reach out to the sun? So well may man in similar manner have access to the sun . . . they are all invisible works, and yet they are natural.

This is certainly not unfettered rationalism but rather science at its best—science that is open to both the knowable and the unknowable. Paracelsus

Dit Boeck gheintituleert Een excellent Tractaet hoemen dye Pocken ende ander Siechten ghenesen sal vèet, is ghevisiteert by heere M. Jan Goosens/Licentiaet inder Godheyt/ van der K. Maiesteyt daer toe ghecommitteert/ende is voorts byder K. Maiesteyt toeghelaten/ gheconsenteert/ende gheoctroyeert Jan Roelants dat hy den tijt van drij Jaren naestcomende tvoorschreven Boeck alleene sal moghen printen ende vercoopen/ende also wordt byder voorschreven Keyserlijcke Maiesteyt verboden allen anderen Printers ende Boeckvercoopers tvoorschreven Boeck nyet nae te printen oft te conterfeyten den voorschreven tijt duerende op dye verbuerte van .XX. Carolus Guldens/ alst blijckt byder Octroyen hem hier op verleent. Ghegeven tot Bruesel Anno M. D. LIII, den sevensten dach in Junio.

Onderteeckent M. Ph. de Lens.

Dit Boeck is ghedeylt in Twee Deelen/waer af deen spreeckt van allen Curen dye tot noch toe ghebruyckt zijn gheweest/soo doort Quicksilver oft anders ende alle sulcken Curen worden oprechtelijck ghecorrigeert ende met seer goeden Consten versien daer hem een yeghelijck wel op betrouwen mach.

Dit Tweede Deel leert hoemen op een seer goede Maniere sal moghen met den Pock houte (Lignum guaiacum gheheeten) ghenesen ende Cureren alle ghebreken der Pocken ende noch vele andere sware ende pijnlijcke Siecten dye hueren oorspronck nyet en hebben vten Pocken maer vt ander ghebreken der natueren/ so dat den genen dye doet ghelijc hier gheleert wordt seer wel te betrouwen/ is elcke sware Sieckte te ghenesen.

A ij Pro-

This small tract on treating the pox was based on the teachings of Paracelsus and was published twelve years after his death. Its frontispiece shows the traditional sickbed scene. *Een excellent tracktaet leerende hoemen alle ghebreken der pocken sal moghen ghensen*, 1553. Paracelsus.

was a great man—courageous, independent, and humble in his own way. His turbulent life demonstrates what powerful chemistry can be made when a revolutionary idea is espoused by a dynamic and forceful personality. Almost single-handedly he tried to drag all of Europe from its medieval lethargy. The bulk of his work was published after his death, and ranks as a considerable corpus. In addition to ten separate works all published before 1590 and all in German, the Library of Congress has his 1616 *Opera* published in Strasbourg and the famous 1658 Geneva *Opera omnia,* regarded as the most complete of the Latin collected editions. This large work contains virtually all of Paracelsus's medical and philosophical writings. Of particular interest is the portrait of Paracelsus by Tintoretto.

The work of Paracelsus did much to further the emerging concept of disease as something organic and not of supernatural origin. From this simple breakthrough, it followed that diseases could be distinguished, described, categorized, and perhaps even cured. No work better exemplifies this scientific state of mind and approach toward disease during the

Renaissance than does that of Girolamo Fracastoro on syphilis. Fracastoro was of a patrician Veronese family and studied medicine at Padua with Nicolaus Copernicus. An eminently typical Renaissance man, he pursued his serious interest in astronomy, geography, mathematics, and the arts as well as medicine. He is best known today for his poem on syphilis, *Syphilis, sive morbis gallicus* (Verona, 1530) which named the disease. Before the publication of this poem, which graphically describes the disease and tells of the young shepherd Syphilus who insulted Apollo and was cursed with the disease, the malady was generally called the "French disease." Naturally enough, the French dubbed it the "Neapolitan disease"—France and Naples each believing it was infected by the other when the French captured Naples in 1495. But Fracastoro's contributions went beyond his famous poem, for in his treatise *De contagione et contagiosis morbis,* published in 1546 in Venice, he offered the first comprehensive explanation of how an infectious disease is spread. The Library has both his poem and his treatise along with his principal astronomical and philosophical works in the *Opera omnia,* published in Venice in 1555. This was the first edition of his collected works. Fracastoro argued that infectious diseases were spread by direct and indirect contact, as well as through the air. His *De contagione* has been described as the first scientifically reasoned statement of the true nature of infection and contagion, which presented the germ theory of disease. Because of this, he is regarded by many as the founder of modern epidemiology.

Much of the progress and many of the hopes of the sixteenth century were set back by the political and economic upheavals of the seventeenth century. Yet while most of Europe was suffering from either economic decline or protracted warfare, England prospered, and at the time of Elizabeth's death at the beginning of the century, such giants as William Gilbert, Francis Bacon, William Shakespeare, and William Harvey were alive. The name Harvey ranks with the best. His major work is rivaled only by that of Vesalius at the head of any list of great medical texts. William Harvey published his *De motu cordis* in 1628, having worked out its thesis twelve years earlier. In producing this book of seventy-two pages, Harvey did a number of astounding and wonderful things—the least of which was his discovery of a major new medical truth. Harvey's discovery of the circulation of the blood—his realization that the same blood moves within a closed circle in our bodies—may seem obvious today, but in 1628 it contradicted traditional Galenic doctrine. Fully aware of the significance of his discovery, and aware too of how strongly it would be opposed, Harvey offered it to the world only when he deemed his arguments to be irrefutable. Herein lies Harvey's real greatness, for his presentation remains a scientific paradigm to this day.

In his book *On the Motion of the Heart and Blood in Animals,* Harvey offered a series of brilliantly conceived and arranged inductive experiments that demonstrated the mechanical and mathematical necessity of his hypothesis. His experiments proved conclusively that the heart was a pump working by muscular force that propelled the blood within a continuous, one-way cycle. Harvey's discovery met with substantial resistance at first

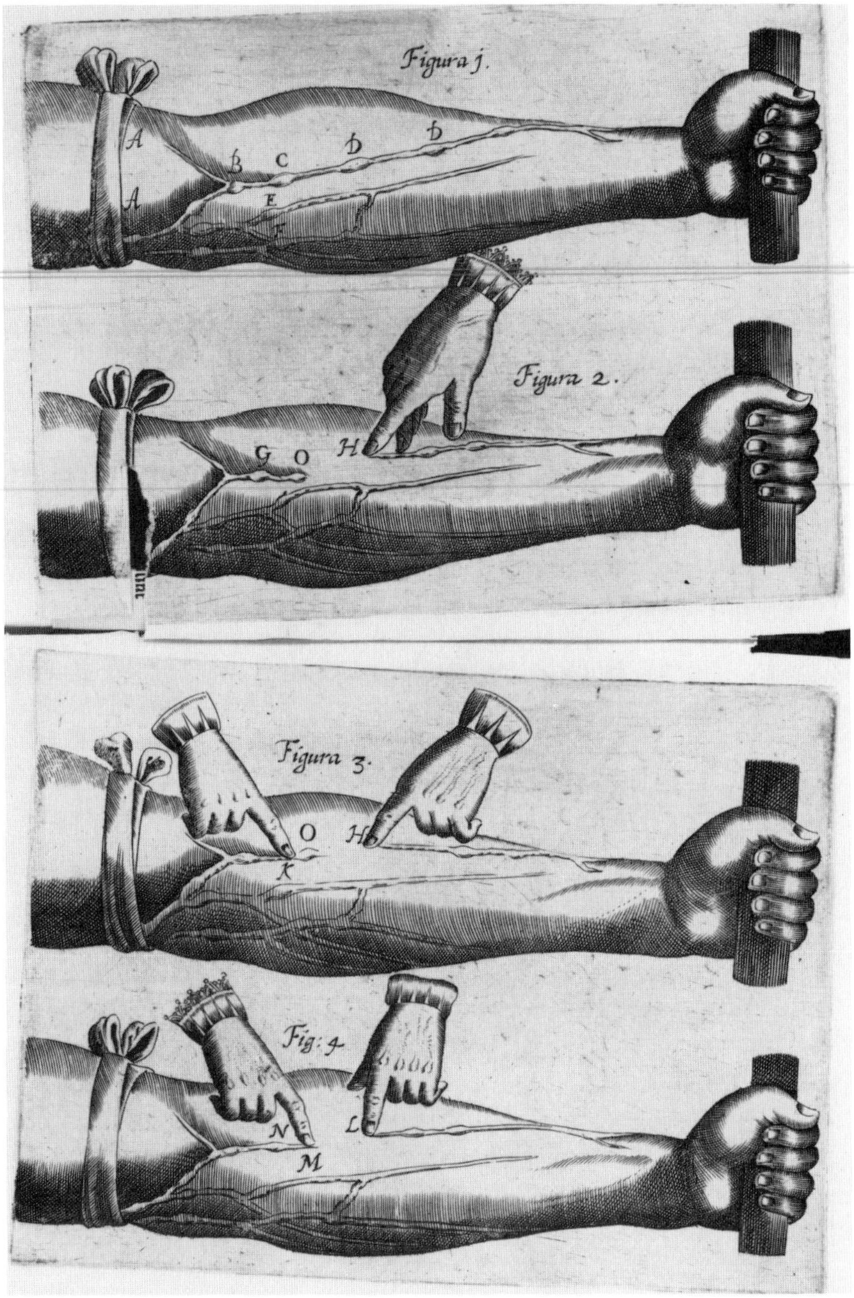

Although several individuals before William Harvey had proposed the notion of the circulation of the blood, none had offered any proof. Not until Harvey's brilliant experiments, based on direct observation and using quantitative methods, was this idea validated by demonstrable results. In this series of experiments, which anyone can easily perform, Harvey showed that blood flowed from the heart in a continuous, one-way cycle—coming from the heart through the arteries to the tissues and returning to the heart via the system of veins. Lacking a microscope, Harvey was unaware of the existence of the capillaries which connected these two systems. Figure 2 is a key here, showing that when the experimenter "milks the vein downward" (from O to H), the one-way valve at O prevents the blood from flowing away from the heart. *De motu cordis et sanguinis in animalibus*, 1643. William Harvey.

but was accepted by most during his lifetime. The Library has a facsimile of the 1628 Frankfurt first edition of *De motu cordis*—the original being very rare, having been printed on poor paper and badly bound.

Two reasons have been offered as to why one of England's greatest scientific books was printed abroad. One is that Frankfurt was the center of the continental book trade, where every semester a book market was held to display the newest publications. Another reason suggested is that Harvey's Frankfurt publisher, William Fitzer, an Englishman, had been recommended by Harvey's friend Robert Fludd. Fludd had been published by Fitzer at Frankfurt and found that by publishing there he not only paid nothing for publication but actually received both free copies and a fee. The earliest copy of Harvey's book in the Library of Congress is the 1643 Latin edition published in Padua. The Library also has the first English edition, published in London in 1653. Together with the 1543 Basel edition of the *Fabrica* of Vesalius, it forms the core of the Library's collections in the history of medicine.

Harvey's link to Vesalius is substantial and real. The continuity of a Paduan medical education joins the two—Harvey studied under Fabricius ab Aquapendente, whose teacher was Gabriello Fallopio, the student of Vesalius. Harvey, the founder of modern physiology, successfully completed the anatomical assault on Galen begun by Vesalius.

Harvey's magnificent work of discovery, demonstration, and exposition towers above all other seventeenth-century medical efforts. Nothing else published during this century of progress is really comparable. Nonetheless, for medicine it was a most active and productive time. Influenced by the mechanical and mathematical orientation of Galileo and Descartes, seventeenth-century medicine is perhaps best characterized by its turn toward the exact sciences. During the early part of the century, a Paduan professor named Santorio Santorio (who was called simply Sanctorius) was one of the first to apply physics or mechanics to medicine. He devised

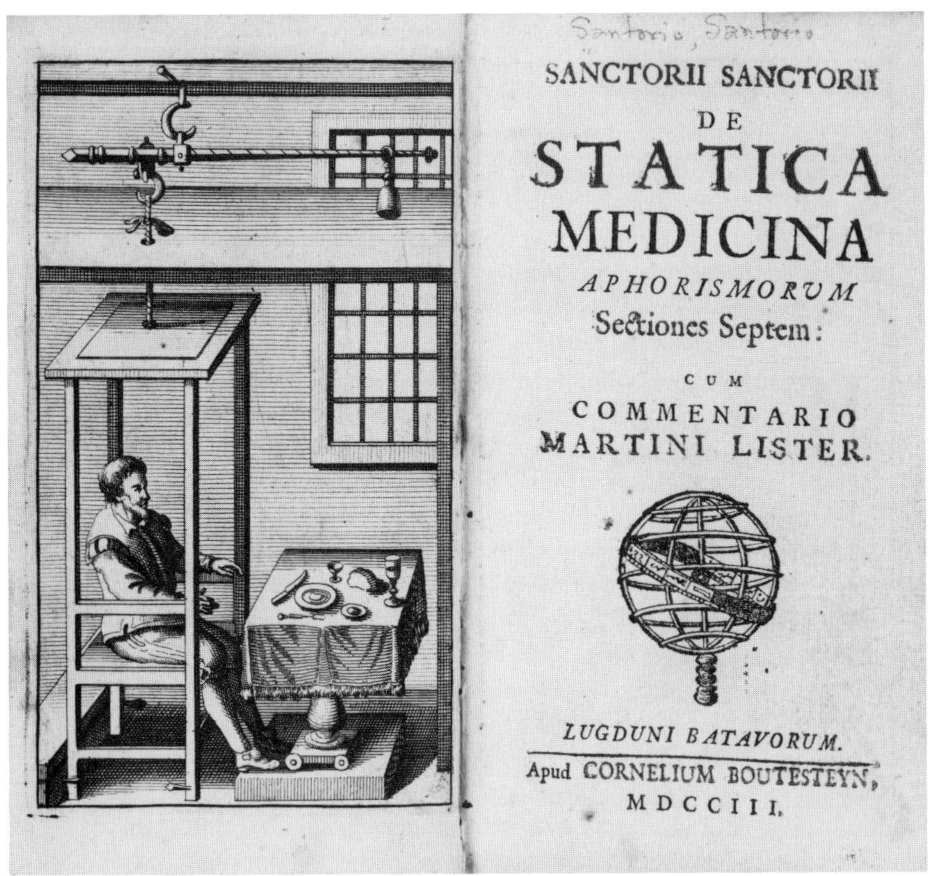

SANCTORII SANCTORII

DE

STATICA MEDICINA

APHORISMORUM

Sectiones Septem:

CUM

COMMENTARIO MARTINI LISTER.

LUGDUNI BATAVORUM.

Apud CORNELIUM BOUTESTEYN, MDCCIII.

The frontispiece to this later edition of Santorio's work shows the author seated in his balance or weighing chair—a platform which also held his bed and work table. With this he would measure his entire intake and output and study the effect of work, rest, and even mood changes on his body. As the inventor of the clinical thermometer and several other medical measuring devices, Santorio was one of the earliest to apply the quantitative approach to medicine and was a forerunner in the study of what we now call metabolism. His book became very popular, and this edition contains commentary by Martin Lister, English zoologist and physician to Queen Anne. *De statica medicina aphorismorum*, 1703. Santorio Santorio.

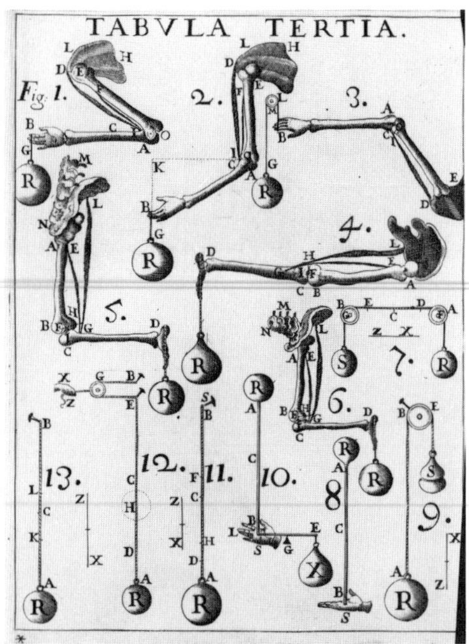

No work so typifies the iatrophysical or mechanistic school of medicine as does Borelli's *De motu animalium.* Following in the mechanical tradition of Santorio, Borelli thought of the body as a machine whose make-up and operation was entirely understandable by mechanical laws and reducible to them. Here he uses the principles of mechanics and statics to demonstrate how the operation of human appendages is similar to that of a pulley system. *De motu animalium,* 1680–81. Giovanni Alfonso Borelli.

scores of instruments and mechanisms too numerous to mention (the foremost being a thermometer to measure body temperature), all toward the end of obtaining measurable data about body functions. His experiments of thirty years are described in a series of aphorisms in his *De statica medicina,* published in Venice in 1614. The Library has the 1676 English translation as well as the 1703 Latin Lyons edition of this important book. The innovative Sanctorius initiated the tradition of experimental medicine, with what became an amazingly successful book.

Continuing the tradition of what came to be known as the iatrophysical school—because it explained body functions and disease in terms of physics rather than chemistry—was a pupil of Galileo's named Giovanni Alfonso Borelli. As a mathematician, Borelli adopted the Cartesian view of physiology, that the body was essentially a mechanism or a machine. This being so, he argued, its physiological functions could be explained by the laws of physics. Toward this end, Borelli set up a laboratory at his home in Pisa and went about his mechanical experiments, studying the physical principles of muscular action and attempting to measure the energy expended by the movement. His famous book on animal motion, *De motu animalium* (Rome, 1680–81), was promiscuous in applying this method, but it did contain many original observations and discoveries, particularly those pertaining to respiration and circulation. The Library has the first edition of Borelli's two volumes, which contain many illustrations showing humans and animals in various positions of muscle exertion.

A somewhat altered but essentially mechanistic view of medicine was the iatrochemical school of this period. Within a mechanistic framework, this school based its explanations of vital phenomena on a chemical interpretation. Two men are known for pioneering work in biochemistry. The first, Jean Baptiste van Helmont, called his own work into disrepute with forays into mysticism and alchemy, which made him a target of the Inquisition. A disciple of Paracelsus, Helmont developed the chemistry of gases and discovered the digestive juices in the stomach and intestine. It was in his most famous work *Ortus medicinae* (Amsterdam, 1648) that he described the first use of the specific gravity of urine. Helmont, however, did little for his scientific reputation with some of his pronouncements. He stated, for instance, that mice could arise spontaneously from spoiled wheat. The Library has the 1652 Amsterdam edition of this work, which was first published four years after Helmont's death by his son.

François de Le Boë, called Sylvius, competes with Helmont as the most influential of the iatrochemists. Unlike the unpredictable Belgian, Sylvius was a dependable scientist and an extremely popular Dutch university professor. In his *Idea praxeos medicae* (Venice, 1672), Sylvius asserted that all physiological phenomena could be explained by reference to chemistry. In this, he abandoned the traditional view that good health depended on a balance of the four humors, and he offered instead the explanation of an acid-alkalai balance. He thus diagnosed and treated all diseases chemically. The Library has the 1679 Amsterdam edition of his *Opera medica,* the collected works of a lecturer who drew students from all over Europe.

The extremes of these systematists who explained everything by reference to a single cause or phenomenon were modified by a late-century trend back to direct involvement with the patient. Thomas Sydenham, a physician who came to be called the English Hippocrates, led medicine away from these dehumanizing extremes with his healthy skepticism of overly theoretical medicine. Sydenham came to the medical profession rather late in his life and thus brought a mature perspective to his medical education. He applied his common sense and brought an independent spirit to what he was taught and decided that most of the prevailing medical practice and theory was nonsense. Returning to the fundamentals of Hippocrates, Sydenham's basic therapeutic belief was that the physician only assisted the life force—nature did the curing. Consequently, he emphasized a need for clinical recognition of particular diseases, recognition that could be obtained only from observation and personal experience. His fame today rests on his firsthand accounts of diseases, and his treatise on gout is considered his masterpiece. The Library has a 1717 London edition of *The Whole Works of That Excellent Practical Physician, Dr. Thomas Sydenham,* which, contrary to its title, does not contain all of Sydenham's writings. The Library's two-volume *Opera medica* (Geneva, 1757) is more complete. Sydenham followed no authority or system in his practice of medicine and aligned himself with no school. He was regarded by his English medical peers as a maverick and an outsider, and in his own time his work had more effect abroad than at home. Sydenham's honesty, his integrity, and, above all, his traditional Hippocratic concern for the patient were constant reminders to physicians not to stray too far from the sickbed. He led by example and by force of character and succeeded in restoring real dignity to the medical profession. His contempt for established theory and practice is well illustrated in his sardonic recommendation of *Don Quixote* as the best practical guidebook for the young physician.

Despite Sydenham's honorable example and proscriptions—guidelines that will always remain valid—the eighteenth century was dominated by the systematists. Continuing and extending the previous century's tendency to construct systems while explaining physiological phenomena, this enlightened age placed emphasis on the measurable and the verifiable—exhibiting a heightened regard for the experimental method. A coincident and contrary trend (which was eventually to succumb) was the impulse toward a type of scientific mysticism akin to the Romantic movement of the time. Thus the eighteenth century, which harbored many a sober rationalist, has been described by some as the golden age of quackery. The progress of other sciences during this time was very rapid, and with the welter of discoveries in allied disciplines so great, it is no wonder that the physician's head was easily turned. The microscopists, such as Malpighi, Leeuwenhoek, and Hooke, had discovered unknown worlds; the chemical discoveries of Black, Lavoisier, and Priestley revealed hitherto unknown elements; and Franklin, Galvani, and Volta tapped unknown forces.

One who happily combined the Hippocratic ideals of Sydenham with the most positivist aspects of the systematists was Hermann Boerhaave, the leading physician of his age. Called "the common teacher of all Europe" by one of his more famous pupils, Boerhaave was an eclectic teacher and

THE

CONTENTS.

a Chap.

The highly original and scholarly physician Bernardino Ramazzini was an observer of the first rank and expounded an entirely new area of medical investigation as well as a new medical discipline—the study of occupational diseases. He recognized the connection between harmful metals and the craftsmen who used them, identifying problems such as lead poisoning in painters, mercury poisoning in gilders and chemists, and diseases of stained-glass workers who handled antinomy. To the questions a doctor should ask his patient he added, "What is your occupation?" *A Treatise on the Diseases of Tradesmen*, 1705. Bernardino Ramazzini.

practitioner, taking what he regarded as the best from each school of thought. His inclination and overall framework was Hippocratic, and it was therefore as a clinician that he excelled. His fame was such that it is said that he had a more direct influence on his contemporaries than any other doctor in history. The Library has two of his more significant works, a 1727 Lyons edition of *Institutiones medicae,* an excellent book on physiology first published in the same city in 1708, and his *Aphorisms,* as published in London in 1724. The former was so popular it was translated into Turkish and Arabic, but Boerhaave's lasting fame rests on his teaching. He instructed generations of European doctors at the patient's sickbed. He provided an example of how best to combine the theoretical and the practical.

The Hippocratic spirit of Sydenham was thus spread in the north of Europe by Boerhaave, while in the south another disciple of Sydenham was demonstrating in a highly original manner his indebtedness to the "English Hippocrates." Bernardino Ramazzini continued and refined the epidemiological studies of Sydenham to the point of focusing specifically on occupational maladies. Occasional references to certain work-related diseases had been made in the historical literature, but it was not until Ramazzini conducted a methodical study of the nexus between occupation and disease that anything remotely scientific or systematic was done. His *De morbis artificum,* first published in Modena in 1700, earned him the deserved title of father of industrial hygiene. In this book, which the Library has in its English translation, *A Treatise on the Diseases of Tradesmen,* published in London in 1705, Ramazzini discussed over fifty occupations and the etiology, treatment, and prevention of their associated diseases. Ramazzini was a most cultivated man and his prose reflects his skill and polish. His masterful essays discuss mercury poisoning in surgeons and lead poisoning in painters, as well as the sciatica of potters and eye troubles of painters. Like Boerhaave, Ramazzini blended his Hippocratic, patient-oriented philosophy and the rigors of the systematists' methods with marvelous results.

No one so typifies the century's emphasis on measurable and observable knowledge as does the Paduan medical professor Giovanni Battista Morgagni. His landmark work, *De sedibus et causis morborum per anatomen indagatis,* contains in exhaustive detail the pathologic findings of a lifetime spent doing autopsies. Published in Venice in 1761 when Morgagni was seventy-nine, his work at once laid the foundations of pathologic anatomy and elevated it to a major branch of medical science. For the first time the connection between the manifestations of a disease and its actual physiological effect—the changes in the diseased organ—were scientifically established.

Morgagni contended correctly that every anatomical alteration in an organ resulted in a change in anatomical function. He further argued that this change could only be recognized by an experienced physician thoroughly familiar with normal anatomy and proper organ function. Morgagni's work was based on nearly seven hundred dissections conducted in a rigorous, careful, routine manner. It is vast in scope and contains a number of brilliant descriptions of new diseases. The Library has the three-volume English translation, *The Seats and Causes of Diseases,* published in London in 1769. Morgagni's classic work is linked to the anatomy of Vesalius and the physiology of Harvey. Through his methodical investigations, he was able to systematize anatomical pathology to the point where he could accurately correlate each condition with its proper clinical symptoms. Morgagni died in his ninetieth year, and was described shortly before that to be as hale as a man of fifty and not in need of spectacles. His *De sedibus* appeared in numerous editions and even today remains alive and useful.

In our survey, the eighteenth century concludes with the publication of what was to become one of the triumphs of empirical research in medicine as well as the notable beginning of preventive medicine. In 1798 Edward Jenner published a thin quarto volume with four colored plates called *An Inquiry into the Causes and Effects of the Variolae Vaccinae.* Based on twenty years of smallpox research, and describing twenty-three experiments, Jenner's *Inquiry* offered the thesis that inoculation with cowpox protects an individual from contracting smallpox. Jenner was a likable, honest country doctor in Gloucestershire who, from the beginning of his practice, learned all he could about smallpox and its transmission. He became fascinated by the local farmers' wisdom that said that dairymaids who had contracted cowpox were immune to the dreaded and more serious smallpox. Fear of the pox had become almost universal. In severe epidemics, one out of three died and those who survived were horribly scarred and disfigured. To Jenner's merit, he studied the pox in a deliberate, scientific manner, carefully proceeding to the fateful experiment of May 1796 in which he vaccinated an eight-year-old boy with cowpox fluid taken from a blister on the hand of an infected girl. Six weeks later, Jenner tried to infect the boy with fluid from human smallpox and could not—the boy was immune. The confident country doctor then repeated these experiments on humans—something that astounds us today—until satisfied he was correct. These and other experiments were described in his 1798 text, which he was forced to publish himself—the Royal Society having politely declined. In the Library of Congress collections are the 1800 second edition, again published in London by the author, and a fascimile of the 1798 first edition. This second edition contained additional information and results that Jenner hoped would help convince some of his more intransigent opponents.

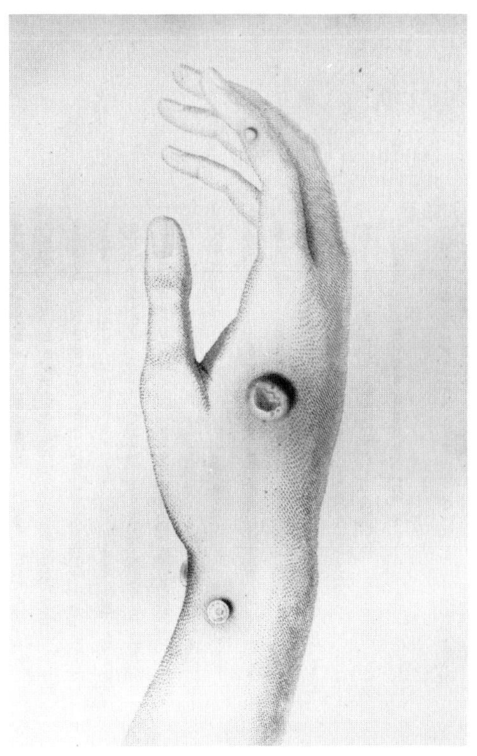

It was from this cowpox blister on the hand of the Gloucestershire milkmaid Sarah Nelmes that Edward Jenner took fluid and injected it into the arm of a healthy eight-year-old boy, giving him cowpox—a mild, transient disease. Six weeks later the boy was inoculated with the dreaded smallpox virus, with no assurance that he would survive the disease. The boy did not contract smallpox, even after receiving a second injection several months later. Although Jenner is rightfully praised as the discoverer of vaccination and the deliverer from smallpox, his risky experiments on humans are chilling to modern sensibilities. *An Inquiry into the Causes and Effects of the Variolae Vaccinae*, 1800. Edward Jenner.

Little persuasion was required, however. Inoculation of all types was not new to continental Europe. Throughout medical history it had had a sort of underground, folkish existence. Once inoculation for smallpox was given a scientific imprimatur by Jenner, it gained immediate acceptance. By 1803, his work had seen many translations, the British royal family had been vaccinated, and Jenner was voted ten thousand pounds by Parliament (he eventually was given twenty thousand more). Jenner's discovery demonstrated what bounties might result from the scientific method. A scourge of mankind was suddenly lifted and Jenner's name was praised worldwide. Yet the humble country doctor probably would have readily admitted that

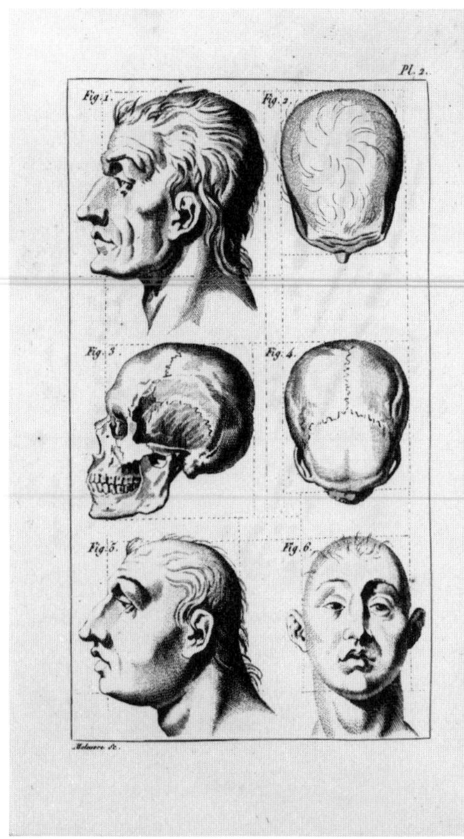

Philippe Pinel was an early reformer in the treatment of the insane, regarding them as sick people rather than as possessed by devils or witchcraft. His work was guided by the idea that mental problems were the result of disease or of pathological changes in the brain. Pinel placed great emphasis on what he called structural defects of the skull and regarded cranial shape as a major factor in a person's mental state. Here he contrasts a roundish skull from a sane person (figs. 3 and 4) with those of two insane people (figs. 1 and 2 and 5 and 6). *Traité médico-philosophique sur l'aliénation mentale*, 1809. Philippe Pinel.

although he had arrived at the "how" of the disease pragmatically, he was at a loss to describe the "why." Not until nearly a century later would the world learn from Louis Pasteur something about the causes of disease.

It was during the eighteenth century that the insane began to be regarded as sick people rather than as criminals or as people possessed by devils. The end of the century saw real reform in the manner in which these unfortunates were treated. These humane changes began in revolutionary France with Philippe Pinel who, while in charge of a Paris insane asylum, directed that the inmates' chains be removed and that they be regarded as sick of mind. The Library has his classic *Traité médico-philsophique sur l'aliénation mentale* (Paris, 1801) in its second edition of 1809, also published in Paris. Pinel's unorthodox views on insanity were embraced by the leading physician of the new United States, Benjamin Rush, who in 1787 took charge of the insane at the Pennsylvania Hospital. Rush is regarded by most as the father of American psychiatry, having produced the first systematic book on the subject in America. Rush argued that the emotions and behavior of the insane suffered as much as the intellect, and he studied the relationship between the body and the mind. His treatise *Medical Inquiries and Observations upon the Diseases of the Mind* was first published in Philadelphia in 1812. The Library of Congress has this work in first edition as well as a modest collection of Rush's papers. In addition to his medical lectures of 1803–4, the Library's manuscript collection includes some of his correspondence with the political luminaries of the time—John Adams, Thomas Jefferson, James Madison, James Monroe, and Thomas Paine. Rush was a signer of the Declaration of Independence.

Both Rush and Pinel were preceded in their pioneering work in the psychological field by an Austrian whom neither would have acknowledged as a real physician. Franz Anton Mesmer began his medical experiments with the latest vogue, electricity and magnetism, and soon made a mystical jump to what he called "animal magnetism." Simply put, he believed in the curative powers of the ancient tradition of the laying-on of hands. Mesmer arrived in Paris in 1778 with a reputation for the unusual and in no time became the darling of French high society. His spa, a glittering hodgepodge of technology and theater, entertained, enthralled, and "cured" its prominent customers with magnetic tubs, hypnotism, and seances. The image of Mesmer in a lilac suit, playing the harmonica and touching his patrons with an iron wand was far removed even from the medicine of his day, and it was not long before his claims for animal magnetism were investigated by the Académie des Sciences. The committee reported unfavorably, but its composition is more interesting than its report. Among the experts were Benjamin Franklin (U.S. ambassador to France), Dr. Joseph Guillotin (proselytizer for painless execution, after whom the guillotine was named), Antoine Lavoisier (pioneer of the new chemistry), Philippe Pinel, and Jean Sylvan Bailly (who later became mayor of Paris). Both Bailly and Lavoisier were to die during the Revolution, beheaded by their colleague's mechanical namesake. Mesmer left Paris in disgrace in 1785, but his 1779 book remained. The Library has a first edition of his *Mémoire sur la découverte du magnétisme animal,* published in Paris and Geneva. Most now agree that Mesmer was a sincere believer in his theories and methods and that the cures he effected were genuine. Through hindsight we recognize the principle underlying his method as the power of suggestion in curing psychosomatic illnesses, and not some unseen force called animal magnetism. In a sense, Mesmer was an unwitting pioneer of psychotherapy.

The nineteenth century begins what is usually called medicine's modern period, and it is within the broad positive connotations of "modern" that many of the reasons for medicine's explosive advancement during this time are found. Modern times connote progress—material and immaterial. The latter entails concepts of freedom that translate into the realities of individuals challenging dogma and separating science from metaphysics. Material advancement made possible such tangible benefits as universities opening to all social classes, expanding industries spawning new technologies, and advances in transportation and communication speeding up the exchange of ideas. Most important of all perhaps was the synergistic effect that produced a greater awareness of the dignity of the individual. In summary, medicine kept pace with the volatile nineteenth-century advances and consequently made great strides.

It is safe to say that it was not until the nineteenth century that medicine was able, in any broad and real way, to help the suffering individual. During this period, technical advances aided the diagnostician as well as the surgeon, and the beginnings of an understanding of the fundamental mechanisms of disease were emerging. All aspects of medicine—from the research laboratory to the operating table—were enjoying the benefits of the rigorous application of the scientific method. By the end of the century, a person's chances were fairly good that his doctor could not only give a name to his medical complaint but probably had an elementary understanding of what it was and how it progressed. With somewhat more luck, the doctor could select the proper treatment and mitigate the symptoms if not cure the disease altogether. The implications of this modest medical advance are most significant. First, it presumes an understanding of the true nature and origin of disease. Second, it assumes that an organized body of standard medical practice is available to guide the physician in his diagnosis and treatment. Last, it presupposes a degree of medical technology hitherto unavailable. Such was the encouraging state of medicine at the end of the nineteenth century.

Among the more dramatic nineteenth-century medical advances were those in the field of human physiology. In 1822, an obscure American army camp surgeon practicing medicine near the Canadian frontier was transformed almost overnight into a specialist on the mechanism of human digestion. The physician, William Beaumont, was called to treat a nineteen-year-old trapper accidentally shot in the stomach. Beaumont's operating skill saved the boy's life but his patient was left with a permanent gastric fistula (an abnormal opening leading to the stomach). To Beaumont's credit, he recognized this unique opportrunity to study the human digestive process in situ, and for the next ten years he conducted hundreds of experiments with the reluctant cooperation of his sometimes not-so-willing patient. Beaumont introduced into his human laboratory various types of food suspended on a silk string and observed the results. From this tedious process, Beaumont was able to describe properly the physiology of digestion, demonstrating the characteristics of gastric motility and describing the appearance and properties of gastric juice (samples of which he sent to scientists throughout the world). He determined that the stomach contained hydrochloric acid and that it broke down food not through maceration or putrefaction but by a chemical process of dissolving. Beaumont conducted careful, detailed studies of all aspects of the digestive process and published in 1833 his classic *Experiments and Observations on the Gastric Juice, and the Physiology of Digestion.* The Library has a first edition of this landmark of American medical literature, which was cheaply printed and bound at Plattsburgh, New York, where Beaumont was stationed. Beaumont's pioneering work in experimental physiology made him a famous man. The cost of both the knowledge gained and Beaumont's fame was paid, however, by the difficult and unnatural life his widely known subject, Alexis St. Martin, was forced to lead. St. Martin was French-Canadian and part Indian. He suffered not only the indignities of having to tour medical colleges as "the man with the window in his stomach," but of having to act as Beaumont's servant as well. The two men's lives became inextricably bound over the years, and when St. Martin was not being chased by Beaumont after running away, he was returning on his own, no longer able to earn a living as a trapper. Ironically, the patient survived the doctor by many years, and died at eighty-two.

Two other giants of nineteenth-century physiology, like Beaumont, also were concerned with the study of digestion. The Frenchman Claude Bernard flourished in mid-century and the Russian Ivan Pavlov produced his great work at century's end. Bernard's fame and accomplishments go well beyond identification with any singular body function, however, and extend to the very heart of the scientific method itself. His most important discoveries began with the study of digestion and metabolism, and they in turn led to his eventual proposal of a general theory of how the organs of the body work. All living organisms, he argued, are characterized by their ability to maintain their internal environment—their "milieu intérieur." Health and life are maintained by the body's ability to adjust internally to changing external conditions and to maintain a proper equilibrium. Bernard was a brilliant experimenter and correlator and an intuitive man as well. He was ahead of his time with his views on the physiological effects of emotion, and he attributed his own indigestion to France's humiliation in the war of 1870. Overall, Bernard is best known as the founder of experimental medicine, having pioneered the analytical technique of artificially producing disease by chemical or physical means. Although Bernard's emphasis on objective experiment advanced medicine considerably, it unfortunately made his personal life rather unpleasant, since his experiments made him the target of many an antivivisectionist. Indeed, Bernard's animal experiments so upset his daughters that they became estranged from him, and his wife even obtained a legal separation. Bernard was regarded much more positively by his scientific peers, however, being called "physiology itself" by one of them. At his death he became the first scientist given a public funeral by France. Many of his particular medical discoveries, such as the glycogenic function of the liver, digestion in the small intestine, and the vasomotor mechanism (nerves which govern the dilation of blood vessels), were published in the *Comptes rendues* of the Académie des Sciences, Paris, of which the Library has a complete set. Although the Library does not have his 1865 *Introduction à l'étude de la médecine expérimentale* in first edition, it does have his two-volume work *Leçons de physiologie expérimentale appliquée à la médecine* (Paris, 1855–56). This treatise also contains his classic work on glycogenesis and experimental diabetes.

Like Bernard, Ivan Petrovitch Pavlov also studied the digestive process and experimented on animals. And like the famous Frenchman, the Russian Pavlov created artificial fistulas in dogs' stomachs, but with a greatly advanced surgical technique. Pavlov's surgical skill was such that he was able to produce permanent gastric and pancreatic fistulas in his dogs without any injury to their nerves or blood supply. Now able to experiment at length, he made impressive gains in the study of the physiology of digestion and contributed significantly to its advance. Pavlov then embarked on one of those singular scientific odysseys of thought that sometimes occur when a mind of genius follows its instincts. Familiar with what he called the "unconditioned" reflex of a hungry dog salivating, Pavlov became intrigued by the possibility of evoking a similar response via a "conditioned" reflex. Thus his famous experiments with bell-ringing and dog salivating demonstrated that repetition of specific stimuli could produce reflexes that have no direct relation to the stimulus. This radical new departure of Pavlov's opened an area of investigation and speculation heretofore wholly ignored and unknown. With the publication in 1897 of his *Lektsii o rabotie glavnykh pishchevaritel'nykh zhelez* (Lectures on the function of the main food-digesting glands) in St. Petersburg, which offered detailed experiments and results of his conditioned-stimulus and selected-response investigations, Pavlov pioneered a new field—the physiology of behavior. Pavlov's 1897 work was quickly translated into German the next year, and this became the version best known outside of Russia. The Library, however, has the original 1897 Russian version. The implications of Pavlov's conditioned reflex discovery as well as its possible application to other nonscientific fields were not lost on Pavlov or his contemporaries. Pavlov continued his work, eventually using his theory of conditioned reflex to explain much of complex human behavior and mental processes. His work has been criticized as being founded on completely mechanistic tenets and as being susceptible to authoritarian misuse. Despite these objections, his work gave great impetus to the fin-de-siècle blossoming of psychology and for the first time indicated that a certain degree of human behavior is explicable by individual conditioned reflexes. As with Sigmund Freud, Pavlov's best work was highly singular and original and exemplifies the creative mind at its intuitive and inductive best.

In this photograph reproduced in a recent edition of his collected works, Ivan Pavlov appears as he was—a determined, energetic, passionate, and very human individual who was respected by all who knew him. Although his conditioned-reflex experiments and later work laid the basis for the scientific study of behavior, Pavlov was no strict behaviorist when it came to human nature. Every year without exception he would holiday with his family, saying "no scientific treatise had a passport to the country," wisely acknowledging and indulging in life's uncomplicated pleasures. *Lektsii o rabotie glavnykh pishchevaritel'nykh zhelez*, 1897. Ivan Petrovich Pavlov.

In addition to these major advances in understanding human physiology, nineteenth-century medicine was perhaps best characterized by revolutionary developments in understanding the nature and mechanism of infectious disease. Not until well into the second half of the century did the knowledge that bacteria both caused disease and functioned as contagious transmissible agents become well known and accepted. Growing evidence of this now incontrovertible fact was mounted first in the field of obstetrics. In 1842–43 Oliver Wendell Holmes, the famous American man of letters, published an article in the *New England Quarterly Journal of Medicine and Surgery* entitled "On the Contagiousness of Puerperal Fever." Holmes was a physician and a professor of anatomy at Harvard University from 1847 to 1862 as well as an eminent essayist. In this article, which the Library has in its collections, Holmes attributed childbed (puerperal) fever to infections that were introduced to the new mother by the hands of her examining doctors, who had touched other infected persons. Holmes stated his case eloquently but offered no empirical proof. His exhortations that physicians wash their hands before and after pelvic examinations were ridiculed by his peers and dismissed. Holmes nevertheless wrote a book years later entitled *Puerperal Fever, as a Private Pestilence,* which restated his case. The Library has a first edition of this work published in Boston in 1855.

Holmes was indeed on to something, but it was left to a Hungarian physician to summon the empirical proof. Ignaz Philipp Semmelweis, an obstetrician at the Vienna Krankenhaus, noted the high mortality rate from puerperal fever among women examined by doctors and medical students, in contrast to the extremely low rate among women attended by midwives only. He then discovered that the doctors and students came to the patients directly from the dissecting morgue. His suspicions were confirmed during the autopsy he performed on his friend Koltetschka. His colleague had died of an infection from a scalpel wound sustained while performing an autopsy on a puerperal fever victim. The dead man's organs exhibited the same changes as those of a puerperal patient. Semmelweis immediately instituted a strict hand-washing policy and subsequently documented a startling decline in puerperal fever deaths. In a short time he had succeeded in virtually eliminating puerperal fever from his maternity wards. His original scientific communication was given the urgent title "Höchst Wichtige Erfahrungen über die Aetiologie der in Gebäranstalten Epidemischen Puerperalfieber" (or, Extremely important experiences concerning the etiology of epidemic childbed fever in lying-in institutions) and appeared in *Zeitschrift der K.K. Gesellschaft der Ärtze in Wien* during 1847–48 and 1849. The Library does not have this journal in its collections. Despite the apparently obvious and spectacular evidence in support of his idea, Semmelweis met with immediate and fierce opposition. His work was refuted by virtually every orthodox obstetrician of his day and he was personally persecuted. After leaving Vienna for Budapest, he published his now-famous treatise *Die Aetiologie, der Begriff und die Prophylaxis des Kindbettfiebers* in Budapest, Vienna, and Leipzig in 1861. This epochal book was basically a mass of barely comprehensible statistics written with hardly any style. Today, an original first edition is very rare; the Library's collections include a facsimile of it. Little notice was taken of the book despite its pioneering premise—that puerperal fever was a contagious form of blood poisoning or septicemia. Semmelweis died at forty-seven, a broken, brooding man. Ironically, he succumbed to septicemia himself while a patient at an insane asylum.

Both Holmes and Semmelweis advocated what is commonly known as sterilization or asepsis. Essentially, this is a preventive measure that keeps germs away from the patient. Pasteur had not yet proposed his germ theory, so neither Holmes or Semmelweis knew exactly why their method worked. It was thus not until Louis Pasteur founded the science of bacteriology with his fermentation researches that the world became aware of the unseen universe of the microorganism and its central role in the cause and transmission of disease.

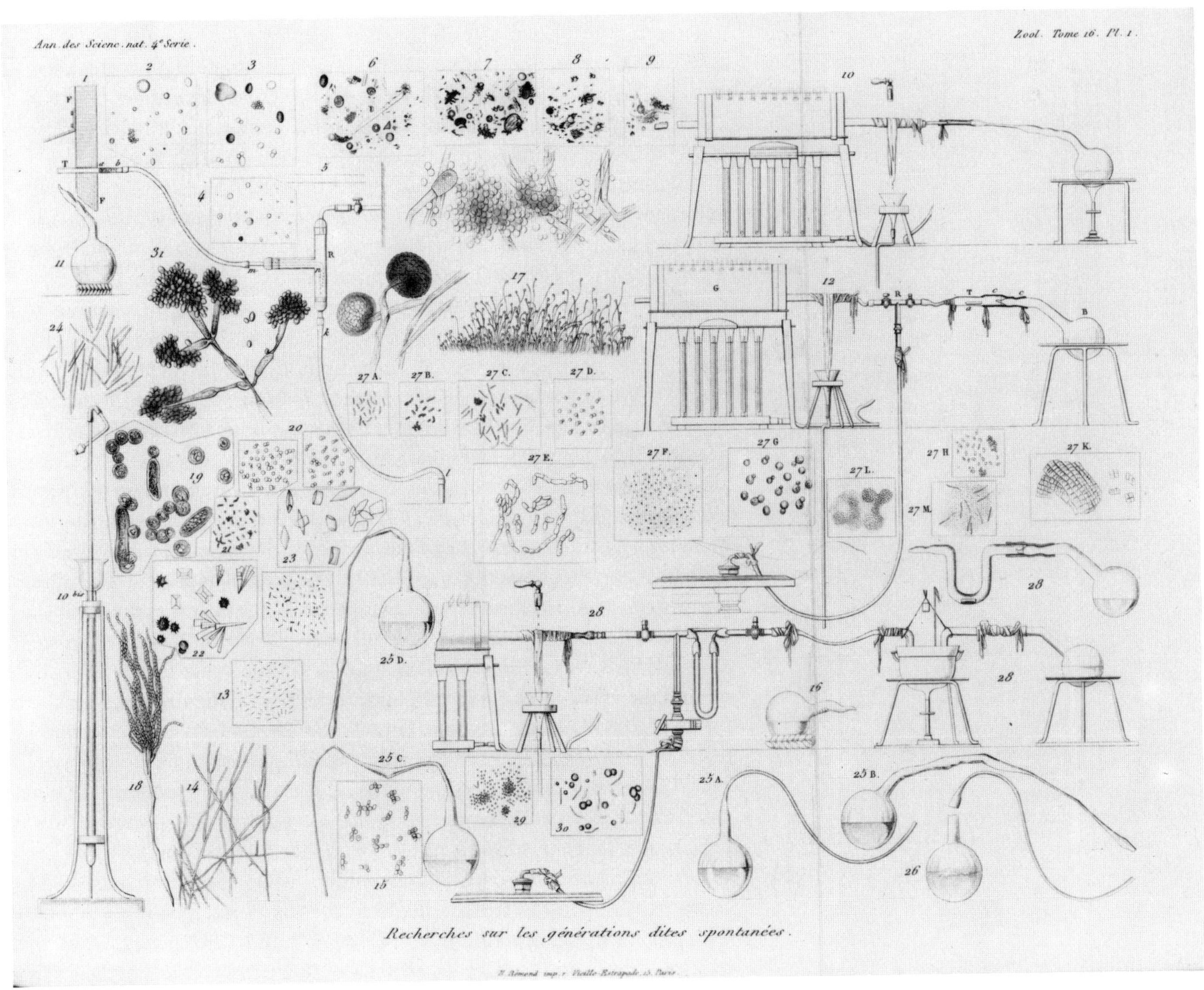

Louis Pasteur possessed a facility for experimentation and a genius that often made the complex appear simple. One of his straightforward experiments, showing that the microbes responsible for decay and fermentation were in the air and not in the decaying matter itself, is illustrated in this plate taken from his 1861 article on spontaneous generation. In the lower right, the long-necked bottle (25A) contains meat broth that was heated, thus destroying all the bacteria in it. Time passed but the broth did not decay, since the bacteria in the air remained in the lower crook of the bottle's glass tube. In bottle 26, the neck is removed, exposing the bottle's contents directly to the air. The broth soon became infested with bacteria. "Mémoire sur les corpuscules . . . ," *Annales des science naturelles*, 1861. Louis Pasteur.

Pasteur was trained as a chemist, and early on discovered that the microscope was his métier. By the age of twenty-six he had made a national name for himself with his microscopic work on the asymmetry of crystals. While in his thirties, he focused on the problem of why his country's wine and beer industry lost so much to spoilage. Investigating the fermentation process, he discovered that bacteria as well as yeast cells were present and concluded that fermentation was more than a purely chemical phenomenon and that it involved a living organism. Furthermore, only the proper organism would provide the desired effect. In a series of careful but simple experiments, Pasteur demonstrated decisively that fermentation is caused by the action of minute living organisms which, left unchecked, would cause spoilage. He recommended a gentle heating at an appropriate point in the process to kill these microscopic organisms. This method became known as "pasteurization." Pasteur made his discoveries known mainly through French scientific journals—primarily the *Comptes rendus* of the Académie des Sciences. The Library has all of Pasteur's articles in that journal as well as those published in the *Annales des sciences naturelles* and the *Annales de chimie et de physique*. The Library also has his books on wine and vinegar, *Études sur le vinaigre* (Paris, 1868), *Études sur le vin* (Paris, 1873), and *Fabrication du vinaigre* (Paris, 1875). Pasteur's studies on milk, wine, and beer were epochal in their significance to medicine. They disproved the traditional notion of spontaneous generation and proved that living organisms cause both fermentation and putrefaction. He revealed a world within a world, discovering what he called "the infinitely great power of the infinitely small."

One who came to appreciate immediately some of the implications of Pasteur's bacterial studies was an English surgeon, Joseph Lister. Like many a conscientious physician, Lister was often disheartened by patient deaths from gangrene or infection, especially after successful surgery. Using Pasteur's demonstration that living microbes can be airborne, Lister hypothesized that postoperative infections could be caused by bacteria in the air. He guessed correctly that suppuration, or the discharge of pus from an infected wound, was similar to Pasteur's putrefaction and could possibly be caused by similar unseen organisms. From that leap, it was a simple matter to discover a chemical to disinfect or combat the growth of these microorganisms. He came upon carbolic acid (phenol), sprayed it on the patient during surgery, cleansed his instruments with it, and even used it to dress wounds, with the result that amputation mortality fell by almost two-thirds. His March 1867 article entitled "On a New Method of Treating Compound Fracture, Abcess, Etc.," published in the *Lancet,* initiated the era of antiseptic surgery. Between March and September of the same year, Lister published six articles in the *Lancet,* further detailing his new system of antisepsis. A complete set of this British medical journal is in the Library of Congress collections.

During this time, Pasteur had saved the French silk industry in 1865 by discovering a tiny parasite that infected the silkworm. This work led him to articulate fully his germ theory of disease and to make his famous discovery that vaccine from germs that were weakened by heating, drying, or chemicals would produce an immunity without provoking any of the disease symptoms. Pasteur's dramatic and risky anthrax and rabies demonstrations in 1881 and 1885 established the validity of using attenuated germs to achieve immunity. Pasteur's germ theory of disease has been called the greatest single medical discovery of all time. Indeed, once Pasteur had revealed the principle, its elaboration became an eventuality. Pasteur was a complex man—a dreamy, even romantic individual who employed the most rigorous analytical methods. Well aware of his contradictory qualities, he used them to his advantage, acknowledging the role of intuition and imagination in his work. "Preconceived ideas are like searchlights which illumine the path of the experimenter and serve him as a guide to interrogate nature," he said. But "imagination must submit to the factual results of the experiments." Pasteur's later work on infectious disease, such as his 1880 article "Sur les maladies virulentes," are also part of the Library's *Comptes rendus* collection.

The great bacteriological beginnings provided by Pasteur came to fruition with the life's work of a German country doctor, Robert Koch. If Pasteur founded modern bacteriology, Koch developed its basic methodological techniques and saw to its final establishment. Koch's forte was his exceptional mastery of the technical aspects of research. He applied this talent to the study of an immediate practical problem—the deadly anthrax bacillus that was killing off his neighbor's animal stock. Through painstaking research in 1876, Koch discovered the anthrax bacillus in the blood and spleen of the dead animals. He then was able to culture the organism, pass it through several mice, and recover the same bacilli at the end of the process. Koch had, for the first time, been able to study and therefore to work out the complete life history and sporulation of a microorganism. As for the anthrax itself, Koch clarified how and when the disease starts and what its duration and cause were. The Library does not have Koch's anthrax paper, which was published in a work by Ferdinand Cohn in 1877, but it does have an 1880 translation, published in London, of Koch's book on *Investigations into the Etiology of Traumatic Infective Diseases,* which first appeared in 1878. This book described the bacteria of six different types of surgical infection and elevated Koch to the front rank of the medical profession. He continued his brilliant investigations and pursued his personal belief that tuberculosis was also an infectious disease. The publication of his paper "Die Aetiologie der Tuberculose" in 1882 capped eight years of intensive research and revealed that Koch had isolated and cultivated the infectious tubercle bacillus.

His tuberculosis paper had a second, far-reaching effect which by itself would have made it famous. In his tuberculosis research, Koch's procedure was so exemplary that it came to provide a model used to this day to prove that a specific organism is responsible for a specific disease. This procedure came to be known as "Koch's postulates." The Library's copy of this paper is found in volume 2 of a work entitled *Mittheilungen aus dem Kaiserlichen Gesundheitsamte,* published in Berlin in 1884. By the time of his death in 1910, Koch had discovered the causes of many other diseases—including cholera, Egyptian ophthalmia, and sleeping sickness—

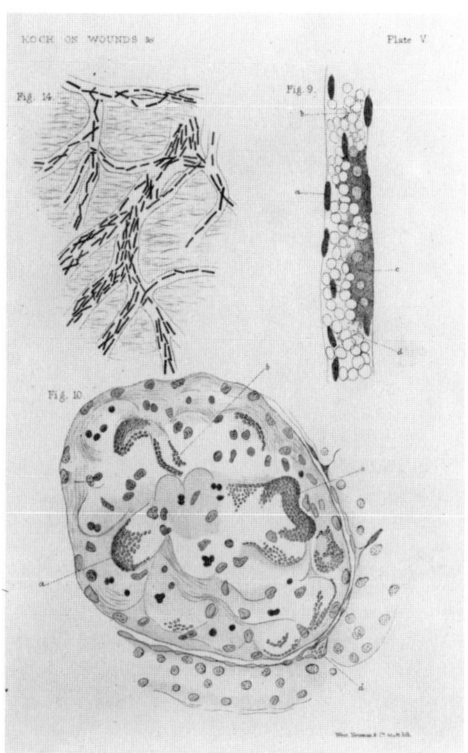

Robert Koch was a painstaking experimenter and a master of technique. Because of this, he was able to identify the organisms responsible for several diseases as well as to study their complete life cycles. This plate is taken from his landmark paper which demonstrated his new techniques in preserving, documenting, and studying bacteria. These illustrations show Koch's own drawings, which he took from his superb photomicrographs. *Investigations into the Etiology of Traumatic Infective Diseases,* 1880. Robert Koch.

58 Dritte Vorlesung.

kann man sie leicht von den homologen (Lobsteins homöoplastischen) dadurch trennen, dass sie von dem Typus desjenigen Theils, in welchem sie entstehen, abweichen. Wenn im Fettgewebe eine Fettgeschwulst oder im Bindegewebe eine Bindegewebs-Geschwulst sich bildet, so ist der Typus der Bildung des Neuen homolog dem Typus der Bildung des Alten. Alle solche Bildungen fallen der gewöhnlichen Bezeichnung nach in den Begriff der Hypertrophien oder, wie ich zur genaueren Unterscheidung vorgeschlagen habe, der Hyperplasien. Hypertrophie in meinem Sinne wäre der Fall, wo einzelne Elemente eine beträchtliche Masse von Stoff in sich aufnehmen und dadurch grösser werden, und wo durch die gleichzeitige Vergrösserung vieler Elemente endlich ein ganzes Organ anschwellen kann. Bei einem dicker werdenden Muskel werden alle Primitivbündel dicker. Eine Leber kann einfach dadurch hypertrophisch werden, dass die einzelnen Leberzellen sich bedeutend vergrössern. In diesem Falle gibt es eine wirkliche Hypertrophie ohne eigentliche Neubildung. Von diesem Vorgange ist wesentlich unterschieden der Fall, wo eine Vergrösserung erfolgt

Fig. 27.

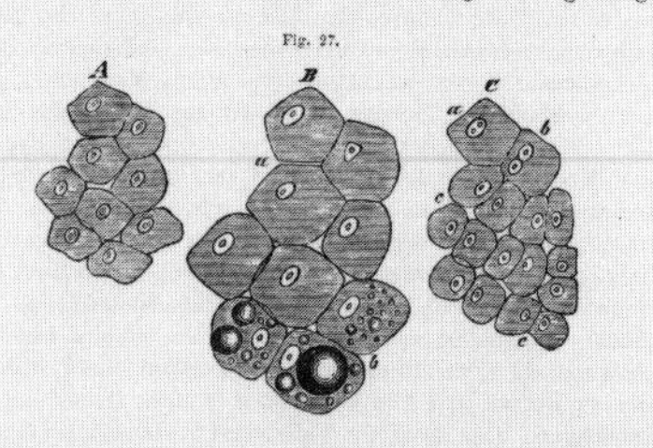

Fig. 27. Schematische Darstellungen von Leberzellen. *A* Einfache physiologische Anordnung derselben. *B* Hypertrophie, *a* einfache, *b* mit Fettaufnahme (fettige Degeneration, Fettleber) *C* Hyperplasie (numerische oder adjunctive Hypertrophie) *a* Zelle mit Kern und getheiltem Kernkörperchen. *b* getheilte Kerne. *c*, *c* getheilte Zellen.

By the mid-nineteenth century, medical research had attained the rigorous standards of modern medicine in the work of Rudolf Virchow. Working at the cellular level, Virchow demonstrated that the structure and appearance of living cells was profoundly altered by disease. In this illustration from his major work, he shows normal liver cells (A) and abnormal liver cells that have increased in size (B), as well as those that have become smaller and multiplied (C). *Die Cellularpathologie*, 1858. Rudolf Ludwig Karl Virchow.

and offered effective preventive measures for others, such as typhoid fever and malaria. The revolutionary work of Pasteur and Koch laid the foundations of bacteriology upon which such twentieth-century greats as Alexander Fleming, Selman Waksman, and Jonas Salk have built their own monumental medical accomplishments.

No mention of nineteenth-century medicine is complete without reference to the name Virchow. One of the greatest pathologists ever, Rudolf Virchow so dominated his time that he was called "the Pope of medicine" by his contemporaries. A man of many talents and interests, he conducted extensive anthropological and archaeological investigations, as well as being a political and social activist. In 1856 he became professor and director of the Institute of Pathological Anatomy of the Charité Hospital at Berlin and remained in that position until his death in 1902. During those years he performed an enormous number of autopsies and conducted a careful microscopic analysis of diseased tissues. In 1858 he published in Berlin, his great work, *Die Cellularpathologie* in which he demonstrated that the structure and appearance of living cells was profoundly changed by disease. The Library has a first edition of this revolutionary work that founded cellular pathology. In his application of cell theory to diseased tissue, Virchow brought to completion the work begun on cells by Robert Hooke over two hundred years before and continued by Marcello Malpighi, Nehemiah Grew, and Robert Brown, then through Matthias Schleiden and Theodor Schwann. Virchow can also be regarded as one of the real founders of scientific medicine—his rigorous work demonstrated that such unscientific theories as spontaneous generation were implicitly false. Virchow did, however, go too far in ascribing all diseases to the cell or to cellular imbalance, and he refused to acknowledge that disease might be caused by invasion from without. Some years later Pasteur's germ theory explained that such invasions occurred. Nonetheless, Virchow's genius gave the science of cellular pathology to medicine and laid the groundwork for later, more fundamental studies of the molecules within the cell.

Revolutionary nineteenth-century discoveries on the nature of disease aside, most laymen and patients would agree that the greatest medical discovery of the time was that of anesthesia. The blocking or relieving of pain made modern surgery possible and transformed surgery from a hurried, torturous procedure to a calm, controlled, scientific process. Though there were many known sleep-inducing drugs with long but spotty histories, ether became the first real successful general anesthetic. By 1830, chemists had discovered ether, nitrous oxide, and chloroform, but no medical applications were made until an American country doctor, Crawford W. Long, performed a successful minor surgical procedure using sulfuric ether. Dr. Long used ether successfully many times afterward but made no effort to publish his results until 1849. By that time, the discovery had been claimed by a dentist, William T. G. Morton, who authorized the esteemed New England surgeon, Henry J. Bigelow, to publish a full account of Morton's public demonstration at Massachusetts General Hospital. Entitled "Insensibility during Surgical Operations Produced by Inhalation," the article appeared in the 1846–47 *Boston Medical and Surgical*

Journal and is regarded as the first published report of this new anesthesia. The Library has this journal as part of its collections. A third contender for title of discoverer was Charles T. Jackson, a chemist and colleague of Morton's who claimed he had instructed Morton in how to use the new gas.

A long, rancorous dispute resulted as to who was the actual discoverer, and the controversy was eventually put to a special committee of the U.S. Senate to decide. The committee was unable to resolve the issue. In its manuscript collections, the Library has the papers of both Dr. Long and Charles Jackson. Long's collection consists of correspondence and legal documents as well as some photographs and offers what he regarded as documentary evidence attesting to his claim of priority. Jackson's collection consists primarily of his letters, in which he discusses his claim and opposes that of Morton. The outcome was sad for all parties. The quiet Dr. Long never received the recognition he deserved; Dr. Morton suffered a mental breakdown and died in poverty; and Charles Jackson went completely insane and died in a mental institution. A final casualty was Horace Wells, Morton's former dental partner, who preceded even Morton in his ether experiments. While he was demonstrating his procedure before Dr. John C. Warren's medical class at Harvard, Wells's patient cried out in pain and Wells was booed and hissed by Warren's students. He soon withdrew from practice and committed suicide in 1848 at the age of thirty-five.

Reference to the mental problems of these individuals is a fitting introduction to the work of one of the most creative and influential figures of the twentieth century, Sigmund Freud. The revolutionary psychological theories of Freud literally created a new medical field that had been heretofore mostly unrecognized and certainly never really explored. Although the scientific credentials of this field are not as firm as most others (traditional modes of scientific demonstration do not always apply), the impact of psychology on virtually every field of knowledge is today both unequivocal and easily demonstrable. Judged by their originality and revolutionary impact (as well as sustained influence), Freud's contributions are as fundamental and as significant as those of Darwin or even Copernicus. It was Freud, the pioneer into the mind of man, who made the world think psychologically and therefore made it more aware of itself. For this modern version of the loss of innocence he has been both praised and condemned—one critic calling him "the greatest killjoy in the history of human thought." Yet once Freud opened the door of psychology it could not be closed. It is safe to say that no other scientist has had such a broad influence, his ideas having penetrated almost every field of knowledge, from art and literature to daily life and speech.

Freud first studied neuroanatomy and neuropathology in Vienna and later went to Paris to work with Jean Martin Charcot, studying the problem of hysteria and the uses of hypnosis. It was from this work begun in 1885 that Freud eventually formulated his comprehensive theory on the determinants of human thought and behavior. His theory first appeared in print in 1895 in *Studien über Hysterie,* published in Leipzig and written in

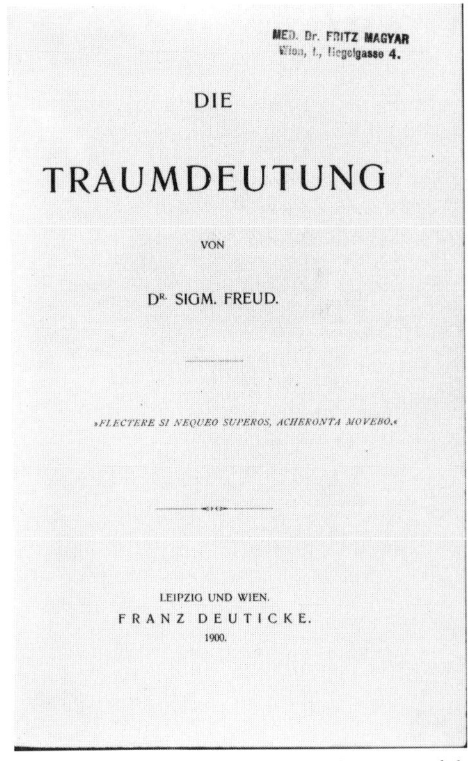

MED. Dr. FRITZ MAGYAR
Wien, I., Hegelgasse 4.

DIE

TRAUMDEUTUNG

VON

DR. SIGM. FREUD.

»*FLECTERE SI NEQUEO SUPEROS, ACHERONTA MOVEBO.*«

LEIPZIG UND WIEN.
FRANZ DEUTICKE.
1900.

It seems somehow fitting that the first year of the twentieth century should witness the publication of Freud's landmark work probing the human mind. Judged by its originality, impact, and influence, his work ranks as high as any in the history of science. *Die Traumdeutung*, 1900. Sigmund Freud.

collaboration with his friend and colleague Joseph Breuer. In this book, which the Library has in first edition, Freud offered his doctrine of psychoanalysis, which stated that human behavior is influenced by unconscious mental processes and conflicts. In 1900 his greatest work, *Die Traumdeutung,* was published in Leipzig and Vienna. The Library has this work in first edition, also. The landmark work contains all the basic components of what became the essence of Freudian psychology—dreams as wish fulfillment, displacement, regression, and the rest. Many of the dreams Freud examined were his own, and he gave detailed accounts and interpretations. His book represents the first attempt at a serious scientific study of the phenomenon of dreams, and Freud always regarded it as his greatest effort.

The Sigmund Freud Collection in the Library of Congress has grown over the years to become the largest collection extant of his papers. This important collection is complemented by the presence of related material belonging to his students and associates. The papers of Alfred Adler and the Freud-Jung letters are perhaps the most notable, with the papers of Siegfried Bernfeld, Rudolf Dreikurs, Arnold Gessell, Maxwell Gitelson, Smith Jelliffe, John Watson, and Edoardo Weiss also included. The major portion of the Freud papers were donated by the Sigmund Freud Archives, Inc., and by Freud's daughter, Anna. In addition to his draft manuscripts and articles, the collection consists of personal and professional correspondence, not all of which is yet open to researchers. It is a sizable collection, numbering over twenty-two thousand items and requiring 22½ linear feet of shelf space. Supporting this collection is a first edition collection of books written by Freud as well as more than fifty books owned by Freud—all with either an inscription to him, his signature, or some other indication of provenance. Taken together, the Library's growing collections in the history of psychology and psychiatry make it one of the major centers of research in that field.

After more than eighty years, much of Freud's work has been changed and modified, yet no amount of change can alter his stature or diminish the fundamental significance of his work. By the force of his genius he generated such an essential reorientation toward the study of the motivations of human behavior that he forced an essential change in our perception of ourselves and of the world around us. With an epic ambitiousness he focused on the most obscure, complex, elusive, and unpredictable subject for study—the mind of man. Alone, he went into unknown territory, and like any brave pioneer, he was able to describe, to designate, and, finally, to offer his personal vision of what he had discovered to the world.

It has been nearly fifty years since Freud died, and the medical world he inhabited no longer exists. In this fairly short span of time, medicine has undergone a sea change in its basic understanding of things and in its ability to diagnose and to treat. Yet despite the many spectacular successes of twentieth-century medicine, few today would suggest that we are approaching any medical nirvana. Besides those numerous diseases that have proven intractable, we witness regularly the emergence of new illnesses or the strengthened permutations of old ones. This combined with our high technology and its sometimes unexpected and often dangerous by-products and effects, as well as the stressful nature of modern lifestyles, all seem to conspire to pose new threats to our overall well-being. So medicine, like its sister disciplines, continues on its scientific treadmill—a frustrating, stimulating, and necessarily endless pursuit of understanding.

The tradition of medicine is linked most intimately with all things human. Unlike most other traditional scientific disciplines, medicine focuses wholly on mankind as its subject—creating the unique scientific situation of an essential identity between the investigator and what he or she is investigating. As a practice, medicine is nearly as old as mankind. As a science, it is one of the latest to mature. Yet whether ancient art or modern science, medicine is concerned primarily with the well-being of the whole human being and cannot help but have a typically humanistic tradition. Despite its magico-religious history with its sometimes altogether wrong-headed and harmful methods (which led one medieval victim to choose as his epitaph, "I died of a surfeit of doctors") or its mechanistic and coldly clinical modern counterpart, the tradition of medicine is essentially that of science both humanized and humane.

5. *Chemistry: Fertile Alchemy*

As an exact science, chemistry is relatively young. Compared to astronomy, its real beginnings as a systematic and unified discipline are found practically in the modern period. Until Lavoisier at the end of the eighteenth-century imposed a unity on chemistry, gave it a common language, synthesized previous work, and defined its legitimate research problems and methods, chemistry was a hodgepodge of all manner and types of science and pseudoscience. Part medicine, part metallurgy, and part guesswork, chemistry was an art whose practitioners would variously treat, distill, melt, vaporize, and even attempt to transmute. If liquid could be transformed into gas and solid into liquid, why not lead into gold, sickness into health, and well-being into eternal youth? By Lavoisier's time, then, chemistry was badly in need of definition and focus.

It is all too easy to blame alchemy for the seemingly retarded development of chemistry. Some might argue that the centuries of alchemical experience took away more than they gave to science. Functioning with no real theory in an almost ad hoc manner, encouraging the intuitive and the subjective, seeking to obscure rather than to clarify and to hide rather than to transmit knowledge, alchemy left a legacy of decidedly unscientific—and perhaps even antiscientific—methods. Were not centuries of human effort wasted by the indulgences of the alchemists?

This judgment can be made, however, only if one assumes that the development of science always proceeds in a rational and linear manner. The history of chemistry illustrates the opposite. Few disciplines have been so dominated by the love of speculation and the desire to do the impossible as chemistry in its alchemical period. The residue of its unscientific and gratuitous explanations tarnished even the best work of the most talented individuals. How else can we explain why Joseph Priestley stubbornly clung to the phlogiston theory in the face of Lavoisier's powerful and correct refutation.

The history of chemistry also shows how difficult the real stuff of chemistry—the knowledge of the composition of substances—can be to discover and to grasp. Was it not easier to study the planets' orbits and to plot their irregularities or to dissect a corpse to learn how the blood moves than to study the air itself? How does one study it, capture it as a subject, or analyze its properties? Alchemy may have been a dead-end path off the main road of chemistry, but its existence alone is not responsible for chemistry's relatively late flowering. Knowledge of the essential constituents of the material world required the solution of a powerfully complex puzzle—one not quickly or easily solved. Historical comparisons are some-

Opposite page:
The alchemist and his assistant in their laboratory are shown here in the chapter 7 frontispiece from Thomas Norton's influential *Ordinall of Alchimy.* Norton's translated work was the longest text in Elias Ashmole's *Theatrum chemicum britannicum*, the largest collection of alchemical poetry in the English language. *Theatrum chemicum britannicum*, 1652. Elias Ashmole.

times revealing and it is interesting to note the comparative levels of development of different disciplines by considering three books in the Library's collections.

The year 1543 is a scientific milestone, marking the real beginning of two modern sciences. In that year in Nuremberg, Nicolaus Copernicus published his *De revolutionibus,* which founded modern astronomy by correctly positing a heliocentric universe. Remarkably, only three hundred miles away in Basel, another revolutionary book was being published in the same year. *De humani corporis fabrica,* written by Andreas Vesalius, was launching modern anatomy and would eventually dispel a millenia of medical confusion with its scientifically accurate description and illustrations of human anatomy. Yet just one year before, in Venice in 1542, a book attributed to the legendary and long-dead Raymund Lully was published. Entitled *De secretis naturae,* this treatise offered the reader instructions detailing miraculous medical cures as well as ways to convert imperfect metals into pure silver and gold. Obviously, this work was the peer of *De revolutionibus* and *De humani corporis* only in chronological terms. But the point is made. While medicine and astronomy were entering their formative periods in the mid-sixteenth century by beginning to experiment and to seek scientific truths for their own sake, chemistry was a backward art, steeped in myth and distraction—and still over one hundred years away from Robert Boyle's healthy skepticism.

The following discussion of major works of chemistry in the collections of the Library of Congress begins with the long heritage of alchemy and attempts to focus on its real and positive contributions. One might ask whether anything could have lasted so long and turned the heads of so many if it were totally useless and not at all practical. At the least, alchemy offered centuries of applied knowledge of the nature and characteristics of materials and their combination. The chapter concludes at the beginning of the twentieth century with the last of the great individual chemists. An underlying theme, linking the early alchemists to these modern chemists, might be the search for the elements of nature—those primary material substances common to all matter.

Alchemy is both the fertile subculture and the grand illusion of chemistry. Typically, it is disparaged as a disreputable pseudoscience—a scientific cul-de-sac, and a sadly laughable one at that. Nonetheless, a significant part of the history of chemistry is undeniably alchemical—chemistry having had a long inductive period—and to slight the contributions of alchemy to the development of chemistry would be to present a distorted picture.

Alchemy has been many things to many people. To some it was a "divine art" whose admixture of astrological symbols, magical charms, and religious ideas would produce gold from base metals. To others, it was a patient and doggedly experimental search into the nature of things. Although practitioners of both schools may have worked with the same materials and performed similar experiments, it was the underlying philosophical concept of each that distinguished the charlatan from the serious experimentalist. The philosophy of the false alchemist was straightforward—his was a singular search for a way to make gold from common metals. Not much philosophy there. The true alchemist, however, was more serious both in his manner and in his objectives. He sought an understanding of the principles of things—trying to discover the essential, invisible force behind the material world.

Whatever the merits of such a search, this goal both assumes and engenders a certain inquisitiveness, a healthy curiosity about the natural world. Inevitably, chemistry was to benefit from these centuries of experimentation and experience with all sorts of metals and materials and combinations thereof. Despite its sometimes irrational premises, fraudulent practitioners, and ludicrous claims, alchemy bequeathed to science a tradition rich in hands-on experience. The alchemist's laboratory was literally a place of labor, the ubiquitous furnace and alembic always at work melting, distilling, sublimating, refining, and evaporating. It is because of this pragmatic, experimental tradition that the rich legacy of alchemy assumes a significance to the history of the chemistry.

Alchemy is commonly said to have been begun by Hermes Trismegistos, the Greek name applied to the Egyptian god Thoth ("Thoth, the thrice great"), the patron of science and learning in general. It was he who supposedly provided the basic document of alchemy, the Emerald Tablet. From its origin in ancient Egyptian mysticism, alchemy followed the mainstream of learning and civilization, flourishing in Alexandria and spreading from there throughout the Greek-speaking world. After Rome fell, the Nestorians and Monophysites, exiled from Byzantium, brought their alchemy to Syria and Persia where it was taken up enthusiastically by the Arabic-speaking peoples. After some centuries, alchemy was spread through the whole of Western Europe by the European scholars who translated the Arab texts. During the Renaissance, alchemy was further stimulated by two contrary impulses—a romantic revival of things ancient and a pragmatic urge to manipulate the natural world. Following this period, alchemy was very much alive and continued to receive varying degrees of serious attention even through the eighteenth century.

Because the legacy of alchemy is such an ancient one, its literature is sizable. Yet as a body of literature it is maddeningly chaotic. The central problem is the secretive nature of both the craft and the craftsman. Unlike the modern scientific method, which values and encourages rapid disclosure of results and methodology, the tradition of alchemy was to obscure and to conceal. As a result, alchemical texts are suffused with a kind of cabalistic symbolism and equivocation that made metaphor their medium. Conveyors of theoretical truths they may not be, but these texts do contain significant empirical information on the nature of materials. Yet by far it is their abstract, equivocal quality that fascinates, intrigues, and attracts. As contemporary evidence of the omnifarious nature of alchemy, there is the convention that views all its texts as being solely allegorical tracts with psychological implications. Alchemy is surely many-sided.

EMBLEMA XIV. *De ſecretis Naturæ.* 65

Hic eſt Draco caudam ſuam devorans.

EPIGRAMMA XIV.

Dira fames Polypos docuit ſua rodere crura,
Humanáque homines ſe nutriiſſe dape.
Dente Draco caudam dum mordet & ingerit alvo,
Magnâ parte ſui fit cibus ipſe ſibi.
Ille domandus erit ferro, fame, carcere, donec
Se voret & revomat, ſe necet & pariat.

I VETE-

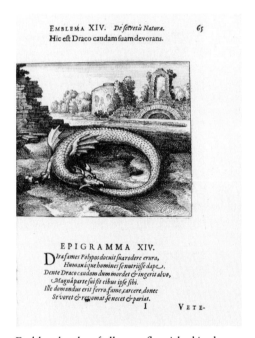

Emblem books of all types flourished in the seventeenth century, and alchemy found their use of symbolic pictures and didactic mottoes especially appealing. This page from Michael Maier's *Atalanta fugiens* typifies the alchemical emblem book with its symbolic "Ouroboros" pictured below the words, "This is the dragon which devours its own tail." The epigram below the illustration tells metaphorically of the unity of all matter and of regeneration of life through death. To the practicing alchemist, the circle and the epigram symbolized the repeated dissolving, evaporating, and distilling he must perform to obtain the desired degree of purity of matter. *Atalanta fugiens,* 1617.

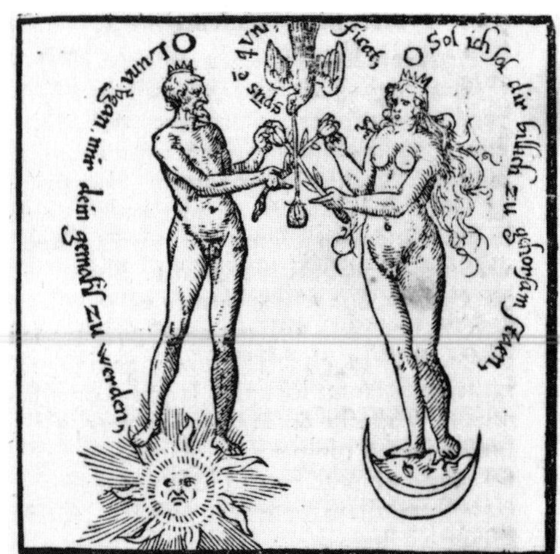

The use of symbolic pictures in alchemical texts was a sort of insider's shorthand. Aside from the ubiquitous dragon or serpent, the image of a marriage was one of the more frequently used symbols. This picture symbolizes the combination of two substances—the union of king and queen, sun and moon, gold and silver. Despite its penchant for the esoteric, nearly all of alchemy's symbolic images are grounded in common notions and relate directly to the realities of everyday life. *Artis auriferae,* 1593.

This simple woodcut is a very crude version of an elegant late medieval manuscript. The story told here is an allegorical rendering of an alchemical process or recipe. Specifically, it illustrates one part of a process (dismemberment signifies purification through separation of parts) which leads ultimately to the generation of the philosopher's stone—that magical tincture responsible for all sorts of chemical miracles. *Aureum vellus,* 1599.

Debates as to their meaning and merits aside, virtually all of the major compilations of alchemical writings are represented in the collections of the Library of Congress. From the sixteenth century to the end of the eighteenth, it was a fairly common practice to gather alchemical books and manuscripts and print them together in collected editions. Nine of these works will be described here. The earliest and one of the most valuable of these collections is *De alchemia,* published in Nuremberg in 1541. The Library's copy of this work is on microfilm. *De alchemia* contains sixteen woodcuts and includes the works of the Arab Djaber, called Geber, and the famous "Tabula smaragdina" (Herme's Emerald Tablet), whose conundrums and equivocations are printed here for the first time. It also contains *Speculū alchemiae,* a work attributed to Roger Bacon that became the original text from which the 1597 English edition of *The Mirror of Alchimy* was made.

Verae alchemiae, a very large work with no illustrations, was published in Basel in 1561. It was edited by the physician Guglielmo Grataroli and includes the works of Roger Bacon, Geber, Arnold of Villanova, and Albertus Magnus, among many others. As a good cross section of alchemical literature, it contains practical knowledge of chemical facts as well as typically extravagant claims of gold-making or boasts of prolonging human life.

Another of the chief collections of the standard alchemical authors is *Artis auriferae,* whose title fancifully tells of the art of making gold. Although this work was first published in Basel in 1572, a more handsome reprinting was done in the same city in 1593, of which the Library has a two-volume copy. It includes the famous *Turba philosophorum* or *Assembly of Philosophers,* which is regarded as the report of a meeting of historical sages and alchemists of classical antiquity who discourse and debate alchemical issues. In 1937, a later edition (1610) of this collected work was offered for sale by a London bookdealer whose catalog noted that "Some of the woodcuts in the work would, from any but the alchemical point of view, be considered highly improper." This Library has this edition on microfilm.

The Library's copy of *Aureum vellus* or *The Golden Fleece* is a rare and interesting work attributed to the mysterious Salomon Trismosin, magician and legendary teacher of Paracelsus. Many doubt whether such a person ever existed. This work opens with Trismosin narrating his many travels through Italy and the East and telling of a tincture with which he made himself young. In veiled language and symbolic illustrations, Trismosin supposedly tells how to prepare the tincture (or philosopher's stone)—the ineffable substance that would effect all manner of chemical miracles. The Library's copy of this collected work was printed in Rorschach on Boden See in 1599 and also contains Trismosin's famous *Splendor solis.* Its pages are darkened and its overall condition indicates that it was heavily used. The book eventually saw many translations. In French, it became well-known as *La Toyson d'or,* published in Paris in 1612 and 1613. Another edition, containing the complete five tractates, was published in Hamburg, 1708–18, under the title *Eroffnete Geheimnisse des Steins der Weisen oder Schats-Kammer du Alchymie.*

The six-volume *Theatrum chemicum* published by Lazarus Zetzner in Strasbourg between 1659 and 1661 contains about two hundred tracts and is the largest and most comprehensive compilation of alchemical works published. This collection is both representative and unique, many of its treatises coming from hitherto unpublished manuscripts. Isaac Newton had this work in his library and is known to have used it often. The library has all six of the small but very thick volumes.

Theatrum chemicum britannicum is similar in name only to the above work. This large volume is a rare collection of strictly English alchemical poetry. It is the work of Elias Ashmole, an English antiquarian and collector, who compiled the thirty-two poetical writings and published them with his notes in 1652 in London. The Library has the 1652 edition. As a source of information on the English medieval alchemical experiences, this work is invaluable. Newton had this work too in his library.

The Library's best copy of *Musaeum hermeticum reformatum* is the 1678 second edition published in Frankfurt. This collection of alchemical writings was first published there in 1625, but this second edition is the more complete. It contains twenty-one separate treatises, each with an engraved title page, and forty-one mostly symbolical illustrations. It was published by the physician and imperial poet Adrian Mynsicht under the pseudonym Henricus Madathanus.

Jean Jacques Manget's *Bibliotheca chemica curiosa,* published in Geneva in 1702, nearly rivals Zetzner's *Theatricum chemicum* as the most complete collection of alchemical texts ever published. Made up of two folio volumes, both of which the Library has in its collections, the work is not only fairly exhaustive but also well organized. Manget, a physician to the king of Prussia, classified the authors according to the subject and nature of their writings. In the first volume there are sixty-nine tracts, and in the second, seventy-one. Manget's compilation is especially significant, since it contains many treatises not found elsewhere. Manget died in his ninety-first year, supposedly without ever having had a day's illness.

The final alchemical work worthy of note is *Bibliotecha chemica,* an alchemical bibliography compiled by Friedrich Roth-Scholtz. Published in Nuremberg in 1719, this useful compilation is written in German, despite having a Latin title. The book does contain gaps, however, since Roth-Scholtz did not live to complete his work. The Library has a recent facsimile of the later 1727–29 edition. Also in the Library's collection is his *Deutsches Theatrum chemicum* published in Nuremberg between 1728 and 1732. This collection of fifty-two different alchemical treatises contains a separate title page for nearly each work. Of these nine major compilations of alchemical literature, most are made up of pre-Renaissance alchemical writings and sixteenth-century medical recipes. In both periods, certain names recur and predominate. The names of Geber the Arab, Roger Bacon, Raymund Lully, Arnold of Villanova, Albertus Magnus, and Basil Valentine became obligatory to the compilers. Many works were attributed to these famous individuals—not all correctly so. Nonetheless, the Library has an impressive collection of their individual efforts. The works of the famous Geber are represented by *Chimia,* published in Lyons

This image depicts the combining of sulfur and mercury—which most alchemists believed to be the two immediate constituents of all metals and minerals. The lion-headed serpent devouring its own eagle head is a variation of the ancient "Ouroboros," which was an emblem of the eternal, cyclic nature of the universe and the unity of all matter. The chemical formula suggested here is not what it appears, however, since sulfur and mercury when combined form only the unmagical cinnabar (mercuric sulfide). The alchemists evaded this reality by arguing that only pure, "sophic," or ideal sulfur and mercury, and not their "vulgar" or ordinary counterparts, would achieve the intended result. *Theatrum chemicum,* 1659–61.

One of the more puzzling works in Manget's collection of alchemical tracts is the *Mutus liber*, a series of fifteen engraved plates with no explanatory text, supposedly depicting the preparation of the philosopher's stone. Here its first plate and frontispiece shows the biblical story of Jacob's ladder, in which God's messengers descend the ladder connecting heaven and earth and promise a sleeping Jacob a bounteous future. These plates are partly symbolic and partly representational and show a man and a woman performing various chemical operations. The book has been attributed to an eighteenth-century French physician named Tollé who used the name Altus or the anagram Saulat. *Bibliotheca chemica curiosa*, 1702. Jean Jacques Manget.

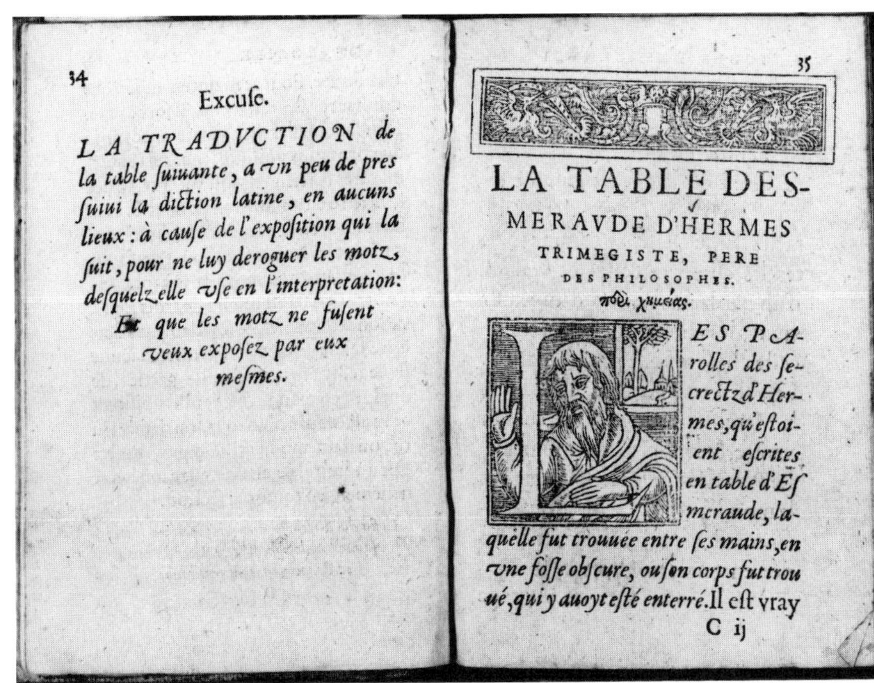
34

Excuſe.

LA TRADVCTION de la table ſuiuante, a vn peu de pres ſuiui la diction latine, en aucuns lieux : à cauſe de l'expoſition qui la ſuit, pour ne luy deroguer les motz, deſquelz elle vſe en l'interpretation: Et que les motz ne fuſent veux expoſez par eux meſmes.

35

LA TABLE DES-MERAVDE D'HERMES TRIMEGISTE, PERE DES PHILOSOPHES.

LES PAROLLES des ſecretz d'Hermes, qu'eſtoient eſcrites en table d'Eſmeraude, laquelle fut trouuee entre ſes mains, en vne foſſe obſcure, ou ſon corps fut troué, qui y auoyt eſté enterré. Il eſt vray

C ij

It is not known for certain that Roger Bacon wrote this book, originally titled *Speculum alchemiae.* It is a short and not particularly distinguished work, treating the origin and composition of metals, and contains some allusions to transmutation as well as obscure metaphysical discussions on the origin of mercury and sulfur. What does distinguish this book is the fact that it is among the first generation of Bacon's works to be printed, his writings having remained in manuscript form since the thirteenth century. Bacon's primary contribution to chemistry, indeed to all of science, was his insistence on the superiority of observation and experiment over mere argument as a method of acquiring knowledge about the natural world. It was opinions like this as well as exaggerated popular accounts of his experiments that resulted in his imprisonment and his being remembered by posterity as *doctor mirabilis,* the "wonderful teacher." *Le miroir d'alquimie,* 1557. Roger Bacon.

in 1668, and by *The Works of Geber,* published in London in 1678. The fascinating Doctor Mirabilis, Roger Bacon, about whom much imaginary lore has been written, is best represented by his famous book *Le Miroir d'alquimie,* published in Lyons in 1557. Bacon's book is almost small enough to fit in the hand, but it is extremely thick. Lully was, like Bacon, a friar. Among works attributed to him in the Library's collection is *De secretis naturae,* published in Venice in 1542. The Library also has the 1541 Strasbourg edition on microfilm. While Lully may have known some secrets of nature, as his book title suggests, they did not prevent him from being stoned to death in Algiers, where he was trying to convert the Muslims to Christianity.

By the fifteenth and sixteenth centuries, alchemy had taken a more practical turn and had entered what has been called its iatrochemical period. Alchemy in service to medicine may be best represented in the Library by the classic *Alchymia* of Libavius. Published in Frankfurt in 1606, this work has been described as the first chemical textbook, in the modern sense of the word. Libavius was the Latinized name of the German scholar Andreas Libau. His book included not only a record of the chemical knowledge of the his time and a summary of much past knowledge but also a unique and very detailed plan of what he considered to be an ideal chemical institute of the future. It is interesting to note that the building he described included not only areas for distilling and crystallizing but also a secret goldmaker's furnace and laboratory. The Library's copy of Libau's rare work is a large book with metal clasps and contains many illustrations of alchemical equipment. The most famous practicing medical alchemist was Paracelsus, but he was considered here in the chapter on medicine.

יְהֹוָה

D.O.M.A

HIPPOCRATES

HERMES

ALCHYMIA
ANDREÆ
LIBAVII, RECOGNI-
TA, EMENDATA, ET
aucta, tum dogmatibus &
experimentis non-
nullis;
TVM COMMENTARIO
Medico Physico Chymico:
QVI EXORNATVS EST VA-
riis Instrumentorum Chymicorum picturis;
partim aliunde translatis, partim
planè nouis:
In gratiam eorum,
QVI ARCANORVM NATVRALIVM
cupidi, ea absq; inuolucris elementariorum &
ænigmaticarum sordium, intueri
gaudent.
PRÆMISSA DEFENSIONE ARTIS:
opposita censuræ Parisianæ:
Cum Gratia & Priuilegio Cæsareo speciali ad decennium.

GALENUS

ARISTOTELES

IN DEO LÆTANDVM

FRANCOFVRTI,
Excudebat Joannes Saurius, impensis Petri Kopffii.
Anno cIↃ. IↃ. VI.

In a style common to many books of its time, this title page is bordered with scenes depicting its subject matter—in this case, alchemy. But the alchemy of Andreas Libavius was primarily pharmacology, or chemistry in service to medicine—accounting for the figures of Hippocrates and Galen on the title page. As a physician, Libavius agreed in principle with Paracelsus in his use of chemical remedies, but also criticized him for being both vague and extravagant in his chemical prescriptions. Libavius presented in his *Alchymia* not only the chemistry of his time but also a summary of past chemical knowledge, all written in plain language. *Alchymia,* 1606. Andreas Libavius.

Three final books require mention before the Boyle period and the early beginnings of modern chemistry, and all three have to do with the practical sides of chemistry. The earliest is a valuable incunabulum by Hieronymus Brunschwig, *Kleines Distillierbuch,* published in Strasbourg in 1500. Basically a medical recipe text, *The Distilling Book*'s special purpose was to apply the technique of distilling with steam to the separation of a plant's medicinal essences from its nonmedical parts. The book contains many skillful illustrations of the period's chemical apparatus, and its woodcuts of plants are striking.

Two other significant early texts are concerned with practical aspects of the study of metals. Georgius Agricola, in his classic work on metals, *De*

Liber de arte distillandi. de simplicibus.
Das buch der rechten kunst
zů distilieren die eintzige ding
von Hieronymo Brunschwygk/bürtig vñ wund artzot der keiserlichẽ frye statt Straßburg.

The chemistry of Brunschwig was more in the tradition of materia medica than alchemy in that it focused primarily on the distillation by steam of a plant's medicinal essences. This frontispiece shows a unique botanical scene where the distilling apparatuses have become part of the garden itself. In addition to his distilled herbal remedies, Brunschwig offered a few unorthodox cures, such as distilled "waters" of ants, frogs, and flies. *Kleines Distillierbuch,* 1500. Hieronymus Brunschwig.

Vannuccio Biringucci ignored both alchemy and iatrochemistry and concentrated rather on metallurgy and chemicals used in warfare. His speciality was the separation of metals from ores, and in this area his work is more complete than Agricola's. The top illustration shows a German cupeling furnace with a brick dome, and the lower shows a cupeling hearth with an iron hood. Both methods exposed the gold or silver ore to a blast of very hot air which oxidized the unwanted metals. *De la pirotechnia,* 1540. Vannuccio Biringucci.

DEL FAR LI CENNERA. 58

RVZZVOLO
VINTVLA

QVEL altro modo che s'hadopera per coprire el cenneraccio, el cappel di ferro mi piace aſſai piu. Perche molto piu ſi puo riſtregnere el fuocho & tenere el bagno caldo, & con eſſo ſi puo affinare el pocho, & laſſi come al maeſtro piace.

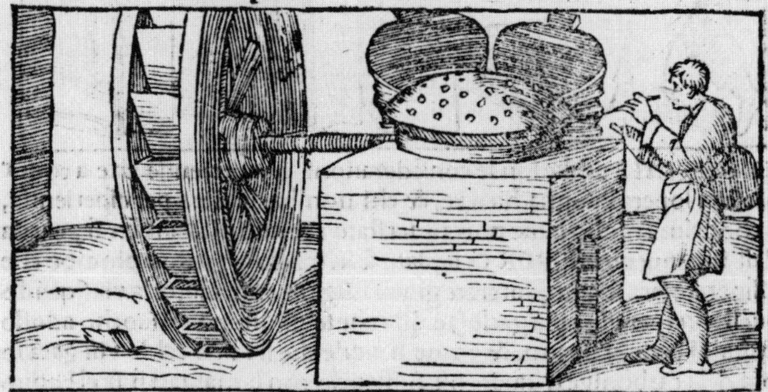

ET COME v'ho detto auanti ſi copreno, anchora quando ſ'affina li cenneracci con certe piaſtre di terra cotta groſſe tre ditta, & larghe mezzo braccio, & longhe quanto el cenneraccio, & queſte mi piacia no molto piu che alcuni de glialtri modi chio habbi veduto adope rare. perche ſacoſtano meglio per tenerla calda ſecondo che laua manchando,

H ii

re metallica, summarized everything connected with the mining industry and metallurgical processes. Agricola's real name was Georg Bauer. His systematic and comprehensive work was published in Basel in 1556, a year after his death, and it contains a magnificent series of 273 large woodcut illustrations. The book is discussed at greater length in the chapter on geology. A smaller work also in the Library's collections is *De la pirotechnia,* which details the processes of sixteenth-century metallurgy. Written by Vannuccio Biringucci and published in Venice in 1540, the book proceeds along the same lines as Agricola's. Written in the vernacular, the text is notable for its clarity and precision. It is intended for the practicing metallurgist and the maker of gunpowder and chemicals used in warfare.

The publication in 1556 of such a systematic and stately folio as Agricola's *De re metallica* exemplifies the sixteenth-century tendency toward a more rational, less intuitive kind of chemistry. Equally important were the beginnings of a movement away from the one-sidedness of applied chemistry. Not until chemistry was completely divorced from medicine would it be established as a separate science. This process was set irrevocably into motion with the publication in 1661 of a little octavo volume that was both anonymous and undedicated. Written by Robert Boyle, the seventh son and fourteenth child of the earl of Cork and Lord High Treasurer of Ireland, the book summoned up a new scientific spirit.

Boyle's book, *The Sceptical Chymist,* first published in London, echoed the emerging experimental science of his time and must have rung in the ears of the Scholastics. Written in English in the form of a dialogue, the text rails against unquestioning adherence to authority and exposes the errors, pretensions, and posturing of his fellow "chemists." Boyle became the ever-questioning skeptic in this book, whose "why's" would only be stilled by demonstrable proof. In this regard, he became the first true exponent of the Baconian method of experimental science, or what was then called the New Philosophy. Although the book became very popular, it did meet with resistance—specifically from the truculent author of the *Leviathan,* Thomas Hobbes, who disparaged what he called "the experimentarian philosophers."

The *Sceptical Chymist* attacked the core of traditional alchemical philosophy concerning the composition of all matter. Specifically, it contended that the elements of Aristotle (earth, air, fire, and water) and the three principles of Paracelsus (mercury, sulfur, and salt) were founded on intuition and bad observation. An element, it claimed, was a real, material substance that could be identified only by experiment. Boyle has been called the catalyst who set off the much-overdue chain reaction in chemistry, and so he was. It was only after Boyle that chemists perceived the larger dimensions of their science and began to put aside their pure empiricism for a grander, more productive theoretical and experimental science. Boyle's *Sceptical Chymist,* of which the Library has the 1668 Rotterdam edition in Latin and the 1680 Oxford edition, began the first revolution in chemistry, without which the second revolution, Lavoisier's a century later, could not have occurred.

Boyle's call for a more systematic chemistry was supported and promulgated by his contemporary Nicolas Lémery. Although Lémery did little original experimental work, his textbook on chemistry, *Cours de chymie,* became the best and most popular work of its time. It was so widely used that Lémery lived to see thirteen editions of it published. Many translations followed its publication in 1675, and for more than fifty years the book served as the most authoritative text in general chemistry. The Library has an English version, *A Course of Chymistry,* published in London in 1680, as well as a battered and well-used French copy pub-

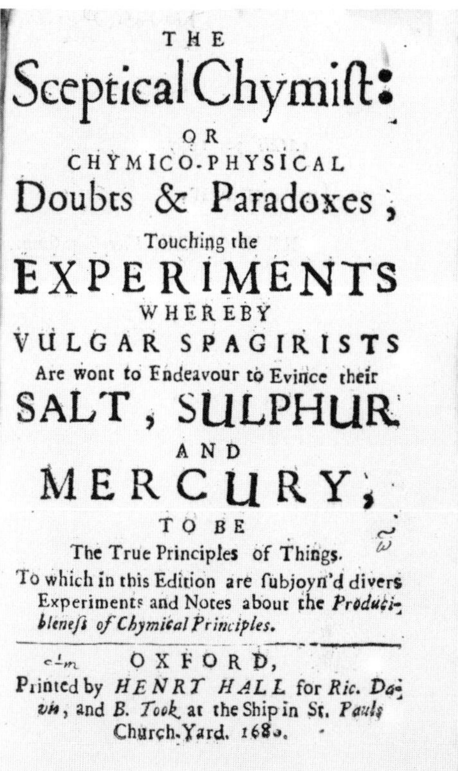
THE
Sceptical Chymiſt:
OR
CHYMICO-PHYSICAL
Doubts & Paradoxes,
Touching the
EXPERIMENTS
WHEREBY
VULGAR SPAGIRISTS
Are wont to Endeavour to Evince their
SALT, SULPHUR
AND
MERCURY,
TO BE
The True Principles of Things.
To which in this Edition are ſubjoyn'd divers Experiments and Notes about the *Producibleneſs of Chymical Principles.*
OXFORD,
Printed by *HENRY HALL* for *Ric. Davis,* and *B. Took* at the Ship in St. *Pauls* Church-Yard. 1680.

The title page of this English version of Boyle's *Sceptical Chymist* draws the battle lines of chemistry straightaway. In this work, Boyle attacks the assumptions and conclusions of the "Vulgar Spagirists" or alchemists whose unproven and confused theories held sway for so long. Espousing the experimental method, Boyle outlined a new concept of what an element is. He rejected the alchemical notion (taken from the Greeks) that all matter is composed of some combination of earth, air, fire, or water and argued instead, "I now mean by Elements . . . certain Primitive and Simple, or perfectly unmingled bodies." Boyle's new concept of elements marked the beginnings of scientific chemistry by wrenching it free from both medicine and alchemy. *The Sceptical Chymist,* 1680. Robert Boyle.

lished in Paris in 1730. Lémery, as an adherent of Boyle, made popular the notion of chemistry as a "demonstrative science" and strongly advocated the experimental method.

This new methodology was turned increasingly toward the matter of fire and combustion—a subject that was to dominate eighteenth-century chemistry. The phlogiston theory of Georg E. Stahl, which purported to explain both why some things burn and why others do not and to describe the nature of combustion, came to dominate eighteenth-century chemistry. Stahl's hypothetical phlogiston, which he named from a Greek word meaning "to set on fire," was described in his *Zufällige Gedanken . . . über den Streit, von dem so genannten Sulphure,* published in Halle in 1718. The Library has this edition, whose title translates *Random Thoughts on the Dispute about the So-called Sulfur.*

The phlogiston theory explained combustion neatly. Combustible objects were rich in the substance and the burning process involved its loss to the air. When the object was fully consumed, it contained no more phlogiston and could no longer burn. Thus a piece of wood or paper possessed phlogiston, but the remaining ash did not. The irony of this false and mistaken theory is that not only was it not an obstacle to the development of chemistry but it actually inspired further research by a large body of its proponents. Such great chemists as Black, Cavendish, Scheele, and Priestley were phlogistonists throughout. The significance of the theory and its saving grace is that it was the first important generalization in chemistry. From a singular and common point of view, a variety of processes, chemical actions, and substances could be comprehensively correlated. The phlogiston theory was by no means bad science—it was simply wrong.

Joseph Black was a rigorous investigator with a high regard for the experimental method. He is best known for his discovery of carbon dioxide, which he called "fixed air," yet his research methods and research focus are equally significant. In his experiments, Black placed great emphasis on the weight proportions of the reacting chemical compounds he tested. His pioneering techniques of quantitative measurement paved the way for the coming revolution in chemistry. Black's discovery of carbon dioxide was included as an appendix to his inaugural dissertation, "De Humore acido a cibis orto et Magnesia alba," presented in 1754 for his medical degree. This was later published as a separate paper, "Experiments upon Magnesia Alba, Quicklime, and Some Other Alcaline Substances." The Library's collections include this rare paper as it first appeared in a 1756 publication of the Philosophical Society of Edinburgh. Apart from this paper, Black published practically nothing in chemistry, focusing mainly on teaching. His lectures were published anonymously in 1770, and a more complete account of them. *Lectures on the Elements of Chemistry,* was issued in Edinburgh in 1803. The Library's collections include the latter two-volume set. Black was a popular lecturer who numbered among his students not only the famous Daniel Rutherford but also Benjamin Rush, who became the first professor of chemistry in America. Black had little difficulty abandoning his phlogiston views once the theories of

Lavoisier became demonstrable. In fact, he welcomed and embraced this new, verifiable information.

Black's contemporary and fellow countryman Henry Cavendish never surrendered to the facts, however. That Cavendish should remain a phlogistonist is especially remarkable, given the fact that his own discoveries gave its opponents a most formidable weapon. But the irony is consistent with this most strange and enigmatic of famous scientists. Reclusive and eccentric, Cavendish ignored every aspect and reality of the world and focused only on his work. Dedicated solely to his solitary experimentation and studies, he almost never spoke, and communicated with his female servants by notes. He ignored his huge inheritance as totally as he disdained the trappings and glories of the world. Preferring not to confront his professors for examinations, he took no degree from Cambridge. Choosing minimal contact with his scientific peers, he published virtually nothing. Despite this seemingly wasted life, Cavendish possessed a protean devotion to his science and wasted little time. In his work is found the proof that water consists of hydrogen and oxygen and that air is a mixture of nitrogen and oxygen in constant proportions. This revelation of the compound nature of what had traditionally been considered basic "elements" was to sound the death knell of the Aristotelian system. Cavendish experimented with electricity, and it was by his passing an electrical discharge through a mixture of ordinary air and what he called "inflammable air" (hydrogen) that he obtained water. These results were documented in two rare papers, "Experiments on Air," written by Cavendish for the Royal Society at the urging of a colleague. The Library has both papers, which were published (one in 1784 and the other in 1785) as part of the society's *Philosophical Transactions*. Cavendish labored at his work with a seemingly inhuman purity of spirit. He cared for neither credit nor acknowledgment. As a result, although his electrical work anticipated most of what was to be discovered in the next fifty years and his air experiments discovered, in addition to hydrogen, the gas we now call argon, such work remained unknown for nearly a century. The unpublished electrical experiments of Cavendish were edited by the equally great James Clerk Maxwell almost a hundred years after they were conducted by Cavendish. The Library has a first edition of Maxwell's *Electrical Researches of Henry Cavendish,* published in Cambridge in 1879.

As the work of Black and Cavendish shows, the study of gases occupied many of the best minds of eighteenth-century chemistry. Joseph Priestley was no exception. Priestley is well known for having discovered oxygen, or what he called "dephlogisticated air." The name he chose for this new gas was indicative of his scientific sympathies—Priestley was a die-hard phlogistonist. Stahl's theory had such an intellectual grip on him that even after he had discovered a new substance, recognized it as such, and determined its properties, he persisted in explaining the phenomenon in terms of the phlogiston theory. This stubborn streak was an essential part of his character, however. As a schoolteacher-preacher of the Unitarian sect and a sympathizer with the revolution in France, Priestley was comfortable as-

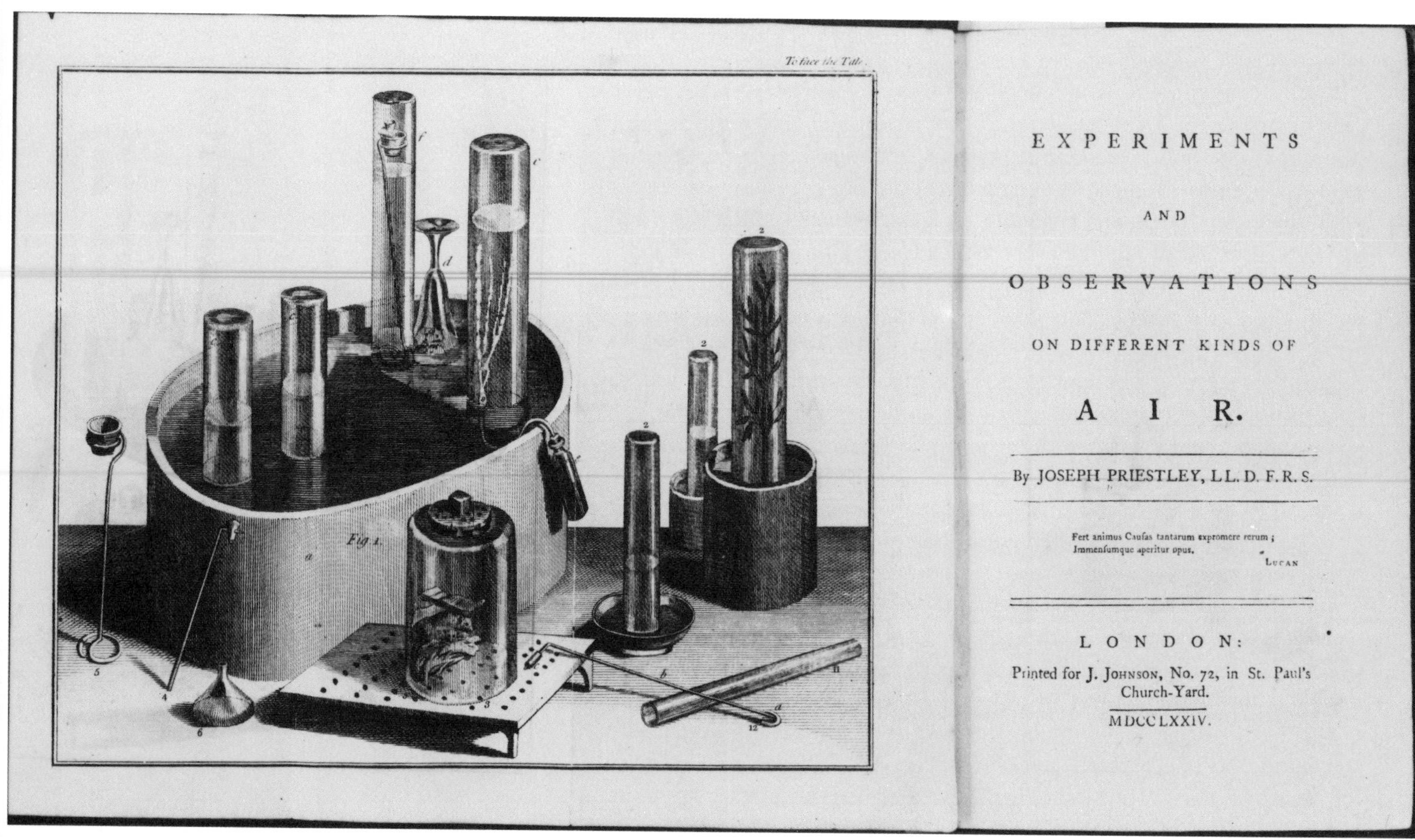

EXPERIMENTS

AND

OBSERVATIONS

ON DIFFERENT KINDS OF

AIR.

By JOSEPH PRIESTLEY, LL. D. F. R. S.

Fert animus Causas tantarum expromere rerum;
Immensumque aperitur opus.
LUCAN

LONDON:

Printed for J. JOHNSON, No. 72, in St. Paul's Church-Yard.

MDCCLXXIV.

In this frontispiece, Priestley shows off his inventive genius by illustrating some of his many experiments. The large oval bowl is his pneumatic trough which held inverted cylindrical jars in water and collected the gases he produced experimentally. Between two of them stands an inverted beer glass holding a mouse that is breathing Priestley's "dephlogisticated air" or oxygen. To the far right, cylinder number 2 illustrates his experiment with plants taking in carbon dioxide in daylight. These and other experiments led Priestley to the inescapable and revolutionary conclusion that air, which had forever been thought to be a simple and indivisible element, was indeed a compound substance. *Experiments and Observations on Different Kinds of Air,* 1774-77. Joseph Priestley.

suming an unpopular and controversial stance in a conservative Britain. His chemical experiments were both ingenious and productive, however, and despite his unpopular religious and political views he was awarded the Royal Society's Copley Medal. Descriptions of the bulk of his chemical experiments are contained in his three-volume *Experiments and Observations on Different Kinds of Air,* published between 1774 and 1777. The Library of Congress collections include the London first edition. In it is found his August 1, 1774, experiment in heating red oxide of mercury, which produced oxygen. Priestley sampled the new gas himself and felt "light and easy" and noticed that a candle burned "with a remarkably vigorous flame" in it. Priestley's scientific reputation and achievements did not guarantee total British toleration of his political views and on July 14, 1791, the second anniversary of the fall of the Bastille, his home and laboratory in Birmingham were destroyed by the townspeople. Priestley then fled to the newly independent United States and settled near the city of his scientific compatriot Benjamin Franklin. It was from Northumberland, near Philadelphia, that Priestley, obdurate to the end, published his last work, *The Doctrine of Phlogiston Established,* in 1800. The Library's first edition copy is a small spare book.

To Karl Wilhelm Scheele is usually credited the independent discovery of oxygen, but Scheele actually discovered it in 1771, some years before

Priestley. A negligent publisher is blamed for the delay in making his discovery known. It was not until 1777 that Scheele's *Chemische Abhandlung von der Luft und Feuer* was first published. The Library's copy was published in 1782 in Leipzig. Scheele was another of those hardy experimentalists of the age for whom work was all-consuming. In the course of his life, Scheele discovered or contributed to discoveries of more new substances of significance than any chemist before or after. He was a brilliant and ingenious investigator whose early death at forty-three may have been hastened by his self-testing, and tasting, of new compounds. At the time of his death, Scheele was still a phlogistonist.

Modern chemistry begins with Antoine Laurent Lavoisier. He has been called the Galileo of chemistry, for he destroyed the traditional but false notions of the old school and placed his discipline on a systematic scientific footing. Lavoisier was a brilliant student from a well-to-do family. He met with quick scientific success and recognition and was elected to the French Academy of Sciences at the age of twenty-three. Early on, he was to appreciate the critical importance of accurate measurement, and it is with his relentless application of the quantitative method that modern chemistry began.

Lavoisier also standardized and gave sense to the language of chemistry, no mean feat given the fancies of its alchemical tradition. In collaboration with other chemists of his day, notably Louis Guyton de Morveau, he published *Méthode de nomenclature chimique* in Paris in 1787, which propounded new principles of chemical terminology. The clarity and logic of his system made it readily acceptable, and it became the basis of modern nomenclature. The Library's copy of this work is a 1787 first edition.

In his most important work, *Traité élémentaire de chimie,* published in Paris in 1789, Lavoisier cleared away the confusion of centuries of alchemical myths and eventually overthrew the theory of phlogiston. Although he was not the first to discover oxygen, he gave it its name and was able to undertand and correctly interpret the experiments of his predecessors. Interpreting and synthesizing the work of Cavendish, Priestley, and other chemists, he offered a clear, unified picture of chemistry as a real, modern science. He demonstrated the principles of the indestructability and conservation of matter and built a rational system of known elements. With this, he established the modern idea of elements as substances that cannot be further decomposed. His two-volume treatise, which the Library has in first edition, is recognized as the first modern chemical textbook. All of its illustrations were done by his wife, Marie-Anne Pierette, who married Lavoisier when she was fourteen.

This famous work appeared the year the French Revolution broke out, and Lavoisier was unfortunately considered by the revolutionaries to be on the wrong side. For years he had been an official at the Ferme Générale, a private consortium that collected indirect taxes for the government; his wife was the rich, beautiful, intelligent daughter of an executive of that firm. He had made many enemies, one in particular being Jean-Paul Marat. On May 8, 1794, the guillotine took off the head of the man who

124 NOMENCLATURE

Noms anciens.	Noms nouveaux.
Gaz méphitique.	Gaz acide carbonique.
Gomme ou mucilage.	Gomme.
Gaz phlogiſtiqué.	Gaz azotique.
Gaz nitreux.	Gaz nitreux.
Gaz phoſphorique de M. Gengembre.	Gaz hydrogène phoſphoré.
Gaz Pruſſien.	Gaz acide pruſſique.
Gaz.	Gaz.
Gilla vitrioli.	Sulfate de zinc.
Gluten de ſroment.	Gluten ou glutineux.
H.	
Hépars.	Sulfures.
Huiles animales.	Huiles volatiles animales.
Huile de chaux.	Muriate calcaire.
Huile de tartre par défaillance.	Potaſſe mêlangée de carbonate de potaſſe en déliqueſcence.
Huile des philoſophes.	Huiles fixes empyreumatiques.
Huile de vitriol.	Acide ſulfurique.
Huile douce du vin.	Huile éthérée.
Huiles empyreumatiques.	Huiles empyreumatiques.

CHIMIQUE. 125

Noms anciens.	Noms nouveaux.
Huiles éthérées.	Huiles volatiles.
Huiles graſſes.	Huiles fixes.
Huiles eſſentielles.	Huiles volatiles.
Huiles par expreſſion.	Huiles fixes.
J.	
Jupiter.	Étain.
K.	
Kermès minéral.	Oxide d'antimoine ſulfuré rouge.
L.	
Laine philoſophique.	Oxide de zinc ſublimé.
Lait de chaux.	Chaux délayée dans l'eau.
Laiton.	Alliage du cuivre & de zinc, ou laiton.
Leſſive des ſavonniers.	Diſſolution de ſoude.
Lignites. (ſels)	Pyro-lignites.
Lilium de Paracelſe.	Alcohol de potaſſe.
Liqueur des cailloux.	Potaſſe ſilicée en liqueur.
Liqueur fumante de Boyle.	Sulfure ammoniacal. — d'ammoniaque.

Antoine Lavoisier was the chief architect in the eighteenth-century reform of chemistry and collaborated in a work that laid out a standardized language for chemistry. This popular work not only eliminated much of the synonymy and confusion of chemical nomenclature but set forth a rational organizing scheme that listed the known elements, metals, and elementary gases, among other things. This sample page shows the archaic names in the left columns and the new, scientific designations on the right. *Méthode de nomenclature chimique,* 1787. Guyton de Morveau.

had laid the foundations of modern chemistry. An apocryphal story recounts that in sentencing Lavoisier, the revolutionary tribunal declared, "The Republic has no need of scientists." National repentance soon followed Lavoisier's execution, however, and in October 1795, the Lycée des Arts unveiled a bust of the scientist, declaring him to have been a victim of tyranny.

Despite his fate, Lavoisier's contributions to chemistry were irrevocable. Following Lavoisier, chemical research was put on a strictly scientific basis. His exact experiments and rigorous use of analytical balance for weight determination at every chemical change paved the way for the atomic theory.

Were it not for one long, almost incidental sentence in John Dalton's first volume of *A New System of Chemical Philosophy,* 1808, the work of this theorist would today have probably been forgotten. But upon that sentence Dalton erected and elaborated his chemical atomic theory. It was Dalton's genius not only to revive the Greek concept of "atomos," or

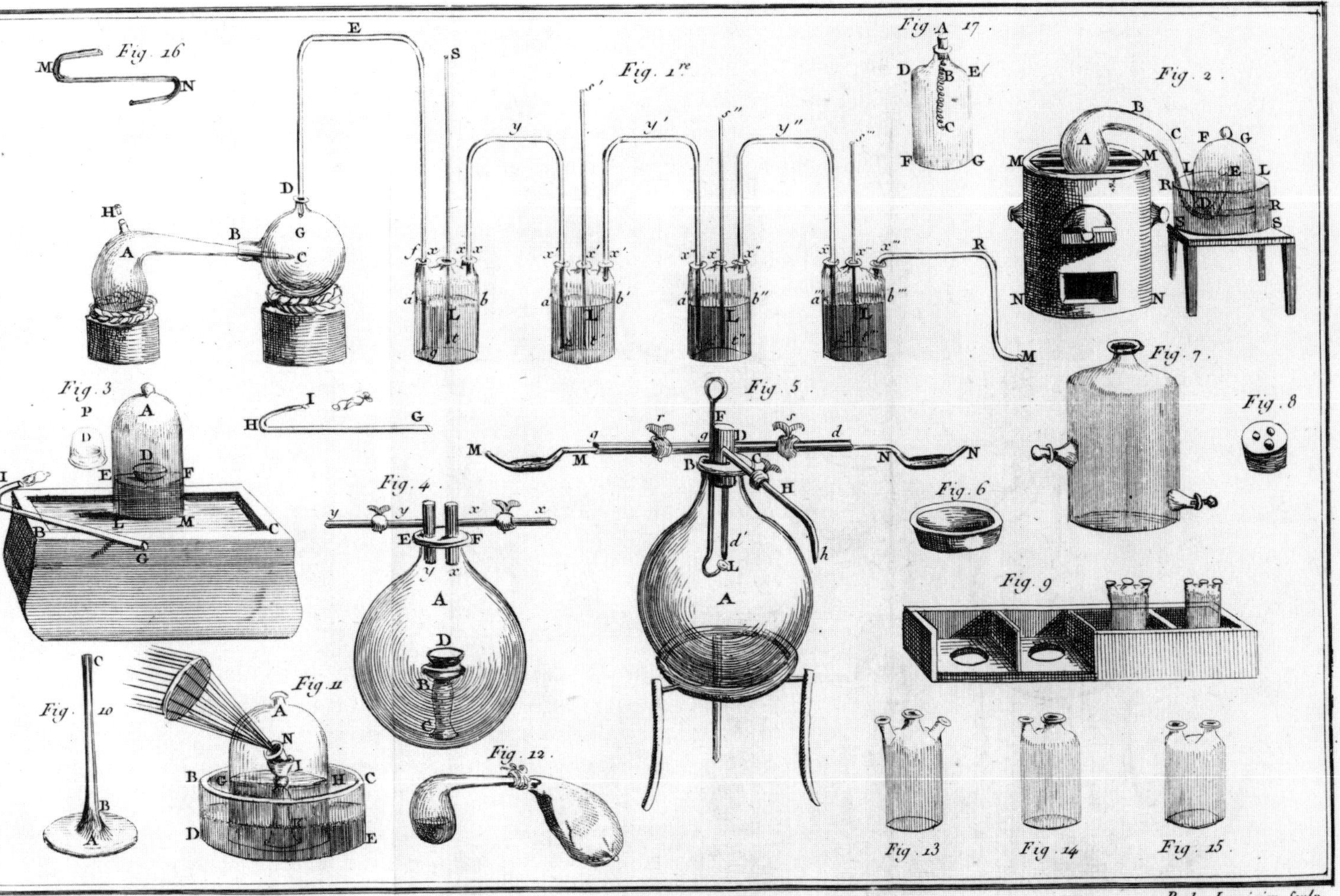

ultimate particles, of Democritus (subscribed to by many, including Boyle and Newton) but to postulate further that: (1) every element is composed of homogeneous atoms whose weight is constant; and (2) chemical compounds are formed by the union of the atoms of different elements in the simplest numerical proportions.

Dalton arrived at this theory in a deductive manner, being more a thinker than a natural experimenter. As is sometimes the case, a very simple theory emerges from the most complex of thoughts. Dalton first formulated his theory in 1803 and published his first volume of *A New System* in 1808. His second volume, published in 1810, contributed substantive additions, as did his third volume, published in 1827. The Library has all three volumes, each printed in Manchester. Dalton not only postulated the concept that a chemical element was quantitatively definable, but he published a table of "relative atomic weights" with symbols. This table indicated proportions by weight of the elements in particular compounds. Dalton's atomic theory has been called a great unification—but it was also

These drawings detail Lavoisier's rigorous methods of determining the subtle weight changes that took place at every chemical reaction. It was through his strict experimental examinations of the reactions involved in the combustion of substances and the calcination of metals that he was able to overthrow finally the phlogistic system and explain definitively the exact role oxygen played in these processes. At the top right (fig. 2) is Lavoisier's apparatus for heating mercury in a confined volume of air—a now classic experiment demonstrating the decomposition and recomposition of air. *Traité élémentaire de chimie,* 1789. Antoine Laurent Lavoisier.

a great clarification for chemistry. Dalton gave to chemistry a theory that worked, a theory that would explain facts, a theory that would eventually make chemistry as truly quantitative as any of the real sciences. Dalton was a simple, modest individual whose Quaker beliefs made him shun the limelight. He became internationally famous despite himself.

Building on the solid scientific foundation erected by Lavoisier and Dalton, chemistry at the beginning of the nineteenth century became a field of rapid growth spurred by the enthusiasm of new discovery. Four individuals, each from a different European nation and all born within three years of one another, best reflect the chemical advances of this exciting period.

Jöns Jakob Berzelius, "that colossal Northman" of Sweden, as a contemporary called him, followed Dalton's path and prepared a list of atomic weights that is regarded as the first reasonably accurate one in history. Berzelius was an absolutely thorough and exact experimenter whose work led to final acceptance of the atomic theory. By 1830 he was considered the foremost chemical authority in the world. His textbook *Lehrbuch der Chemie,* first published in 1803, went through five editions before his death. The Library has both the 1825–31 Dresden second edition in four volumes and the Dresden fourth edition, 1835–41, in ten volumes, both translated from the Swedish into German by Friedrich Wöhler.

Two others, a Frenchman and an Italian, made significant separate advances in pneumatic chemistry. Joseph Louis Gay-Lussac, an adventurous French chemist who had traveled with Alexander von Humboldt and made a balloon ascension with Jean Biot, announced his law of combining volumes in 1808. Through his extensive investigations, Gay-Lussac demonstrated that, like the two volumes of hydrogen and the one volume of oxygen required to form water, similar volumetric relations existed between all gases that chemically combine with one another. Gay-Lussac is represented in the Library by his *Recherches physico-chimiques* (Paris, 1811). Gay-Lussac made other important contributions to chemistry, but it remained for Amadeo Avogadro to demonstrate the link between Gay-Lussac's law of combining volumes and Dalton's atomic theory. In an essay published in *Journal de physique, de chimie, d'histoire naturelle et des arts* in 1811, which is in the collections of the Library of Congress, Avogadro offered the bold hypothesis that equal volumes of all gases and vapors contain the same number of "ultimate molecules" at the same pressure and temperature. Avogadro's hypothesis offered insight into the properties common to different substances, but like many simple truths, his ideas were ignored and rejected by his peers. Not until nearly half a century later would his theory become recognized as one of the cornerstones of chemistry.

As quiet and self-effacing as was the aristocratic Avogadro, so the Englishman Humphry Davy was brash and arrogant. Davy's name is associated with an impressive number of developments—not all limited to chemistry. But his work linking the new field of electricity with chemical investigation enabled him to discover several new elements. Among these

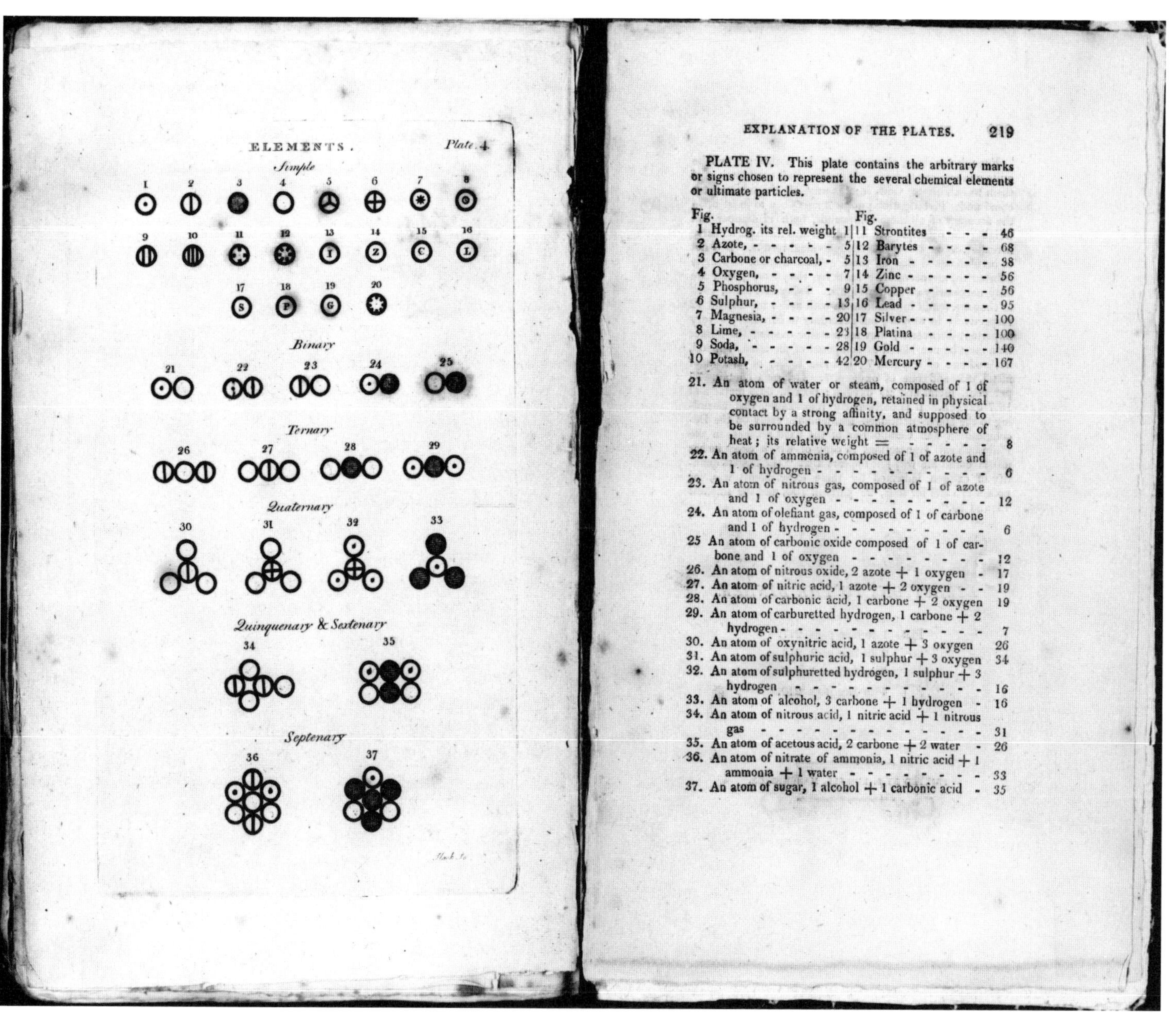

EXPLANATION OF THE PLATES. 219

PLATE IV. This plate contains the arbitrary marks or signs chosen to represent the several chemical elements or ultimate particles.

Fig.			Fig.		
1	Hydrog. its rel. weight	1	11	Strontites	46
2	Azote,	5	12	Barytes	68
3	Carbone or charcoal,	5	13	Iron	38
4	Oxygen,	7	14	Zinc	56
5	Phosphorus,	9	15	Copper	56
6	Sulphur,	13	16	Lead	95
7	Magnesia,	20	17	Silver	100
8	Lime,	23	18	Platina	100
9	Soda,	28	19	Gold	140
10	Potash,	42	20	Mercury	167

21. An atom of water or steam, composed of 1 of oxygen and 1 of hydrogen, retained in physical contact by a strong affinity, and supposed to be surrounded by a common atmosphere of heat; its relative weight = - - - - - - 8
22. An atom of ammonia, composed of 1 of azote and 1 of hydrogen - - - - - - - - - - - - 6
23. An atom of nitrous gas, composed of 1 of azote and 1 of oxygen - - - - - - - - - - - 12
24. An atom of olefiant gas, composed of 1 of carbone and 1 of hydrogen - - - - - - - - - - 6
25 An atom of carbonic oxide composed of 1 of carbone and 1 of oxygen - - - - - - - - - 12
26. An atom of nitrous oxide, 2 azote + 1 oxygen - 17
27. An atom of nitric acid, 1 azote + 2 oxygen - - 19
28. An atom of carbonic acid, 1 carbone + 2 oxygen 19
29. An atom of carburetted hydrogen, 1 carbone + 2 hydrogen - - - - - - - - - - - - - - 7
30. An atom of oxynitric acid, 1 azote + 3 oxygen 26
31. An atom of sulphuric acid, 1 sulphur + 3 oxygen 34
32. An atom of sulphuretted hydrogen, 1 sulphur + 3 hydrogen - - - - - - - - - - - - - 16
33. An atom of alcohol, 3 carbone + 1 hydrogen - 16
34. An atom of nitrous acid, 1 nitric acid + 1 nitrous gas - - - - - - - - - - - - - - 31
35. An atom of acetous acid, 2 carbone + 2 water - 26
36. An atom of nitrate of ammonia, 1 nitric acid + 1 ammonia + 1 water - - - - - - - - 33
37. An atom of sugar, 1 alcohol + 1 carbonic acid - 35

discoveries were potassium, sodium, barium, calcium, and boron. His discovery of nitrous oxide led to its use as the first chemical anesthetic. The Library has a first edition (1812) of his text *Elements of Chemical Philosophy,* published in Philadelphia. His most important papers were published in *Philosophical Transactions.* Davy had the not-uncommon habit of his day of experimenting on himself and was often disabled by a noxious sniff or taste. In 1812 he damaged his eyes in a nitrogen trichloride explosion. Of all his many discoveries, it is often said that his discovery and tutelage of the young Michael Faraday was his greatest.

Nearing the middle of the nineteenth century, chemistry had matured to the point of specialization. Work was beginning in the new field of organic

Beginning with the notion that a chemical element was something quantifiable, Dalton was the first to prepare a table of atomic weights—thus making chemistry a truly quantitative science. This table was based on Dalton's idea that each element is composed of a certain number of atoms with a constant weight and that chemical compounds are formed by the union of the atoms of different elements, always in the simplest numerical proportions. *A New System of Chemical Philosophy,* vol. 1, 1808. John Dalton.

chemistry, and two German chemists, Justus von Liebig and Friedrich Wöhler, pioneered this fertile field. Their collaboration began after Wöhler's wife died tragically young. The twenty-eight-year-old Liebig then asked the thirty-one-year-old Wöhler to join him in an intensive research effort. The two became steadfast friends—the patient, quiet Wöhler counterbalancing the aggressive, hot-tempered Liebig. The fruitful outcome of their collaboration was their confirmation of the idea that in organic chemistry groups of elements were more significant building blocks than the individual elements themselves. This was a radical concept. In inorganic chemistry, the opposite was true.

Three of Liebig's books in the Library's collection are noteworthy. *Anleitung zur Analyse organischer Körper,* published in Brunswick in 1837, detailed the constitution of organic compounds and described the modern method of chemical analysis. His *Die organische Chemie in ihrer Anwendung auf Physiologie und Pathologie,* published in Brunswick in 1842, is regarded as the first formal treatise on organic chemistry as applied to physiology and to pathology. Liebig also introduced the use of mineral fertilizers, and the Library has the first American edition (published in Cambridge and Boston in 1841) of his famous work *Organic Chemistry in Its Applications to Agriculture and Physiology.* Wöhler is represented most significantly in the Library's collections by his article "Ueber künstliche Bildung des Harnstoffs," which describes the first synthesis of an organic compound. Wöhler synthesized urea from ammonium cyanate by simple heating. His article was published in *Annalen der Physik und Chemie* in 1828.

No science is all experimentation, and chemistry is no exception. Occasionally great strides are made by one individual pondering a problem, playing intellectual games, or simply intuiting—seeing the inner organization of nature, understanding the connections between apparent randomness. In 1869 an unknown Russian professor of chemistry, Dmitri Mendeleev, published his first table of periodic elements. It was both a work of real science and the product of true inspiration. For some time, Mendeleev had brought his genius to the problem of the atomic weight of each element. What was the underlying, essential relationship between the different elements? How did the property that made them different or alike (their atomic weight) flow from that single constant? It is said that Mendeleev had a passion for the elements, and it must have been so, for he kept the problem constantly before him. On a card he wrote an element with its known atomic weight. At one element per card, he had sixty-three cards, which he shuffled in a game his friends called patience. He would arrange his cards in all manner of groupings and rows, horizontal and vertical, searching for a correct, systematic sequence. He was playing an abstract, quantitative game of chemistry, and he soon perceived its hidden rules. In 1869 Mendeleev published in the *Journal of the Russian Chemical Society* a table showing the classification and organization of the elements that forms the foundation for the modern periodic table. The Library has this periodical in its microfilm collections. With this seventeen-page paper, titled "Sootnoshenīe svoĭstv s atomnym viesom elementov," Mendeleev

— 70 —

но въ ней, мнѣ кажется, уже ясно выражается примѣнимость выставляемаго мною начала ко всей совокупности элементовъ, пай которыхъ извѣстенъ съ достовѣрностію. На этотъ разъ я и желалъ преимущественно найдти общую систему элементовъ. Вотъ этотъ опытъ:

			Ti=50	Zr=90	?=180.
			V=51	Nb=94	Ta=182.
			Cr=52	Mo=96	W=186.
			Mn=55	Rh=104,4	Pt=197,4
			Fe=56	Ru=104,4	Ir=198.
			Ni=Co=59	Pl=106,6	Os=199.
H=1			Cu=63,4	Ag=108	Hg=200.
	Be=9,4	Mg=24	Zn=65,2	Cd=112	
	B=11	Al=27,4	?=68	Ur=116	Au=197?
	C=12	Si=28	?=70	Sn=118	
	N=14	P=31	As=75	Sb=122	Bi=210
	O=16	S=32	Se=79,4	Te=128?	
	F=19	Cl=35,5	Br=80	I=127	
Li=7	Na=23	K=39	Rb=85,4	Cs=133	Tl=204
		Ca=40	Sr=87,6	Ba=137	Pb=207.
		?=45	Ce=92		
		?Er=56	La=94		
		?Yt=60	Di=95		
		?In=75,6	Th=118?		

а потому приходится въ разныхъ рядахъ имѣть различное измѣненіе разностей, чего нѣтъ въ главныхъ числахъ предлагаемой таблицы. Или же придется предполагать при составленіи системы очень много недостающихъ членовъ. То и другое мало выгодно. Мнѣ кажется притомъ, наиболѣе естественнымъ составить кубическую систему (предлагаемая есть плоскостная), но и попытки для ея образованія не повели къ надлежащимъ результатамъ. Слѣдующія двѣ попытки могутъ показать то разнообразіе сопоставленій, какое возможно при допущеніи основнаго начала, высказаннаго въ этой статьѣ.

Li	Na	K	Cu	Rb	Ag	Cs	—	Tl
7	23	39	63,4	85,4	108	133		204
Be	Mg	Ca	Zn	Sr	Cd	Ba	—	Pb
B	Al	—	—	—	Ur	—	—	Bi?
C	Si	Ti	—	Zr	Sn	—	—	—
N	P	V	As	Nb	Sb	—	Ta	—
O	S	—	Se	—	Te	—	W	—
F	Cl	—	Br	—	J	—	—	—
19	35,5	58	80	190	127	160	190	220.

In his first periodic table, published in 1869, Dimitri Mendeleev demonstrated what became known as the periodic law—that the properties of chemical elements show a recurring pattern when arranged in the order of increasing atomic weight. In this table, the vertical rows contain chemically related substances. Using his table, Mendeleev was able not only to observe a number of previously unsuspected analogies among the elements but to interpret the obvious gaps as undiscovered elements—elements whose existence he then posited and whose properties he described. The periodic table has proved to be an effective system of organization for chemistry—serving to classify and clarify—and a productive conceptual device as well, predicting both new elements and unknown relationships. "Sootnoshenie svoistv s atomnym viesom elementov," *Zhurnal Russkago khimicheskago obshchestva,* 1869. Dimitri Mendeleev.

pronounced that the properties of the elements are a periodic function of their atomic weights. In effect, this meant that if the elements were assembled by order of their atomic weights, the distinct groupings or families of related substances could be observed. Such a fundamental discovery seemed certainly sufficient for a thirty-five-year-old scientist, but Mendeleev then went further and made a truly inspired leap of thought. In using his newly discovered mathematical key to arrange the elements, he perceived that there were, logically, elements yet undiscovered that were missing from his table. Still at the theoretical level, he then interpreted the gaps *as* gaps, and left room in his table for the unknown elements. Also, he was able to predict the characteristics of these unknown elements in terms of their atomic weights and specific gravities. Mendeleev lived long enough to see many of his predictions verified by experiment, when new elements such as germanium were discovered. In addition to his famous text *Osnovy khimīi* (or *The Principles of Chemistry*) (St. Petersburg, 1895), the Library also has a copy of his 1865 St. Petersburg dissertation, *O soedinenīi spirta S Vodoiu* (On the compounds of alcohol with water). The glowing genius of Mendeleev bestows on the humble inductive process a subtlety heretofore unseen.

While Mendeleev's story shows that scientific discovery can have a beautiful symmetry, the career of the American mathematical physicist Josiah Willard Gibbs shows that science can also be powerfully complicated and even unapproachable. Gibbs worked with the known principles of thermodynamics and applied them mathematically to chemical reactions. In a series of papers published between 1874 and 1878 in the *Transactions of the Connecticut Academy of Arts and Sciences,* Gibbs laid out his interpretation of what actually occurred during a chemical reaction. Several of his findings became rules of chemistry, most notably the "phase rule." Gibbs's work incisively described the relations between chemical, electrical, and thermal energies. But sadly, the bulk of his findings were over the heads of most of the scientific community. Recognition of his great work came only after his peers and his science caught up with him. In 1901, Gibbs was awarded the Royal Society's Copley Medal, in recognition of his having laid the foundation of physical chemistry. In the Library's collections is Gibbs's work as it appeared in *Connecticut Transactions.*

Gibbs's receipt of the Royal Society's award demonstrated that the community of science is ideally an international one. But the active joint venture of an Englishman, Lord Rayleigh, and a Scot, William Ramsay, showed science off at its cooperative best. Rayleigh was a physicist intrigued by atomic weights, and he set out to measure the densities of gases as accurately as possible. In doing this, he was puzzled by the nitrogen he obtained from air, which was always specifically heavier than that obtained from chemical compounds. In 1892 he published a letter in *Nature,* presenting his problem to the scientific world and requesting suggestions.

Ramsey was teaching in London and took up the curious problem. The two eventually worked together, identifying the leftover bubble of gas as a new gas—one that was more dense than nitrogen and that made up about 1 percent of the atmosphere. This brought to mind the leftover gas of Cavendish, a century earlier. The new gas would not combine with another element and was so named argon by the pair, from a Greek word for "inert." Rayleigh and Ramsay published their results in 1895 in

Philosophical Transactions of the Royal Society. Their article was entitled "Argon, a New Constituent of the Atmosphere." The Library has this edition as well as the 1896 version printed in the Smithsonian Contributions to Knowledge series. Ramsay continued this work and by 1898 had discovered four more new gases—helium, neon, krypton, and xenon. By 1920, both Rayleigh and Ramsay were dead, and chemistry as a separate theoretical discipline given to mostly individual, personal investigation was fading away.

Chemistry in the twentieth century has become more specialized, more applied, and more interdisciplinary. Several twentieth-century chemists—now biochemists, physical chemists, and photochemists—are represented in the Library's manuscript collections. Svante Arrhenius, considered one of the founders of modern physical chemistry, is represented in the Library not only by his paper on electrolytic dissociation, "Försök att beräkna dissociationen hos i vatten lösta kroppar," in *Öfversigt af Kongl. Vetenskaps-Akademiens Fordhandlingar,* 1887, but by his correspondence with the American physiologist Jacques Loeb during the years 1904–23.

Among other notable modern chemists whose papers are deposited in the Library of Congress are Irving Langmuir and Albert Szent-Gyorgi, both Nobel Prize laureates. Langmuir's collection is sizable, consisting of letters from such contemporaries as Niels Bohr and Vannevar Bush, as well as copies of his speeches and articles. Most significant are his experimental notebooks for the years 1894–1957, which contain the data that led to his development of the gas-filled incandescent lamp, the high vacuum power tube, and atomic hydrogen welding. Szent-Gyorgi's collection is more limited and consists only of his correspondence with the famous physicist George Gamow.

The tradition of chemistry is as much characterized by its unscientific alchemical past as botany is by its old herbal ways. Although alchemy contributed valuable hands-on experience to chemistry with its emphasis on applied knowledge, it detracted from those contributions with its wasteful attempts at doing the impossible. Combining myth with fact and empiricism with wild speculation, the alchemist might be described as one who sought the limits of the possible. The contradictory nature of his craft led to such typically unproductive efforts as searching for a method of transmuting base metals into gold, but alchemy also served to cultivate a healthy and ultimately productive inquiry into the essential principles of nature. When this fertile alchemical tradition was given scientific order and purpose and was focused on its proper subject, its yield was bountiful. Once it overcame the limits and distractions of a disorderly craft, it attained both purpose and progress.

Albertus
Plini⁹

moneri non pōt. dicit Albertus. Horizon aūt quia vbicūq; locoꝝ variaꝫ accessus et recessus/in maribus variari necesse est. Sunt aūt qui motibus oceani alias ab his que posite sunt rationes afferunt. Aiunt eñ in mari abyssum ꝓfundissimā: de qua rupti sunt fontes abyssi ꝛc̄. Iuxta quā et cauerne nō parue: in q̄bus de spiramine aquarū venti concipiunꝫ: qui aquas maris per patentes terrarū cauernas in abyssum attrahunt: et ea exundante rursus magno impetu repellunt. Hos aūt ventos spiritus procellarū intelligunt: qui ab influxu lune/et ne vacuū fit in natura frequentius mouenꝫ. Alias causas breuitatis gratia transeo.

De terre motu. Cap. XVII. Ma.

Ex his etiā tremor terre ꝓuenit. Cum eñ in visceribus terre vapores obstructi exeundi locū non inueniunt. in die per radios solares calefiunt et subtilianꝫ: rursus aūt per nocturnū frigus ingrossanꝫ: et vtroq; modo moti exitū querentes/ latera terre et cauernarū concutiunt/ nonnunq̄ etiam rumpunt: et terram tremere faciunt. sepius autem in nocte.

Colles fiunt.

Diluuiū particulare

Pestis sequitur terremotum

Q; si exitū nō inueniūt: nōnunq̄ terrā ad modū monticuli aut collis eleuāt: quā si rūpūt/cineres ꝛ lapides eijciūt. Foueas magnasq; voragines causant q̄ si circa flumina eueniūt: ipsa ad distantiā aliquā tota absorbēt. Si aūt in alueo extiterint hi inclusi spūs aquā extra terminos ripaꝝ eijciētes/ ꝑticulare diluuiū efficiūt. Dis. Sane. Sz ob quā cām nōnunq̄ pestis terremotū comitaꝫ? Ma. Spūs hi inclusi si venenosi fuerint/exeūtes aerē corrūpūt et inficiunt: ex q̄ pestilentiā ꝑticularē saltē in locis vbi hec accidūt seq̄ necesse est. Dis. Fortasse illi⁹ ecclia meminit in bñdictiōe salis et aq̄/cū dicit: Nō sit ibi spūs pestilēs et aura corrūpēs. Ma. Nimiꝝ: cū et ꝑtāte tenebraꝝ his spiritib⁹ multotiēs vtanꝫ: aut vt hoies ledāt in corpe: aut sensus alienēt et visionib⁹ decipiāt. Ab his nōnunq̄ oculi et vult⁹ eoꝝ q̄ subterraneos specus aut antra ingrediunꝫ/immutanꝫ: maxime aūt eoꝝ qui a demonibus responsa ac-

6. *Geology: The Secret in the Stone*

Few natural objects appear less capable of telling a story than does a rock. Dense and mute, it sits unmoving—the essence of stolidity and passivity. It can be tossed playfully or hurled purposively. It can be used as a tool or chipped and shaped to match man's image. It can be literally supportive, or it can be destructive by virtue of the use made of it. All such uses and more were manifested early in man's history. But the recognition that a stone, any stone, has a past and contains within it a story of both essential changes interminably wrought as well as unimaginable violence and evidence of physical extremes did not occur to man until relatively recently. The individual rock, of course, particularizes the story of the earth. And geology is the scientific study of that story.

The metaphor of the earth as a book to be read, with its rocks and strata as the alphabet, is especially useful, for not until it was so regarded could there be any real progress in geology. All the answers of geology rest in the earth itself—the secret is in the stone—and they await only the knowing and observant eye. For centuries however, that eye was distracted, if not beclouded, by the belief that Scripture had provided a workable and true theory of the earth. Generations of students began and ended their studies at the dogmatic reference point of Genesis. Tensions between religious dogma and the emerging science of geology inevitably increased as new and unsettling discoveries were made. By the mid-eighteenth century, empirical evidence was beginning to erode the historically solid base of the given. A turnabout did not happen suddenly, however, and this accounts for the relative youth of geology as a science.

Ironically, the dominance of geology by religion reached its apex during the rational, mechanistic, post-Newtonian era near the end of the seventeenth century. At that time, it was not unusual to discover many a geological text written by a clergyman—especially in England. The Library of Congress has several such books in its collections. Their authors struggled laboriously during that age of reason to offer an empirical explanation of the particulars of Genesis. It was in just such a religio-rational milieu that Archbishop Ussher soberly proclaimed, in his *Annals of the World,* that the world began during October of 4004 B.C.

Such a book represents a severely misguided, albeit charming, attempt at the scientific method. At the core of that method is a single imperative—that the truth be sought honestly and be faced openly. But the truth of geology befuddles the mind and contradicts the senses. Even more upsetting are its implications. Fish fossils atop mountains, bones of extinct elephants beneath the streets of Paris, massive glaciers that seem to be

Opposite page:
The devastation and terror of an earthquake is captured in this simple woodcut, "De terre motu." The idea that such a catastrophic happening was the result of a real, physical dynamism inherent in the earth itself—an outward sign of its ceaseless evolution—was not accepted until the nineteenth century. *Margarita philosophica,* 1515. Gregor Reisch.

moving—such phenomena are more than curiosities and demand explanation. Attempting that explanation unfolds a timeless tale of constant, almost imperceptible change marked sporadically by sudden, violent cataclysms. It also reveals that the beginnings of the earth preceded man's own existence and that ages upon ages of elementary change have occurred which took no heed of man's presence once he did arrive. This revolutionary truth of geology, first stated in full by James Hutton, eventually had the same shattering impact as the Copernican revolution and the later Darwinian revolution. The physical world that man had formerly regarded as his private garden—a safe, secure, stable place designed to shelter him—was shown to be anything but that. The most cultured and cosmopolitan city in the world—Paris—was built upon the hardened overflow of terrible volcanoes. Those imposing symbols of stability and imperturbability, the mountains, had once been the bottom of the sea. Entire continents were on the move. Decidedly, the world was a much different place than it appeared.

Yet it is appearances that the geologist studies. To the untrained or the uninterested, a pile of rocks is simply that. But the geologist sees the order in apparent chaos and the chaos in what seems order. To him, a cut in the earth becomes a slice of the past, a map of time, a story told with gravel and fossils. Geology may be the most directly observational of the sciences and the one that puts the most participatory demands on its students. The books and maps described below were written and drafted by scientists and scholars. But more than most scientists, these individuals demonstrated a robust, active involvement with their science. Jean Guettard trod 1,800 miles to make a mineralogical map of France; William Smith spent a lifetime walking his native England; Louis Agassiz nearly died descending into the ice water of a glacier well; and Alfred Wegener did succumb during the winter of an Arctic expedition.

These men of talent and even genius did not pursue the story of the earth for sport. They and all their fellows recognized the significance of geology. They saw that it provoked great conflicts between revealed and empirical truths and dealt with issues that impinged directly on the nature of man and his role on earth. For them, geology involved contemplation of immense destructive forces, yet it also required an attitude and a perspective of a certain calm timelessness. Finally, they knew that to learn the lessons of geology, we must stoop, dig, and above all, go from ourselves to the earth. For, as Cuvier said over 150 years ago, "You have thus found the theory of the earth in the earth itself."

Although some early notions about the earth itself were at times remarkably accurate, those ideas usually focused on a singular geological aspect or phenomenon and existed in isolation, apart from any overall theory. Early investigators were more likely to be geographers than geologists, simply because a sense of place necessarily preceded any serious or prolonged contemplation of the origin and nature of things. A man usually first asks where he is and then proceeds to ask why and how he is there.

Nonetheless, some very early insights were achieved. The Greek Theophrastus, in his treatise on stones, *De Lapidibus,* offered accurate and practical information on minerals he had obtained from miners and jewelers. On the other hand, he contributed the false but persisting notion that fossils were not the remains of real animals but rather were the result of some plastic virtue within the earth. Only fragmentary parts of his treatise survive, and the Library has the original Greek version, *Peri lithōn*, contained in volume 2 of the 1495–98 Venice edition of Aristotle's *Opera*, as well as a much later French version, *Traité des pierres* (Paris, 1754).

Another Greek, Eratosthenes, offered a more solid contribution. In the third century before Christ, he announced his virtually correct calculations of the size of the earth itself—figuring its circumference to be a little over 25,000 miles. Both his method and his mindset were decidedly scientific. Noting that at the summer solstice the sun was directly overhead where he stood in Alexandria and that at the same time it was seven degrees from zenith at Syene in southern Egypt, he realized it was possible to calculate the circumference of the earth using the actual distance between the two places. His calculation hinged on two rather bold assumptions—that the earth was round and that it had an equal curvature to its surface. He also made the bold hypothesis that the presence of sea shells in the Libyan desert indicated the previous extension of the ocean to that area. Little of his work has survived, but the Library has a first edition of the only published collection of all of Eratosthenes's fragments, *Eratosthenica*, edited by Gottfried Bernhardy in Berlin in 1822. The significance to geology of these Greek forerunners is found not so much in their substantive contributions as in their basic assumptions. That is, they attributed geologic phenomena (earthquakes, volcanoes) to natural causes instead of seeking supernatural explanations. Such a view of the natural world is imperative to the scientific method.

A keen eye also helps, and the widely traveled Greek geographer Strabo saw much of the world. Strabo based his work on Eratosthenes and consequently believed that only a small portion of the earth was known to man. He not only postulated the existence of unknown land masses but taught that the elevations and subsidences of the sea might affect whole continents. Strabo viewed the earth as a dynamic and changeable place. His powers of observation were such that he recognized the long-dormant Vesuvius as a volcanic mountain. His main work, *Geographica*, is represented in the Library by a 1472 folio edition published in Venice. The Library also has his *De situ orbis* (Venice, 1510), a geographical encyclopedia containing much of the knowledge resulting from his travels. A large book with very dense type, it offers descriptions of habits and customs of various countries. A Venice 1502 edition is also in the collections.

The geological writings of Albertus Magnus (Albert, Count von Böllstadt), a thirteenth-century Dominican monk, are worthy of mention primarily because they serve both as a summary of geological knowledge up to his time and as a guide to the next two centuries. The Library's earliest copy of his *De mineralibus* is the 1491 Pavia edition of a work first published in book form nearly fifteen years earlier in Padua. It was written about 1260. One of the most comprehensive works of its kind, *De mineralibus* is divided into five books dealing with stones, minerals, and metals.

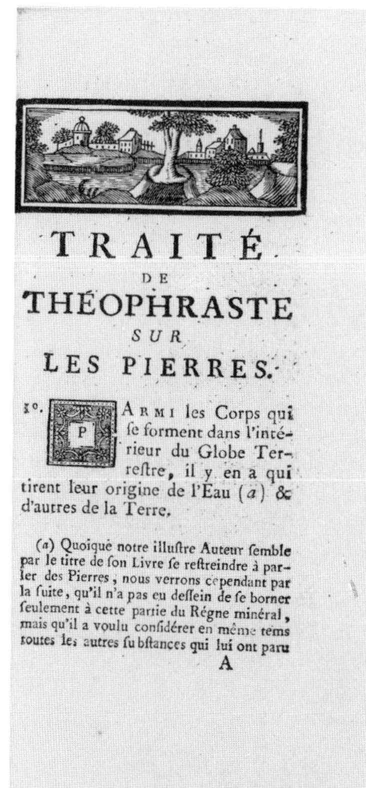

TRAITÉ
DE
THÉOPHRASTE
SUR
LES PIERRES.

1°. PARMI les Corps qui ſe forment dans l'intérieur du Globe Terreſtre, il y en a qui tirent leur origine de l'Eau (*a*) & d'autres de la Terre.

(*a*) Quoique notre illuſtre Auteur ſemble par le titre de ſon Livre ſe reſtreindre à parler des Pierres, nous verrons cependant par la ſuite, qu'il n'a pas eu deſſein de ſe borner ſeulement à cette partie du Régne minéral, mais qu'il a voulu conſidérer en même tems toutes les autres ſubſtances qui lui ont paru

A

The only separate Greek work on rocks and minerals to have survived is a short work by Theophrastus, pupil of Aristotle. Its size, organization, and style indicate that it was part of a larger work, now lost. Its emphasis is on the uses and applications of minerals, and it remained an authoritative practical work for eighteen hundred years. This page is from an eighteenth-century French edition. *Traité des pierres,* 1754. Theophrastus.

In this treatise on geography, the widely traveled Greek Strabo offered several geological observations. He sketched an earth with fires at its center, thus accounting for earthquakes and volcanoes as pressure-releasers. He also noted the transporting action of water and the elevation and subsidence of land areas. To Strabo, the earth was a changeable and very dynamic place. *De situ orbis,* 1510. Strabo.

By 1569 *De mineralibus* had gone into seven editions and had been translated into Italian. The Library also has the 1569 Cologne edition. The size of this tiny, worn edition belies the amount of information it contains.

Georg Bauer, who Latinized his name to Agricola, has been called both the father of mineralogy and the father of metallurgy. The Library has two of his most significant works, both first edition copies. Agricola was a physician who spent his life in mining regions and became interested in mineralogy and its connection with medicine. As his interest broadened, he produced two books, both published in 1546 in Basel. *De ortu & causis subterraneorum,* which the Library has in first edition, made significant contributions to physical geology and is considered the first work in that field. In it, Agricola recognized and was the first to describe the interplay of water and wind in shaping the landscape. He also discussed the origin of ore deposits and the eroding action of water. There are no illustrations in the book. Agricola's other work, *De natura fossilium,* which the Library also has in first edition, bound with *De ortu & causis subterraneorum,* earned him his place in the field of mineralogy. Here he described and classified about eighty different minerals, some of which had never been described before. This work provided a new scientific classification system—one that ordered minerals according to their physical properties.

His crowning work on metals, *De re metallica,* was published in Basel in 1556, a year after his death. The treatise consists of twelve books and summarizes every aspect of mining and metallurgical processes. Agricola placed great emphasis on observation and stated in his preface: "That which I have neither seen, nor carefully considered after reading or hearing of, I have not written about." The book contains a series of 273 large woodcut illustrations by Hans Rudolf Manuel Deutsch. Most illustrate machines and processes, of which Agricola noted, "I have not only

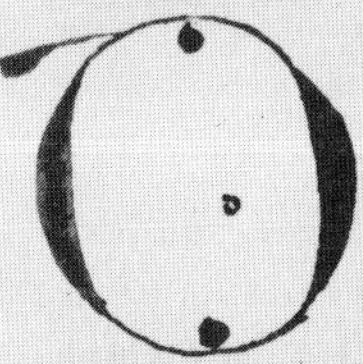

As one of the leading scholars of the thirteenth century, Albertus Magnus offered a theory on the origin of minerals and stones. The "mineralizing virtue" of the earth came from the planets and the stars, he said, which radiated their mineralizing influences on a receptive earth. This theory was basically an elaboration of Aristotle's ideas. *De mineralibus,* 1491. Albertus Magnus.

described them but have also hired illustrators to delineate their forms, lest descriptions which are conveyed by words should either not be understood by men of our own times, or should cause difficulty to posterity . . . ." The Library's first edition copy is in excellent condition. *De re metallica* took Agricola twenty years to write and another five years to have printed. Much of this classic is devoted to a refutation of ancient beliefs and practices. Agricola abhorred speculation not based on observation. His work also for the first time revealed industrial techniques and processes that previously had been family-held secrets. The book saw immediate translation into German and Italian and was subsequently reissued in ten editions. It was first translated into English by Herbert C. Hoover and his wife, Lou Henry Hoover, in 1912. The Library's first edition copy of the Hoovers' translation was published in London. Agricola obviously intended his masterpiece to be both a textbook for his time and a gift to posterity. In both goals he succeeded admirably.

Two other sixteenth-century works focus on discrete segments of geology—metallurgy and fossils. The first, by Lazarus Ercker, reflects the influence of Agricola. Entitled *Beschreibung aller fürnemisten mineralischen Ertzt unnd Bergkwercks Arten,* Ercker's treatise on ores and assaying amplifies Agricola's description and classifications. First published in Prague in 1574, his work was basically a text for the practicing assayer. It proved so accurate and so useful that it continued to be consulted for the next two centuries. The Library's copy is a Frankfurt second edition, published in 1598. It is a large but thin, water-stained book whose many illustrations depict mostly mining scenes.

A contemporary of Ercker, the Frenchman Bernard Palissy published in 1580 *Discours admirables,* in which he discussed the origins of petrified wood and the occurrence of fossil fish and mollusks. Palissy was an ardent French Huguenot as well as a famous potter. Appointed "inventor of rustic pottery to the king and queen mother" by Catherine de Medici during her years of attempted reconciliation with the Protestants, he and his sons built a pottery grotto for the queen in the garden of the Tuileries. In his book, Palissy pointed out that he had found many fossil conchylia that were identical to living mollusks and therefore concluded that the areas where he had found such fossils must have at one time been submerged. Such a view was contrary to what was taught by the Scholastics, such as Albertus Magnus, who denied that fossils were petrified remains of once-living animals. Palissy was the first man in France to express such a radical view, and it offered his religious enemies further proof of his heresy. Palissy, who is represented in the Library by a 1777 Paris edition of his works, *Oeuvres de Bernard Palissy,* edited by Faujas de Saint-Fond, died in the dungeons of the Bastille in 1590. The Library's copy of the 1777 edition is inscribed to "Monsieur Franklin" by its publisher, Ruault.

Palissy's head-on clash with the religious orthodoxy of his time was one of the more dramatic incidents in the characteristically antagonistic relationship of geology with the Christian faith. For, more than most scientific disciplines, geology in its subject matter impinged directly upon church dogma, specifically that regarding the subject of creation. The church believed it had received a definitive outline in Genesis and therefore condoned only those theories that fitted within the narrow confines of that tale. One such theory, that of Archbishop James Ussher, is a famous example of the unscientific method. In his *Annals of the World,* of which the Library has both the Latin first edition (London, 1650) and the English version (London, 1658), Ussher accepted the Bible's days of creation as literal units and began his calculations. He then set the earth's beginning on "the entrance of the night preceding the twenty third day of October, 4004 B.C." Both theologians and the faithful readily accepted his offering

Opposite page:
Agricola's extensive knowledge of all aspects of mining and metallurgy enabled him not only to amass a personal fortune but to contribute a classic work to the field he loved so well. His *De re metallica* was the product of a lifetime of practical experience and reflected Agricola's scientific bias for the empirical and the demonstrable. It is a most straightforward work, offering details on even the smallest technical aspects of mines and mining. The illustrations, which took the designer Blasius Weffring three years to complete, are similarly detailed and show mostly machines and processes. This illustration from Book 8 on how to extract metals demonstrates one sluicing method for obtaining the small black stones from which tin is made. *De re metallica,* 1556. Georg Agricola.

270 DE RE METALLICA

ſculo ſurſum uerſus batillo ligneo agitatos & conuerſos lauant: ut arena reliqua ab eis ſecernatur. Poſtea ſemper ad eundem laborem redeunt, donec eos materia metallica deficiat, uel riui in foſſas agendas deduci non poſsint.

Riuus A. *Foſſa* B. *Ligo* C. *Ceſpites* D. *Furca ſepticornis* E. *Batillum ferreum* F. *Lacuſculus* G. *Alter lacuſculus ei ſubiectus* H. *Batillum paruum ligneum* I.

I

Kurtzer Eingang in die folgenden fünff Bücher.

Was Probirn sey?

DAs Probiren ist gar ein herrliche alte nützliche Kunst / etwa vor gar langen zeiten / durch die Alchimiam, wie auch alle andere feuwer arbeiten erfunden / durch welche man nicht allein eines jeden Ertztes vnnd Berckart natur / vnnd was es für Metalln in sich hab / auff den rechten halt / wieviel ein Centen desselben mehr oder wenigers gewichts halte / erlernen vnnd erfaren kan / sondern es lernet

Wo von die Kunst deß probirn herkomb / vnnd warzu sie nutzt.

auch diese Kunst ein jedes Metall in sich selbst zu erforschen / ob ein zusatz darbey / was vnd wie viel derselbig sey / vnnd dann wie die Metall von solcher vermischung oder zusatz / auch andern eingeleibten Metallen in viel weg zu separirn vnd zu reinigen / vnd endlich für fein / rein / vnd aller zusätz ledig zuvrtheilen seyn / Derhalben auch diese Kunst den Bercklleuten die Berckwerck suchen / vnd nutz davon erlangen wöllen /

A sehr

This woodcut view of an assay laboratory is from Lazarus Ercker's practical treatise on ores and assaying. Stimulated by Agricola's work on mining, Ercker (who was superintendent of mines to Rudolph II of Austria) sought to bring some intellectual order to the specific problems of testing and assaying ores. His comprehensive manual was instructive, useful, and accurate and remained in use for two centuries. *Beschreibung aller fürnemisten mineralischen Ertzt unnd Bergkwercks Arten,* 1598. Lazarus Ercker.

and scribbled the date in many a Bible. It then became unthinkable that the earth was more than a few thousand years old.

For the next century or more, geological fact would compete with speculation and fancy as many a well-intentioned individual attempted to reconcile emerging discoveries and new ideas with the old accepted beliefs—usually to the disservice of both. Nils Steensen, known as Steno, experienced the heights of both worlds. Once a professor of anatomy at Padua, Steno was a wealthy Dane who took up the study of fossils. In his *De solido,* or *Dissertation on a Solid Body,* published in Florence in 1669,

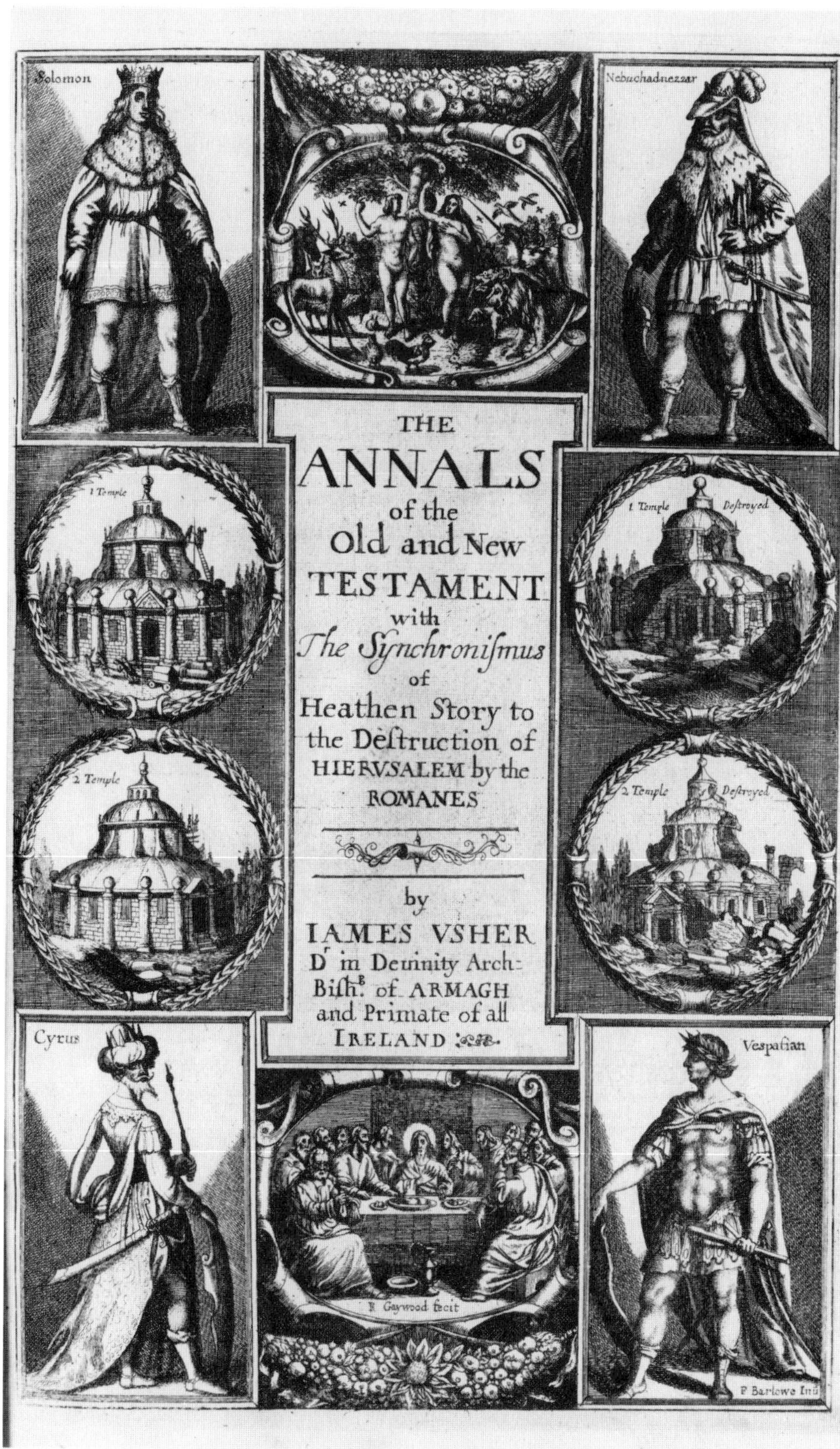

THE
ANNALS
of the
Old and New
TESTAMENT
with
The Synchronismus
of
Heathen Story to
the Destruction of
HIERUSALEM by the
ROMANES

by
IAMES USHER
D^r in Devinity Arch:
Bish^p of ARMAGH
and Primate of all
IRELAND

What became known as scriptural geology is partially represented by this work of Archbishop James Ussher of Ireland. Ussher interpreted the Old Testament and calculated that the earth's creation took place during the year 4004 B.C. He also dated the Great Flood at 2349 B.C. Such specificity from a respected religious source lent enormous credibility to the popular notion that the earth was geologically young. *The Annals of the World,* 1658. James Ussher.

a work that has become extremely rare, Steno offered a major contribution to the advance of geology. Having studied the composition of the earth's crust in the Tuscany region, he was the first to recognize that the strata of the earth contains the chronological record of its geological history. *De solido,* of which the Library has the first English translation, by John G. Winter (New York, 1916), contains Steno's famous diagram showing six successive types of stratification. Steno also made clear the true origin of fossils, as the remains of once-living things. In this theory he reflected the ideas of the Italian school, led by Fallopio, Fracastoro, and Moro.

Steno's fossil work was accepted by few, and he turned increasingly to the religious life. He had always been a devout Lutheran, but in 1667 he became a Catholic and gave up medicine and geology. He was eventually to become a bishop in northern Germany. In his final years he was described as practicing an extreme and unbalanced asceticism, which eventually undermined his health. Thus Steno, whose scientific genius laid out some of the principles of modern geology (and whose medical accomplishments were equally brilliant), stopped his career in mid-step and strode off in an entirely different direction. His lasting work, *De solido,* is a small treatise, almost an outline, for it was intended as an introduction to a larger work—a work he never began.

As the organic origin of fossils became increasingly evident toward the end of the seventeenth century, the struggle to reconcile the emerging story of the earth with that of received doctrine continued. Three books, all published during the last twenty years of that century, purported to explain fossils and much else in terms of the great flood. Fossils were dismissed simply as the vestiges of creation that became interred after the flood of Noah. And theories of the earth's origin were more complicated and fanciful. The first of these books was the Reverend Thomas Burnet's work, whose complete title is characteristic of the age: *The Sacred Theory of the Earth, Containing an Account of the Original of the Earth, and of All the General Changes Which it Hath Already Undergone, or Is to Undergo, Till the Consummation of All Things.* First published in London in 1681, this work is represented in the Library by the London edition of 1726, as well as the 1699 Latin edition. Burnet offered a theory of the earth that explained the scriptural account of both creation and the flood by natural processes. An angry God had baked our earthly paradise and had rent open great fissures, from which issued the waters of the deluge. The final result was the shattered and ruined earth of our time—a chaos of mountains and plains. Burnet's book garnered considerable attention and was a particular favorite of King Charles II of England. First published in Latin, with only twenty-five copies issued, his work saw a later, larger English version.

In 1695, John Woodward, professor of physick in Gresham College, published in London his *Essay towards a Natural History of the Earth.* Woodward best exemplifies the orthodox religious views of his academic contemporaries vis-a-vis geology. As a man of science, Woodward tried to accommodate all observed phenomena with Scripture. He was a diluvialist to the core and argued that the earth's crust was "taken all to pieces" by

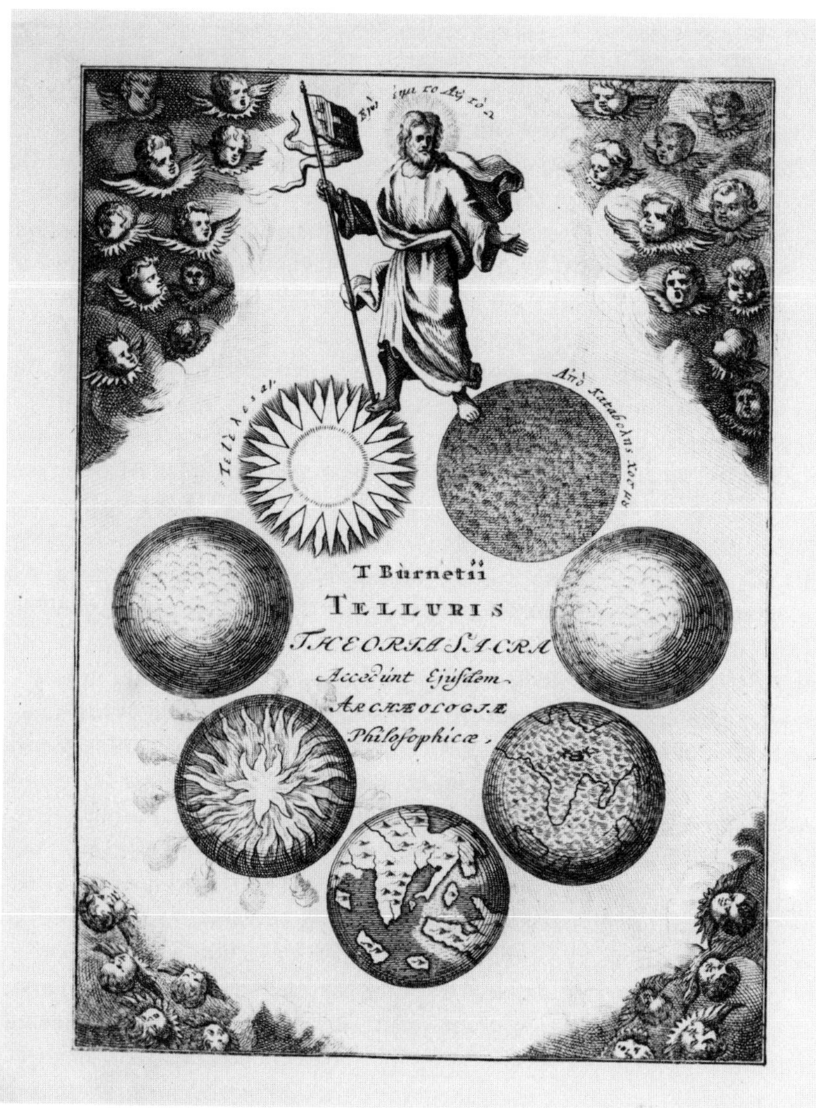

A unique blend of science, scripture, and Platonism, Thomas Burnet's explanation of the formation of the earth's features made for a very readable and popular book. His theory described a once perfectly habitable earth which became the irregular place of today after the great flood had done its rough work. Although the book was a popular success, it was criticized by both theologians and scientists. The former said Burnet had been too liberal or allegorical in interpreting scripture and had overemphasized the role of natural forces, whereas some scientiests refuted his use of Cartesian mechanics with their own Newtonian version. Despite his labored explanations and selective interpretations, Burnet's basic emphasis on the shaping power of natural forces was an important contribution to the future science of geology. This frontispiece depicts some of the major changes the earth has endured over time and predicts a universal conflagration that will end the world. *Telluris theoria sacra*, 1699. Thomas Burnet.

the deluge and eventually resettled into the strata we see today. Woodward's work had a wide circulation, and the Library has a first edition copy, a small unillustrated book.

A year after Woodward's publication, another book appeared in London, written by Isaac Newton's successor to the chair of mathematics in Cambridge. William Whiston was recommended for the position by Newton himself, and his book's title suggests a confident state of mind. Entitled *A New Theory of the Earth, from Its Original, to the Consummation of All Things, Wherein the Creation of the World in Six Days, the Universal Deluge, and the General Conflagration, as Laid Down in the Holy Scriptures, Are Shewn to Be Perfectly Agreeable to Reason and Philosophy,* Whiston's book attributed the flood to the close passing of a great comet that unleashed an incessant deluge upon the earth. Not to be outdone by Burnet, with whom he disagreed, Whiston dated his cloudburst on paradise at some time on the eighteenth of November, 2349 B.C. Whiston's ideas seem a remarkable fancy, but it is to his credit that he was one of the first to suggest a liberal interpretation of the six "days" of creation, and he allowed that the earth was much older than most believed. The Library has his work in first edition.

Burnet, Woodward, and Whiston all represent what might be called the religious school of geology. Each offered a major theory of the earth within the context of Scripture. Each made an earnest attempt to explain a received and traditionally accepted doctrine in light of the known laws of physical science. All three men offered plausible but hypothetical causes for natural phenomena. In doing so, none added anything to the science of geology and they may even have retarded its progress. With hindsight, we can say that their work demonstrates that although there is a place in science for speculation, imagination, and even romance, when these aspects preempt the inductive process and become an end in themselves, the results will probably be unsound.

Geologic thinking in the eighteenth century showed signs of increasing scientific maturity. If Steno was the solitary high-water mark of the seventeenth century, the eighteenth saw several outstanding individual contributions. That century of progress began, however, with the taint of scientific hoax and farce. A gullible Würzburg professor, Johannes Bartholomew Beringer, published *Lithographiae Wirceburgensis* in 1726 in Würzburg—a treatise revealing his paleontological discoveries in the nearby hills. Beringer's amazing stones, which he so passionately described in his *Würzburg Lithography,* had in fact been hand-carved and were part of a malicious academic conspiracy to discredit the doctor. It is not surprising that the Library does not have a copy of Beringer's 1726 first edition since, upon discovering his own name carved among the "discovered" rocks subsequent to publication, the hapless Beringer set out to buy back all the copies. The Library does have a 1963 translation of this infamous work.

By mid-century two Frenchmen, Jean Étienne Guettard and Nicolas Desmarest, not only had contributed to a growing body of geological knowledge but had made invaluable methodological advances. Desmarest summed up the new methodology with his cryptic "Go and see." The accomplishments of both men underscored how essential reliance on field observation and direct examination was to the science of geology. Instead of relying on speculation and supposition, these men both walked across the same region of central France and built their theories on firsthand experience. Guettard preceded Desmarest in his work and is credited with creating the first true geologic maps. His maps of France were the result of his discovery that rocks and minerals were not scattered randomly but were arranged in bands that ran on for miles. Guettard patiently traveled the 1,800 miles required to produce his maps. The Library has the results of his first mineralogical map as published in 1746 in the *Mémoires de l'Academie royale des sciences,* entitled "Mémoire et carte mineralogique sur la nature & la situation des terreins qui traversent la France & l'Angleterre." Guettard also correctly identified the Auvergne region of central France as volcanic in origin, although to his time no volcanoes had ever been known in that part of France. Most of his findings were

Opposite page:
The irony of Jean Guettard's work is that although he eschewed scrupulously any attempt to theorize, he gave to geology the beginnings of a "big idea" by ignoring political boundaries and viewing the earth as an integrated whole. Guettard was the first to plot geological data on a map. Using symbols for rock formations, mineral deposits, and fossils that he discovered in certain regions of France, he noticed that three horseshoe-shaped bands came to an abrupt end at the English Channel. Further researches indicated that indeed, a mineralogical map of the adjoining southeast English counties revealed the continuance of these same bands. Here is Guettard's map of one of his three "Bandes" (made of clay and calcium carbonate) connecting France and England. "Mémoire et carte minéralogique sur la nature & la situation des terreins qui traversent la France & l'Angleterre," *Mémoires de l'Academie royale des sciences,* 1746. Jean Étienne Guettard.

Mem. de l'Ac. R[le] des Sc. 1746. Page 392. Pl. 32.
Carte Minéralogique,
Où l'on voit la Nature et la Situation des terreins qui traversent la France et l'Angleterre.
Dressée sur les Observations et pour un Memoire de Mr Guettard de l'Acad. des Scienc[es]
Par Philippe Buache de la même Academie.
1746
Explication des Caracteres.
MER DU NORD
LA MANCHE
OCEAN
Bretagne
Picardie
Champagne
Normandie
Suite de l'Explication des Caracteres.

published in the Royal Academy of Sciences *Mémoires.*

Nicolas Desmarest explored the same region of Auvergne that Guettard did and also made field studies in Italy, from Padua to Naples. Focusing his formidable analytical powers on Guettard's volcanic evidence, Desmarest made a bold jump in theoretical as well as observational geology. He interpreted correctly the lava flows of Auvergne as belonging to different epochs—one older than another. Like Guettard, he too published mostly in scientific journals. His most famous paper, "Mémoire sur la determination de trois époques de la nature par les produits des volcans . . .", appeared in the 1806 *Mémoires de l'Institut des sciences, lettres, et arts: Sciences mathématiques et physiques,* which is part of the Library's collections. Desmarest implicitly offered to geology the original concept that landscapes undergo ceaseless change which was eventually to become a basic tenet of geologic thought.

The idea that much of Europe was built on the remains of once-active volcanoes had implications which far surpassed the scope of geology. If this were true, might not old fires be rekindled? In the Age of Reason, these new ideas threatened the stability of an orderly world, a world that was purposeful and knowable. Volcanoes and earthquakes—the embodimet of violence and irrationality—were the antithesis of the rationalists' natural paradigm. Certainly Desmarest's ideas must be false.

To minds so disposed, the explanations of Abraham Gottlob Werner came as a welcome balm. Volcanoes were very much the exception, he argued, and played only a minor role in the earth's geologic history. Water and not fire precipitated the formation of nearly all rocks, he said, and in 1787 by a single sentence—"I hold that no basalt is volcanic, but that all of these rocks . . . are of aqueous origin"—he set off a raging controversy. Werner was a born teacher and lecturer, to whom his enthusiastic students were devoted. His followers and the advocates of his water theory were called "Neptunists," and those opposed were called Vulcanists or Plutonists. At issue was more than simply the origin and nature of certain rocks. More importantly, questions of theory and methodology were at stake. Significant issues, such as how the earth itself was to be studied and how theory and facts were to be tested, applied, and judged were contingent on which school of thought prevailed.

Werner's style has been called "geology by dictum" by one historian, and so it was. Basing his sweeping generalizations on field experience limited to his native Saxony and Bohemia, Werner applied what he thought to be true to the entire globe. He would brook no contrary opinion nor would he consider the merits of any contradiction. He became as immovable in his theory and in his self-assurance as one of his great rocks. Werner has been accused of causing the science of geology to retrogress, and that judgment may be harsh. Indeed, others claim it was he who first elevated geology to the rank of a real science. But his refusal to go wherever the truth might lead, regardless of ego, taints his work. Werner is represented in the Library's collections by *A Treatise on the External Characters of Fossils,* (Dublin, 1805). This is the first English translation of his first publication, *Von den äusselichen Kennzeichen der*

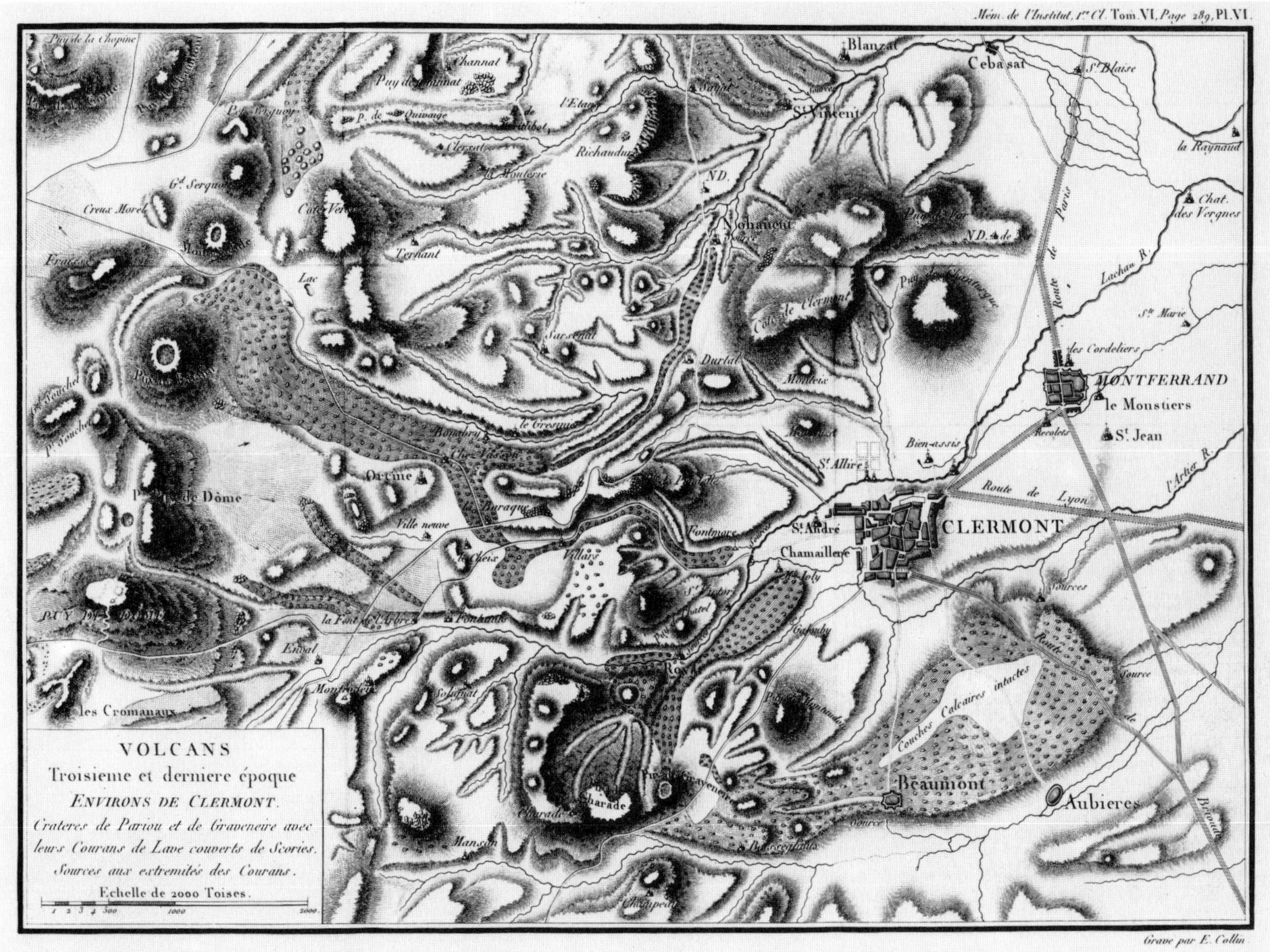

Fossilien, dated 1774. Written during his student days at Leipzig, this book brought him instant fame by introducing entirely new methods in the description of minerals. Despite his great capacity for work, Werner had an antipathy to writing and consequently published very little.

An important contribution to the literature of volcanic geology was *Recherches sur les volcans éteints du Vivarais et du Velay* by Barthélémy Faujas de Saint-Fond. A first edition of his splendid folio, published in Grenoble in 1778 and lavishly illustrated with engravings of old volcanoes, is in the Library's collections. Faujas's folio demonstrates rather convincingly the association of columnar lavas with volcanic cones, that is, that both are volcanic in origin, but his work had no immediate influence on Werner or his school.

The school of thought in opposition to Werner's Neptunists—the Vulcanists—argued that the earth had internal heat and that fire as well as water had been a major factor in the formation of geological strata. This group's outstanding figure was the famous Scotsman James Hutton. Hut-

Unlike Guettard, Nicolas Demarest concentrated on a fairly small geographical region of France, but he investigated it most intensively. Focusing on the Auvergne region of central France, Desmarest became famous for correctly identifying the basalt or black rock of that region as volcanic in origin. Tracing the lava flows of ancient volcanoes and mapping them in great detail, he proposed that the irregular landscape of that area was the result of volcanoes active during vastly different ages—indicating ceaseless change over long periods of time. Demarest was constantly revising his geological maps of the Auvergne region and this one is from his expanded version of 1805. "Mémoire sur la détermination de trois époques de la nature par les produits des volcans," *Mémoires de l'Institut des sciences, lettres, et arts,* 1806. Nicolas Desmarest.

Independently of Desmarest, Barthélémy Faujas de St. Fond offered the same idea that basalt was not of aqueous origin (formed by or under water) but was rather the product of volcanic activity. He investigated the ancient volcanoes of east-central France and discovered the same columnar pillars of lava that had so fascinated Desmarest. Unlike the precise geologic maps of Desmarest, however, Faujas de St. Fond's work contained dramatic illustrations of his discoveries, such as this romantic scene on a volcanic hill in Velay. ***Recherches sur les volcans éteints du Vivarais et du Velay,*** 1778. Barthélémy Faujas de St. Fond.

ton had been trained in medicine but never practiced and became instead a gentleman farmer with an avid interest in geology. His accomplishments in that field were to go far beyond the confines of the Neptunist-Vulcanist debate and were to lay the groundwork for modern geology. His "Theory of the Earth," published in 1788 in the inaugural volume of *Transactions of the Royal Society of Edinburgh,* is the result of a lifetime of careful, inductive study of the earth. This first volume is in the Library's collections. In it, Hutton offered the radical idea that the earth's terrain is the product of a continuous process brought about by natural forces. It was these calm, incremental, yet inexorable forces of physical nature, he argued, rather than any supernatural intervention, that shaped and would continue to shape the earth's surface. Hutton's theory was named "uniformitarianism," so called because it viewed the geologic evolution of the earth as a slow, continuous transformation—one marked only rarely by sudden, catastrophic change. Hutton realized that the geological gradualism he preached necessitated enormously long epochs and that he was in effect replacing the essentially teleological earth science of his time with one whose methods could "find no vestige of a beginning—no prospect of an end." Apart from the rigorous and careful inductive methods that so distinguished his work, Hutton displayed a brave intuitiveness—a willing-

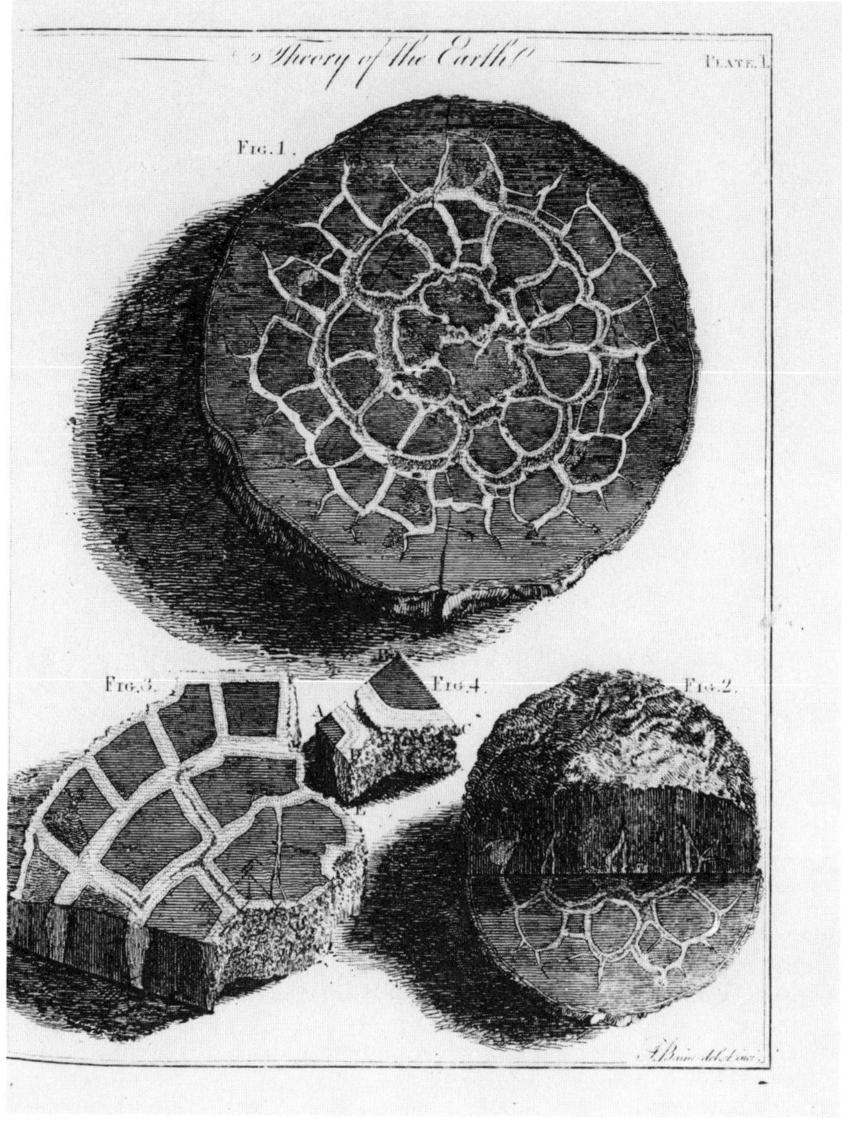

The idea that natural processes have been constantly at work shaping the earth over enormously long periods of time became one of the most fundamental principles of geology. Called "uniformitarianism" when first expounded by James Hutton, this theory argued that most rocks were sedimentary in origin, having been transformed into rocks by the internal heat of the earth. In this illustration from his article, Hutton demonstrates how a horizontal slice of iron-stone achieved its distinctive septarian characteristics, "by means of fusion, or by congelation from a state of simple fluidity and expansion." "Theory of the Earth," *Transactions of the Royal Society of Edinburgh,* 1788. James Hutton.

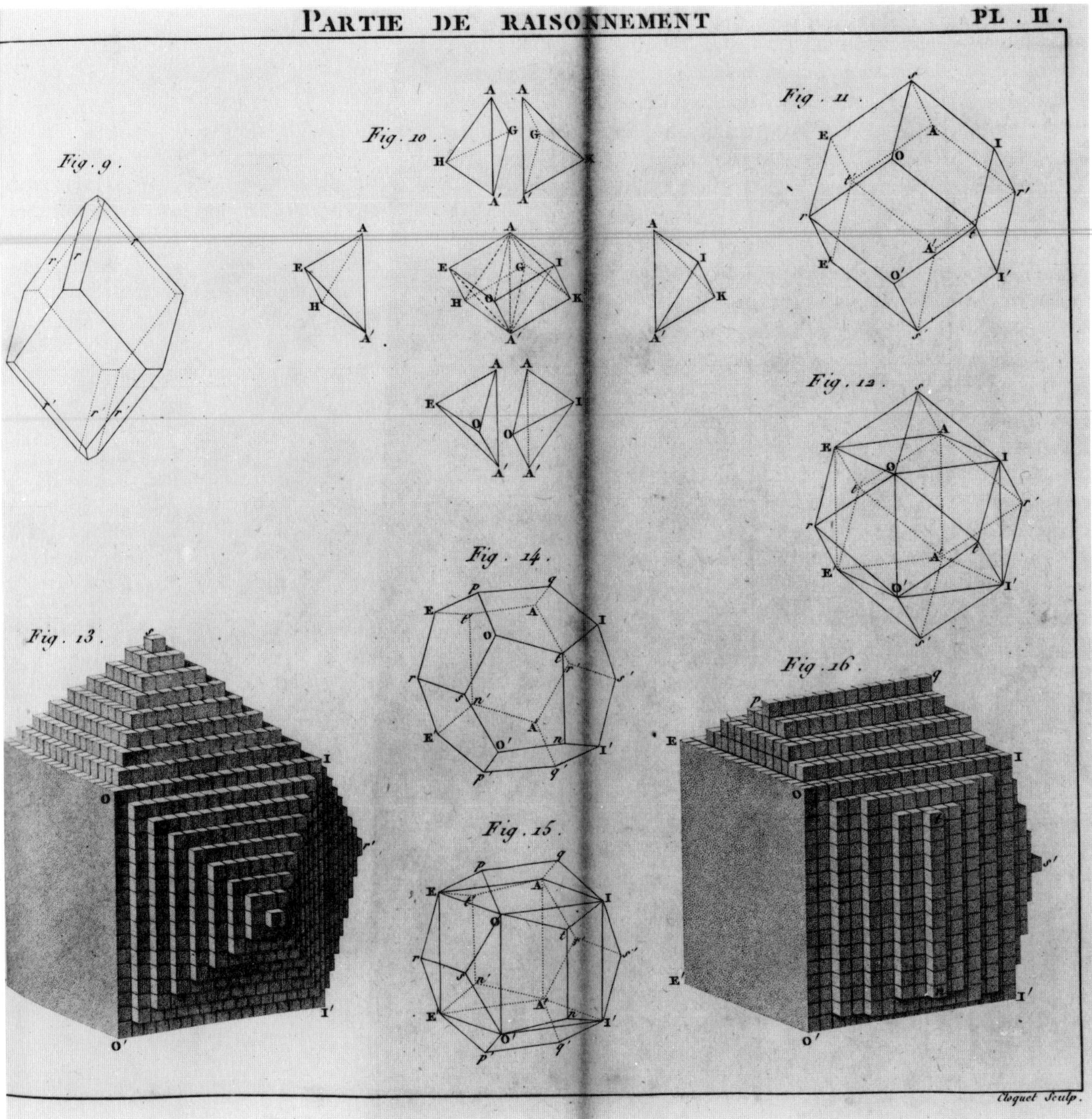

By the nineteenth century, the science of geology had attained a new level of sophistication and specialization with the founding of the subdiscipline of crystallography by René Just Haüy. Haüy's laws of crystal symmetry had their beginnings in an accident—after dropping a piece of calcite, Haüy noticed that the pieces were all rhombohedral. Upon research, he eventually discovered that the ultimate particles forming a particular mineral all have the same shape. This geometrical definition of mineral species brought the rigor and precision of mathematics to one branch of geology and laid the groundwork for modern crystallography. *Traité de minéralogie*, 1801. René Just Haüy.

ness to conceive of nature on a grand scale. He did not seek easy solutions or expect any. "The raising up of a continent from the bottom of the sea is an idea too great to be conceived easily. . . . In like manner destruction of the land is an idea that does not easily enter into the mind of man in its totality." Hutton's words contain implicity the substance and style of modern geology. Hutton was a calm, assured, gentlemanly scientist with no axe to grind and no constituency to impress. Deliberate in everything, he was over sixty years old when, at the urging of colleagues, he first published his "Theory of the Earth." He had presented his theory in 1785 in a paper read to the Royal Society of Edinburgh, which published it three years later. In 1795 he wrote an expanded version in two volumes.

The passions of the Neptunist-Vulcanist debate cooled as the merit of Hutton's theory eventually became evident. His ideas were not immediately well-received, however, and actually took some time to become accepted. This was in part because of the peculiar and difficult writing style of Hutton himself, which must have discouraged and perhaps even misled many readers. Fortunately, like Charles Darwin, Hutton had his Huxley. John Playfair, a gifted mathematician with a penchant for geology, was a close friend and colleague of Hutton's. It was through his literary skill that the world came best to know the real principles of Huttonian theory once they had been endorsed and thus made acceptable by Charles Lyell. Five years after Hutton's death, Playfair published his *Illustrations of the Huttonian Theory of the Earth* in Edinburgh in 1802. Playfair's logical and lucid presentation and development written in what has been called some of the most luminous scientific prose ever composed made for a prize whose wrapping equaled its contents in brilliance. The Library of Congress collections contain a first edition of Playfair's masterpiece.

This turn-of-the-century period of active and fertile debate has been called the heroic age of geology. The new science was only beginning to coalesce and to stake out its territory as the nineteenth century began. Indeed, the name geology itself had only recently come into general use. To the Swiss scientist and mountaineer Horace B. de Saussure goes the distinction of having first used the term in his writings without qualification or apology. Born into a rich, patrician family, Saussure was a brilliant student who devoted his life to the study of mountains. No theoretician, he sought first and foremost to observe and to describe accurately. His best work, *Voyage dans les Alpes* (Neuchâtel, 1779–90), recounts in four volumes his many geological excursions through his native mountains and was the first book to describe the beauty and delight of the high Alps. Until Saussure, the high regions were regarded fearfully as the "montagnes maudits"—desolate and dangerous tracts. His straightforward yet elegant work is represented in the Library collections by a first edition copy.

While the robust Saussure was clambering over the rocks he was studying, a quiet priest, René Just Haüy, was contemplating his rocks in a different manner. Where Saussure conducted his surveys in the mountains, examining rocks in their natural state, Haüy did no field research but made his excursions in Paris—visiting various mineralogical collections.

Whereas Saussure distrusted theory, Haüy excelled at theoretical mineralogy. Haüy's initial treatise, *Essai d'une théorie sur la structure des crystaux* (Paris, 1784), joined mathematics and mineralogy and laid the foundation of the mathematical theory of crystal structure. In this work, which the Library has in first edition, Haüy offered his geometrical law of crystallization. Having accidentally dropped a piece of calcite, Haüy observed the fragments all to be rhombohedral in shape. Further studies revealed that the variations in forms of crystals of the same substance could be explained mathematically. In his laws of crystal symmetry, Haüy brought evidence and hypothesis together in a unified theory. The science of crystals—that is, of the ordered state of solid matter—had begun. The Library also has a first edition copy of Haüy's five-volume masterpiece, *Traité de minéralogie* (Paris, 1801), which contains eighty-six plates illustrating the intricate structure of crystals. This work gained scientific fame throughout Europe for the abbé who, a short time before, had barely survived the French Revolution's September 1792 clerical massacres. Haüy was arrested that year for refusing to take the oath required by the revolution's civil constitution of the clergy. It was only through the intervention of some influential friends that he was released.

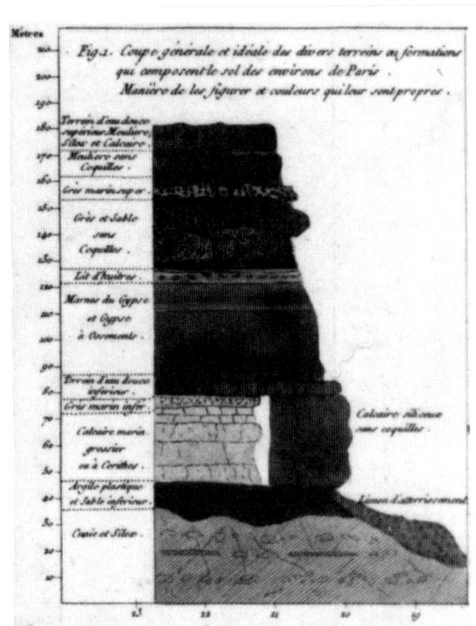

From *Essai sur la géographie minéralogique des environs de Paris,* 1811. Georges Cuvier and Alexander Brongniart. See p. 112.

One who seemed to soar above the wrenching politics of early nineteenth-century France was the magnificent Georges L. C. F. D. Cuvier. Cuvier was a prodigious child, and by the time he was thirty-four, he occupied three of the most distinguished scientific posts of France and became unquestionably the most eminent European scientist of his time. Primarily a zoologist and an astonishingly good comparative anatomist, Cuvier discovered in 1796 ancient elephant bones in the soil of Paris. His questions about their entombment and preservation led him in turn to geology. At this point he struck an alliance with Alexandre Brongniart, a distinguished naturalist who had succeeded Haüy as professor of mineralogy at the Museum of Natural History. Together they prowled Guettard's old territory, the Paris Basin. For four years the two men studied the area, searching for evidence of a link between successive strata or layers and the fossils they contained. In 1808 they published their initial findings in a memoir and in 1811 elaborated and refined them. The later work, published in Paris as *Essai sur la géographie minéralogique des environs de Paris,* is in the Library's collections. Here Cuvier and Brongniart offered a table of stratigraphical succession that demonstrated how fossils could be used to determine a correct geological sequence or chronology. In effect, they placed paleontology on an accurate, scientific basis. The Library also has their 1811 geologic map of the Paris Basin, which represents the distribution of the basin's different strata and their structural relationships. Cuvier also produced in 1812 a work which earned him the title of founder of vertebrate paleontology. In his four-volume *Recherches sur les ossemens fossiles,* published in Paris, of which the Library has a facsimile of the first edition copy as well as the third edition (1825), Cuvier applied his enormous deductive skills to the skeletons of fossil vertebrates. He not only gave accurate, reconstructed accounts of the habits of these prehistoric animals but also showed that many fossils were of animal types that no

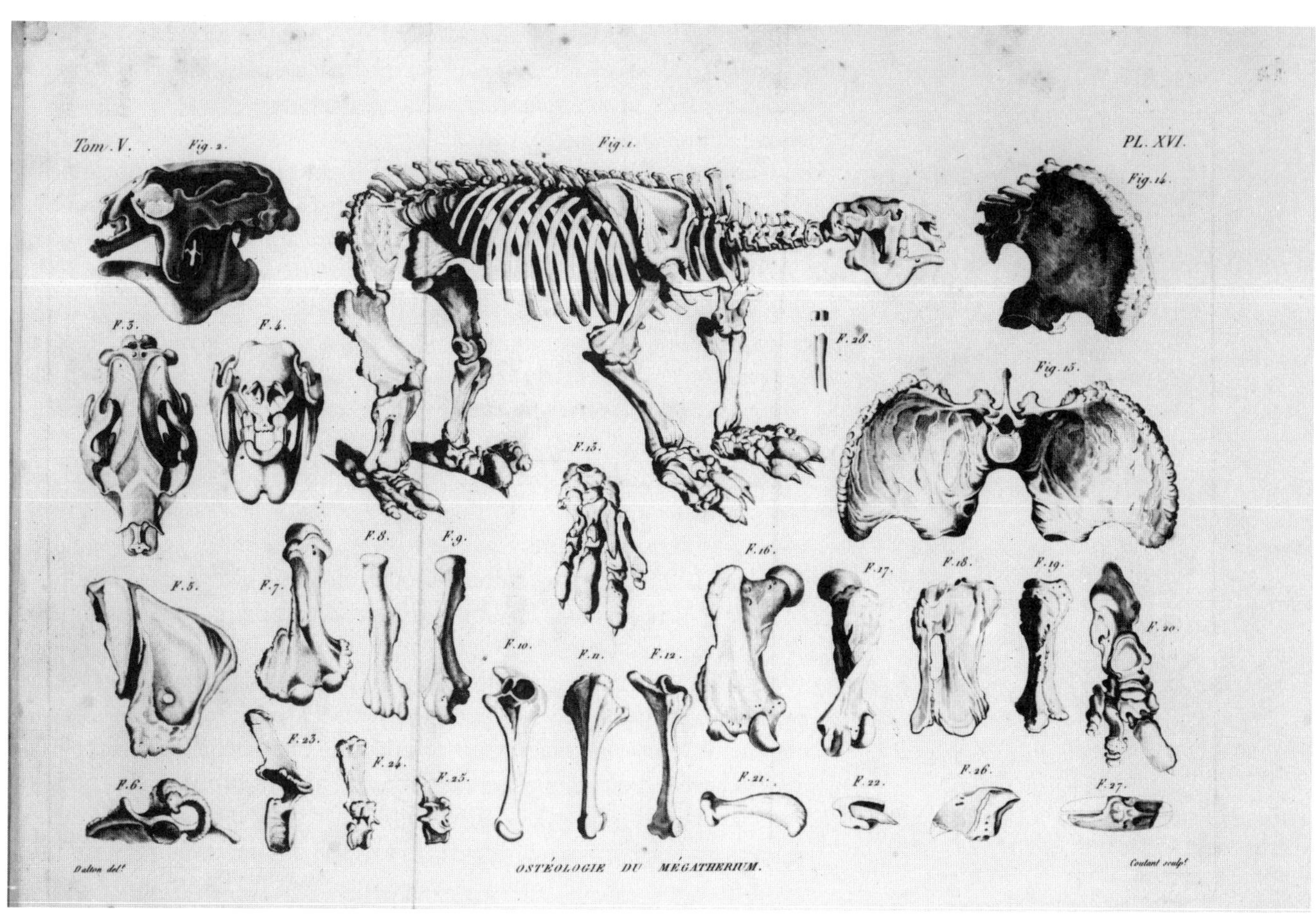

Cuvier coined the name "Megatherium," meaning literally huge beast, and applied it to very large animals known only by their fossil remains. A master of comparative anatomy, he was able to demonstrate that this skeleton of a sloth discovered in South America belonged to a species distinct from any living sloths and was thus the remains of an extinct species. This was proof that species actually could become extinct. *Recherches sur les ossemens fossiles*, 1825. Georges Cuvier.

longer existed. To Cuvier, this record of extinctions indicated a series of catastrophes or revolutions that the earth had suffered. Therefore, to his 1812 *Research on Fossil Remains* he prefixed a preliminary discourse of his theory of the earth, in which he attempted to reconcile his geological findings with biblical accounts. This eloquent but erroneous work was later published separately as *Discours sur les révolutions de la surface du globe*, of which the Library has a Paris, 1826, first edition copy. Cuvier's later geological work does not detract from his native genius and his brilliant career. Singlehandedly, he defined and revivified the sciences of zoology and comparative anatomy. A latecomer to geology, he nonetheless contributed a great deal to the new science before his unsuccessful attempts at an overall theory. Cuvier was the scientific lord of his time—an eloquent, imperious man with a great shock of red hair and piercing blue eyes. His head literally was outsized, and at his death an autopsy revealed his brain weighed 1,850 grams, 500 grams above the average.

No scientist of Cuvier's time could have been more contrapuntal to the great man himself than was the Englishman William Smith. The son of a yeoman farmer, Smith not only made stratigraphical discoveries identical to those of Cuvier, but his accomplishments preceded Cuvier's by some years.

Detail and key from *A Delineation of the Strata of England and Wales with Part of Scotland,* 1815. William Smith. See pp. 110–11.

Both men were born the same year, 1769, but the similarity ended there. In contrast to the aristocratic Cuvier, the broadly built Smith very much looked the part of the honest, staid farmer. He was a man of little formal education and much native intelligence, whose singlemindedness and methodical purposefulness produced one of the great classics in geological literature. Smith's medium was the geologic map, and it was his famous eight-foot nine-inch-by-six-foot two-inch map entitled *A Delineation of the Strata of England and Wales with Part of Scotland,* published in London in 1815, that formally laid down the principles of paleontological stratigraphy. The Library of Congress collections contain a copy of this remarkable work of geological cartography.

Smith was a self-taught surveyor whose job enabled him to closely observe the terrain of his native England. Early on, Smith recognized the importance of fossils as a dating device, realizing that each stratum had its own characteristic form of fossils. This was an entirely new observation. Although Smith was writing in 1799, he did not publish until 1815. Cuvier and Brongniart made public their identical discovery in 1811.

Smith's magnificent 1815 map of England is the first attempt to represent on a large scale the geological relations of such an extensive tract of ground. Smith ingeniously colored the bottom of each separate layer a more intense shade than its upper areas, which made his map easily intelligible. It became the model for all subsequent geological maps and proved a vital geological principle, stated by Smith: "The same species of fossils are found in the same strata, even at a wide distance." Smith's enthusiasm, insight, and determination enabled him to discover the record of nature in her rocks. To that end he sacrificed his personal life and any financial security he may have had. Very late in life he received the honor due him and was called by his peers "the father of English geology." Unlike the multitalented Cuvier, Smith knew only one great thing—but he knew it better than any man. He confined himself to his particular passion—the empirical observation of his country—and left all speculation and theories of the earth to others.

Only occasionally in the history of science is a book produced that is truly revolutionary. Even more rare is a book that deliberately and self-consciously effects a revolution in thought. Such a book is Charles Lyell's *Principles of Geology.*

In 1827, when Charles Lyell had reached thirty years of age, he had already rejected Neptunism, embraced the neglected arguments of Hutton, and decided he would write a book "to establish the principle of reasoning" in geology. He then dedicated three years of extensive travel and fieldwork to the gathering of geologic evidence to support these principles and, in 1830, produced in London the first of three volumes designed to effect a complete reorientation of geological thinking. The Library has all three volumes in first edition.

Actually, Lyell's "principles" were those of another Scotsman, James Hutton, and Lyell acknowledged his debt. Despite Playfair's eloquent presentation, Hutton's principle of uniformitarianism had not convinced many, and it was not until Lyell took hold of it that the principle really came alive. Lyell had a rare interpretive faculty and was as good at writing as Hutton had been bad. He wrote deliberately for the general reader and criticized those who cloaked their ideas in overly technical language. It was his goal to convince everyone, not only his fellow scientists, that there existed a continuity of life and a natural process in the physical world. He drew out Hutton's premise to its logical conclusion and supported that argument with a wealth of evidence and clear reasoning. The subtitle of his *Principles of Geology* pointedly tells his purpose: *An Attempt to Explain the Former Changes of the Earth's Surface, by Reference to Causes Now in Operation.* Lyell scorned theological explanations of natural phenomena and criticized and denounced unscientific and unsupportable hypotheses. He was consistent in his uniformity: "No causes whatever from the earliest time . . . to the present, ever acted, but these now acting; and they have never acted with different degrees of energy from that which they now exert."

Other geologists may have contributed more original ideas than Lyell—certainly Hutton did—but none had perceived or communicated the essential unity of the mature science and so wonderfully expressed it in a body of principles. Nor had anyone captured geology in its proper scale. Lyell provided the feeling and perspective of timelessness so necessary to an understanding of the physical history of the earth. Like a prophet, he turned heads and opened eyes.

His book was an immediate sensation and went through twelve editions in Lyell's lifetime. During the forty-five years between those editions, Lyell made annual field trips, and each new edition was improved, modified, and embellished. Although he wrote other books, *Principles of Geology* was his life's work. No tribute to the book's influence and worth can improve upon that made by Charles Darwin, when he said, "I always feel as if my books came half out of Lyell's brain." When Darwin sailed in 1831 on the *Beagle,* he had the first volume of Lyell's *Principles of Geology* with him. The second volume reached him in Montevideo. Darwin and Lyell later became close friends.

Roderick I. Murchison was also a close friend of Lyell's and accompanied him on his field travels through France and Italy. Murchison had come late to geology, having been lured away from a leisured life of dedicated fox hunting by the chemist Humphrey Davy. At thirty-two years of age, Murchison plunged into his geological studies with the vigor and enthusiasm he brought to everything. With all the instincts of the natural hunter, Murchison sought to resolve one of the more intractable geological problems of his time. William Smith's stratigraphical work had given order to the secondary rocks of England, but those older rocks below what Murchison called the "interminable greywacke" had resisted all ordering attempts. Through his dogged field surveys, Murchison eventually observed an order in the apparent chaos of rock formations. His great 800-page work, *The Silurian System,* (London, 1839), established for the first time the stratigraphic sequence of early Paleozoic rocks. As with Smith's discovery, Murchison's findings could be applied to geological formations elsewhere in the world. His *Silurian System,* of which the Library has a first edition copy, enabled geologists to accurately trace the earth's history backward across an increasingly long span of time. Following the custom that new stratigraphical names be chosen from the geographical region explored, Murchison chose to name his system after the Silures, an old British tribe that in Roman times inhabited the areas he explored.

Murchison's great effort is but one of a number of highly significant geological works published during the first half of the nineteenth century. This prodigious period was characterized not only by great discoveries but by the full emergence of the geologist as observer or field-worker. No individual was more vigorous and daring in his never-ending field research than the robust Jean Louis Rodolphe Agassiz. Although trained in medicine, Agassiz was very much interested in zoology, and by the time he was twenty-six had published the first of five volumes on fossil fish. This huge work became a classic and alone would have established his scientific fame. But the glaciers of his native Switzerland pulled him away from these studies, and he began to ponder the origin and nature of their looming magnificence. Agassiz spent the better part of five summers climbing the Alps and investigating the remains of old glaciers and the actions of present ones. The results of his investigations marked the beginning of modern glacial geology. In his *Études sur les glaciers* (Neuchâtel, 1840), which the Library has in first edition, Agassiz put forth his theory of the Ice Age. The earth, he argued, had once suffered a great drop in temperature, with the result that all of Switzerland was covered by a vast sheet of ice. His 1840 visits to Scotland and Ireland revealed the grooved and polished bedrock and the lodes of pulverized rocks he had come to recognize as evidence of glacial movement. By analogy, then, he generalized his theory to most of Europe, North Asia, and North America. His *Études* is a large book accompanied by an atlas of thirty-two plates with drawings of glaciers. In 1846 Agassiz was invited to lecture in the United States and two years later was appointed professor of zoology at Harvard University. He remained in America until his death, becoming one of the foremost personalities and scientists of the United States. Agassiz opposed the Darwinian interpretation of evolution by natural selection, despite having provided both Lyell and Darwin with a great deal of supportive information.

At this time in America, another eminent geologist also was struggling with the dilemma of Darwin. New York-born James Dwight Dana studied and later taught at Yale College and eventually became a reluctant convert to Darwinism. His famous text, *System of Mineralogy* (New Haven, 1837), became the authoritative reference text on minerals throughout the world. The Library of Congress has a first edition copy of this standard work, which was published when Dana was only twenty-four, containing the brilliant young man's original contributions in mathematical crystallography and classification as well as all that was known about minerals.

After returning from the Wilkes Expedition of 1838–42, Dana turned his attention to tectonics, one of the major problem areas of geology. He soon became the principal exponent, if not the originator, of the contracting-and-cooling-earth hypothesis. He sought to apply that concept

A summer vacation in the Alps allowed Louis Agassiz several weeks' time to explore the mountains in that region and led to his startling conclusions that not only had an Ice Age occurred several thousand years before but that glaciers actually move. In support of his glacial theory, Agassiz showed glaciers in different positions and different elevations. Here he shows the middle part of the Zermatt glacier. What appears to be a dried river bed is actually the moraine or the accumulation of earth and stones carried and finally deposited by the glacier when it reached its melt line. The rocks on the right where the sheep and shepherdess stand are polished as a result of glacial erosion. Agassiz's ideas concerning the transporting power of glaciers and the existence of a Pleistocene glacial period met with twenty-five years of resistance before they were accepted. *Études sur les glaciers*, 1840. Louis Agassiz.

to mountain-building, pointing out that folds found in the Appalachian mountains can be explained only by a horizontal force produced by contraction. In 1863, Dana published his *Manual of Geology* in Philadelphia, a work which likewise became a standard authority and solidified his worldwide reputation. The Library has a first edition of this work. Though a man of frail health in his later years, Dana nonetheless lived to be eighty-two.

Dana wrote of Appalachian mountains being created and growing in great upfolds from powerful lateral thrusts and Agassiz described and

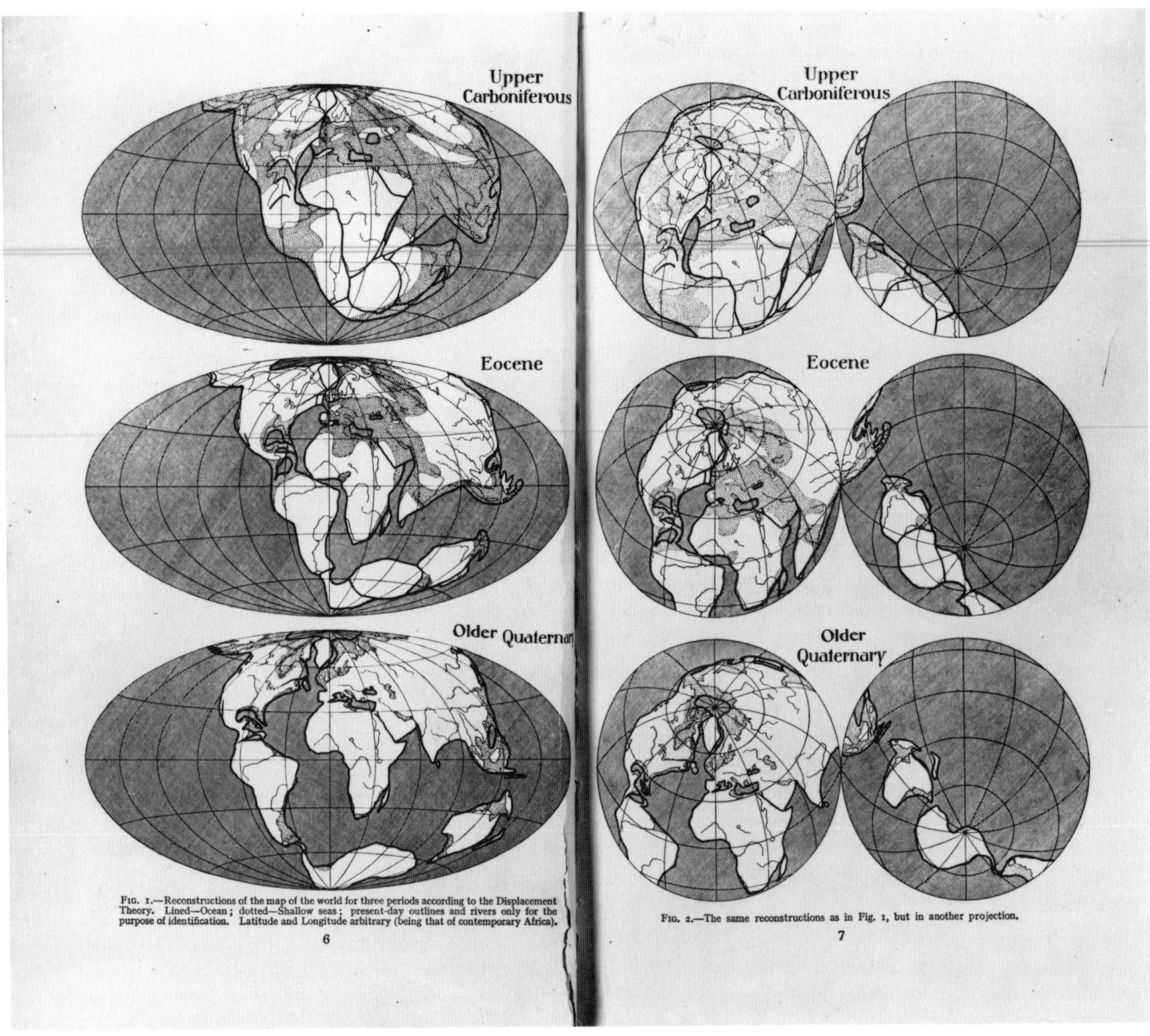

Upper Carboniferous
Eocene
Older Quatern
Fig. 1.—Reconstructions of the map of the world for three periods according to the Displacement Theory. Lined—Ocean; dotted—Shallow seas; present-day outlines and rivers only for the purpose of identification. Latitude and Longitude arbitrary (being that of contemporary Africa).
6
Upper Carboniferous
Eocene
Older Quaternary
Fig. 2.—The same reconstructions as in Fig. 1, but in another projection.
7

Although the puzzle-like fit of the coastlines of South America and Africa had been noted for some time, none offered so revolutionary an explanation as did Alfred Wegener. The continents had been displaced horizontally in great blocks over an enormous period of time, he said, and indeed were still in motion. Wegener envisioned a single, original, great landmass, in the distant geologic past, a Carboniferous continent, which he called Pangaea, that began to drift apart in a gradual and permanent manner. In this illustration, Wegener depicts the global break-up of the continents. *The Origin of Continents and Oceans*, 1924. Alfred Wegener.

measured the actual yearly movement of the Aar glacier in Switzerland but a German meteorologist, Alfred Wegener, wrote of entire continents in motion. His theory of continental drift, or "continental displacement" as it was first translated from the German, had a bold, grand sweep to it. First published in Brunswick in 1915, Wegener's book, *Die Entstehung der Kontinente und Ozeane,* offered the rather revolutionary thesis that the continents were once part of a single landmass, which slowly separated and gradually drifted apart. Although the idea of a single, original landmass was not new, no one had offered anything like Wegener's drift theory to explain our contemporary separateness. Well over a century before, Alexander von Humboldt noted that not only were the coastlines of Africa and South America parallel and made for a "puzzle-like" fit but the rocks of both coasts were similar. Von Humboldt, however, postulated that erosion from marine currents had split the continents. Still others had argued for a quick, catastrophic splitting. No one had offered the startling idea of gradual, continuous drift. Wegener's book, of which the Library has both the 1922 Brunswick edition and the London 1924 English translation of the third edition, argued that the continents were still in motion. This seemed to defy both logic and the senses. And although the geological and biological evidence compiled by Wegener continued to mount, the theory was at a loss to account for a convincing mechanism or mechanical explanation. At first, Wegener argued that the lack of an explanation did not keep the theory from being a fact. "The Newton of the drift theory has not yet appeared," he said. His subsequent attempts to supply a mechanism were shown to be untenable—detracting from his central concept. Being an expert on Greenland, Wegener made several expeditions to that icy region; on the fourth expedition he died at the age of fifty. But the man was ultimately vindicated. Most earth scientists today agree that substantial movement and distortion of the earth's outer layers has occurred over time. Current research in plate tectonics—in which the earth's surface is considered to be divided into a series of moving slabs or plates that are formed at mid-ocean ridges and that return to the earth's interior at convergent plate boundaries—has its origins in Wegener's theory. Wegener's missing mechanism is now believed to be mainly the convection currents resulting from the earth's molten core. Wegener's vindication further underscores the importance of James Hutton's fertile ideas of dynamic gradualism. Two hundred years after he first proposed his theory, Hutton's basic concepts are still in force.

Perhaps no scientific discipline has a tradition so entwined with religion as geology. Although other disciplines occasionally raised issues or ideas that had religious implications or in some way caused a stir (zoology's theories of the origins of man, for instance) none was dominated by religious dogma both at its very beginnings and throughout most of its history. This conflict of "truths"—revealed and empirical—was beginning to be played out only a century and a half ago. The idea that the answers to the great questions of geology lay within the earth itself was a concept that directly countered geology's religious tradition. Yet unlike most other disciplines, which could often incorporate much of their nonscientific or prescientific past, geology had to reject the bulk of its heritage to mature. The notion of a dynamic, still-changing planet is at the core of geology's new tradition. Although such an idea is today unremarkable, it became so only after a monumental struggle.

TV PVS A·
RITH
ME
TICAE·
BOETIVS
RAS
TAGO

7. *Mathematics: Things Are Numbers*

Mathematics stands apart from other scientific disciplines in that its subject does not immediately present itself. Astronomy investigates the stars and the planets, and zoology studies animals—but what is the prime concern of mathematics? To some it is simple counting with all its permutations—mathematics is, in this view, a tool to number reality. Others regard it as an idea, a concept whose purity and abstractness transcend any need for application. Thus we have mathematics as bookkeeping or mathematics as truth, mathematics as existing independently from the mind of man or mathematics as the intellectual creation of mankind.

It may be correct to say that in the traditional sense, mathematics has no real subject at all—none which can be found in the world of concrete reality. There is an "otherness" about mathematics by which it functions or exists apart from what we regard as the real world. The abstract idea of two or "twoness" that we may entertain in our minds is not contingent on the actuality of there being two of anything before our eyes. Indeed, once man made the mental leap and took from the particular those universal elements that could be applied to another situation, he ascended to the level of abstraction and universality. The recognition that a number is an idea entertained by the mind, separate and distinct from any physical objects or reality, was a giant step in man's intellectual growth.

Yet as pure and abstract as mathematics might be, both its origins and its development have been intimately linked to the practical needs of men and women. Mathematics as a system probably originated as a practical tool to help with ancient man's agricultural and engineering needs. The first counting system was probably used to number animals or crops. Since then, man has skillfully enlarged his mathematical arsenal from simple arithmetic and geometry to algebra, analytic geomety, and calculus.

So mathematics has a two-fold, seemingly contradictory, nature. It is both abstract and concrete, ideal and practical. From the beginning, however, mathematics was made up of the interplay of these opposites. It is a creation of the mind steeped in pragmatism—a useful intelligible idea. As a language universal in space and time, it has been spoken both by the pharoahs who measured the Nile delta and by the astronauts who plotted the grids of the moon. The fabulously fertile mix of the ideal and the real in mathematics makes for a discipline with scores of branches and even more avenues of application. It is a realm of knowledge that contains both the raw simplicity of basic arithmetic and the elegance and aloofness of logic. But its real nature is found in neither extreme. Mathematics is not solely the handmaiden of science. This demeans its significance and dimin-

Opposite page:
This illustration from an early sixteenth-century encyclopedic work shows the Form or Figure of Arithmetic, possibly the Muse of Arithmetic. She is standing between Pythagoras, doing ancient counter reckoning, and Boethius, who is using the modern system of Arabic numerals. *Margarita philosophica*, 1503. Gregor Reisch.

ishes its role. Neither is mathematics supreme among the disciplines, with all others subordinate to it. This overemphasizes its importance. It is safe to say that during the nineteenth century, mathematics became a realm of knowledge by itself, neither contingent upon any other discipline nor necessarily superior to any.

The litany of discoveries which follows in this chapter tells the story of how mathematics grew up. It recounts the intellectual journey leading to the nineteenth-century heresy of non-Euclidean geometry, which liberated mathematics from its self-imposed restraints. From then on, mathematics needed only to be self-consistent to be valid and workable. It had attained the special status of being a true intellectual achievement.

At the heart of this lofty achievement of man's mind is a phenomenon that links it with the most ordinary of our endeavors—the universal human impulse for order. Whether this human need is called the search for truth, for beauty, or for any of the traditional universals, it describes a uniquely human pursuit. Put this way, our mathematics becomes a search for symmetry, for unity, for connections and similarities. We search through nature's panoply, sifting with our mathematics for her hidden order, seeking always to know more, to understand better.

The following includes some of the best of these attempts and accomplishments. Emphasis is placed more on the intellectual, mathematical achievement itself than on its practical applications. In a sense, the other chapters in this book deal with those.

Nothing remains from the first great age of mathematics but clay tablets. Dating as far back as 2000 B.C., these tablets quietly record the contributions of Babylonia, whose ancient capital is part of modern Iraq. From Babylon came both an idea and a method; the idea being the notion of abstraction, the method being the invention of a mathematical notation system. With the decline of Babylon, this tradition passed to the ascendant Egyptians and then on to the Greeks.

Greek mathematics flourished in the third century B.C. with the works of Euclid, Archimedes, and Apollonius. All three came out of the mathematics school in Alexandria. Euclid is the first of this triumverate, commonly regarded as the greatest mathematicians of antiquity. His *Elementa geometria* is a systematic exposition of the leading propositions of elementary geometry and also contains a theory of numbers. Written in thirteen "books," Euclid's *Elements* is a codification of all Greek mathematical knowledge since Pythagoras. It is his unique axiomatic arrangement, organization, and presentation of two and a half centuries of work that made this treatise the most successful textbook of all time. Since it was first printed in Venice in 1482, the *Elements* has seen over one thousand editions. The Library's first edition copy has diagrams set in a wide margin close to the theorems to which they relate. It is regarded as the first substantial book printed containing geometrical figures.

Opposite page:
This beautiful page is from the first printed edition of Euclid's *Elements*. Euclid's systemic exposition of the leading propostions of elementary geometry and his unique axiomatic arrangement and presentation made this work the most successful textbook of all time. This late fifteenth-century version of Euclid is a revision by Campanus of Novara (thirteenth century) of an Arabic version of the Greek text made by Adelard of Bath (twelfth century). The creative use of diagrams by the printer, Ratdolt, to illustrate textual theorems produced a satisfying balance of text and illustrations. *Elementa geometria*, 1482. Euclides.

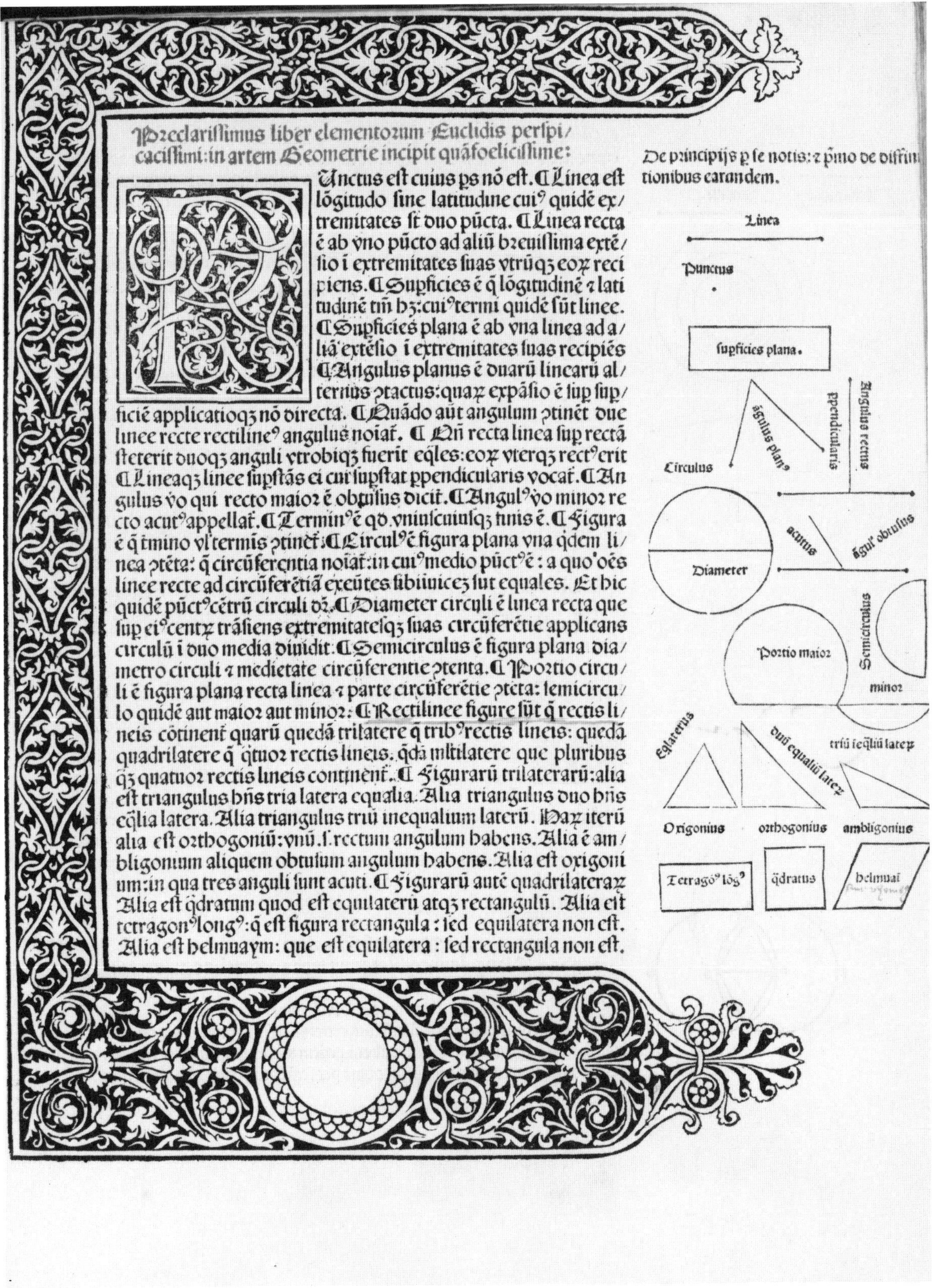

Preclarissimus liber elementorum Euclidis perspi/
cacissimi: in artem Geometrie incipit quã foelicissime:

Punctus est cuius ps nõ est. ¶ Linea est
lõgitudo sine latitudine cui⁹ quidẽ ex/
tremitates sũt duo pũcta. ¶ Linea recta
ẽ ab vno pũcto ad aliũ breuissima extẽ/
sio i extremitates suas vtrũqz eoꝝ reci
piens. ¶ Supficies ẽ q̃ lõgitudinẽ ⁊ lati
tudinẽ tñ hz: cui⁹ termi quidẽ sũt linee.
¶ Supficies plana ẽ ab vna linea ad a/
liã extẽsio i extremitates suas recipiẽs
¶ Angulus planus ẽ duarũ linearũ al/
ternus ꝯtactus: quaꝝ expãsio ẽ sup sup/
ficiẽ applicatioqz nõ directa. ¶ Quãdo aũt angulum ꝯtinẽt due
linee recte rectiline⁹ angulus noiat. ¶ Qñ recta linea sup rectã
steterit duoqz anguli vtrobiqz fuerit eq̃les: eoꝝ vterqz rect⁹ erit
¶ Lineaqz linee supstãs ei cui supstat ppendicularis vocat. ¶ An
gulus vo qui recto maior ẽ obtusus dicit. ¶ Angul⁹ vo minor re
cto acut⁹ appellat. ¶ Termin⁹ ẽ qd vniuscuiusqz finis ẽ. ¶ Figura
ẽ q̃ tmino vl termis ꝯtinet: ¶ Circul⁹ ẽ figura plana vna q̃dem li/
nea ꝯtẽta: q̃ circũferentia noiat: in cui⁹ medio pũct⁹ ẽ: a quo oẽs
linee recte ad circũferẽtiã exeũtes sibiinuicez sũt equales. Et hic
quidẽ pũct⁹ cẽtrũ circuli dz. ¶ Diameter circuli ẽ linea recta que
sup ei⁹ centꝝ trãsiens extremitatesqz suas circũferẽtie applicans
circulũ i duo media diuidit. ¶ Semicirculus ẽ figura plana dia/
metro circuli ⁊ medietate circũferentie ꝯtenta. ¶ Portio circu/
li ẽ figura plana recta linea ⁊ parte circũferẽtie ꝯtẽta: semicircu/
lo quidẽ aut maior aut minor. ¶ Rectilinee figure sũt q̃ rectis li/
neis cõtinent quarũ quedã trilatere q̃ trib⁹ rectis lineis: quedã
quadrilatere q̃ q̃tuor rectis lineis. q̃dã multilatere que pluribus
q̃z quatuor rectis lineis continent. ¶ Figurarũ trilaterarũ: alia
est triangulus hñs tria latera equalia. Alia triangulus duo hñs
eq̃lia latera. Alia triangulus triũ inequalium laterũ. Haꝝ iterũ
alia est orthogoniũ: vnũ .s. rectum angulum habens. Alia ẽ am/
bligonium aliquem obtusum angulum habens. Alia est oxigoni
um: in qua tres anguli sunt acuti. ¶ Figurarũ autẽ quadrilateraꝝ
Alia est q̃dratum quod est equilaterũ atqz rectangulũ. Alia est
tetragon⁹ long⁹: q̃ est figura rectangula: sed equilatera non est.
Alia est helmuaym: que est equilatera: sed rectangula non est.

De principijs p se notis: ⁊ pmo de diffini
tionibus earundem.

Over the years, the name Euclid became literally synonymous with geometry, primarily because of the enduring quality of his axioms. Presented in a brief, logical, even elegant manner, these axioms used the deductive method so typical of Greek thinking. First, Euclid would offer an obvious or self-evident postulate and from that he would lead the reader through a series of convincing steps, eventually to accept the truth of a final proposition. Although little of the *Elements* is original with Euclid, as a work of synthesis it stands alone. And as with few large works from classical antiquity, it has survived intact, escaping any major textual corruptions. Little is known of Euclid the man, nor is it known exactly when he was born or died. His writings reveal nothing of a personal, temporal, or historical nature, being purely and totally mathematical. This singlemindedness may have bored some young readers, but its purity and elegance led Bertrand Russell to describe his first experience of Euclid to be as shattering as first love. Since Euclid's *Elements* was the first substantial source of mathematical knowledge and one that was used by all succeeding generations, its overall concept of mathematics and its particular notion of proof set the course for all subsequent mathematical thinking.

Archimedes too studied in Alexandria, where his teacher had been a pupil of Euclid. He was an aristocrat, said to have been related to the king of Syracuse. As is not the case with Euclid, much is known of his life and deeds, and stories about him abound. Together with Newton and Gauss, he is regarded by many as one of the greatest mathematicians the world has known. His penetrating intellect and powerful genius make him one of the intellectual giants of the West and, indeed, a modern among ancients. It is said that he and Newton would have understood each other perfectly. Archimedes seems to have possessed the type of genius that Leonardo da Vinci came to typify—the lofty intellect roaming free. Both had a great breadth of interests, an ability to apply an intricate idea concretely and ingeniously, and a disregard for convention. The achievements of Archimedes in pure and applied mathematics have become a part of our cultural lore as well as part of the history of science. The story of his running naked through the streets of Syracuse screaming "Eureka!" at his sudden discovery of the first law of hydrostatics attests to his appealing and transcendant personality. And as the name Euclid is synonymous with geometry, so "the principle of Archimedes" is known by any school child to be the law of hydrostatics. Archimedes also contributed significantly to theoretical mechanics and discovered the fundamental principles relating to the lever. Apart from mechanics, his mathematical work included finding the volume of a sphere, the calculation of pi (the ratio of the circumference to the diameter of a circle), and a new scheme for representing large numbers in verbal language.

Unlike Euclid, who wrote in a systematic manner for students, Archimedes wrote brilliant individual essays aimed at the most educated mathematicians of his day. They are all masterpieces of mathematical exposition; they are highly original; and they are presented in an economical and skillfully rigorous manner. His writings are represented in the

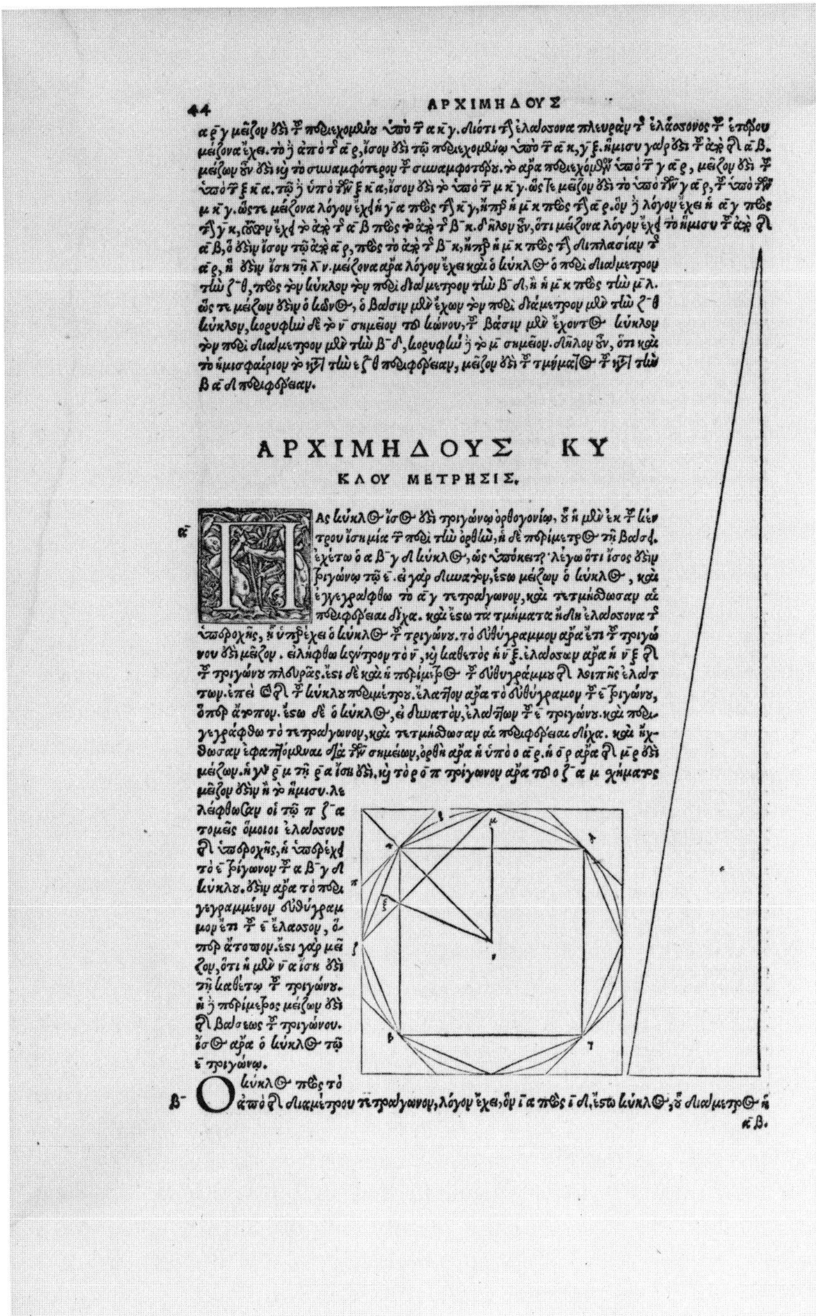

44 ΑΡΧΙΜΗΔΟΥΣ

ΑΡΧΙΜΗΔΟΥΣ ΚΥΚΛΟΥ ΜΕΤΡΗΣΙΣ.

The publication in 1544 of the first complete edition of Archimedes's works was crucial to the scientific revolution of the next century. It was the availability of this work that enabled Galileo, Descartes, and Newton to extend the frontiers of not only mathematics but all the exact sciences. Archimedes aimed his brilliant mathematical essays at the most educated mathematicians of his day, and the rigor, economy, and originality of these essays must have made them a joy to rediscover nearly two thousand years later. *Opera, quae quidem extant, omina*, 1544. Archimedes.

Library's collections by a 1544 collection *Opera, quae quidem extant, omnia,* printed in Basel. Apart from one small tract published in 1503 and another imperfect edition in 1543, this is the first complete edition of Archimedes's works. The text is in both Greek and Latin and is accompanied by many diagrams. It was this publication in 1544 that influenced and inspired the mathematical work of Galileo, Descartes, and Newton.

Among the many stories of the life and accomplishments of Archimedes, one of the more colorful is his assistance in the defense of Syracuse against

The mathematical advances in curved line geometry made by Apollonius during the third century B.C. became suddenly applicable and then indispensable to astronomy once it was understood that the planets followed paths that were not always circular. Up to the beginning of the eighteenth century, Apollonius of Perga was only partially known to the West—the first four books of his *Conics* being known in Greek, the next three existing in Arabic, and his last book being lost. In 1705, the remarkable Edmund Halley undertook the task of preparing a complete and definitive Latin translation of his entire work. Working from Arabic manuscripts with only a Latin key, Halley produced a version that was not only accurate but matched Apollonius's style of expression. Halley's accomplishment gains even more luster with his "restoration" of the lost eighth book, which he reconstructed from the comments of Pappus (ca. A.D. 300), who possessed the Greek text. *Conicorum*, 1710. Apollonius Pergaeus.

the Roman general Marcellus during the second Punic War. As the story is told, Archimedes devised all sorts of cunning contrivances to repel the invader—using catapults, grappling cranes, and burning-glasses to destroy ships. Because of this, it took the Romans three years to take the city—and then it was from the rear. Archimedes died in his seventy-fifth year in a manner, described by Plutarch, that makes him a martyr to science. While drawing geometrical figures in the sand, he was challenged by one of the Roman soldiers who had taken the city and was slain after imperiously ordering, "Do not disturb my circles." The Romans later erected an elaborate tomb to Archimedes, on which was engraved, according to his wishes, the figure of a sphere inscribed in a cylinder, commemorating one of his famous theorems.

The third great mathematician of antiquity was Apollonius of Perga. He too learned mathematics from Euclid's successors. His chief work is *Conicorum*, a systematic treatise on conic sections—those curves derived from slicing a cone. Made up of eight books, seven of which are extant, the *Conics* is a monumental work described by one historian as having created the genre of an exhaustive monograph on a particular topic in mathematics. Indeed, this work not only included all that was previously known about conic sections but its comprehensive and highly original extension of the knowledge of these curves made them a closed subject to thinkers for centuries to come. His work at once became accepted as the standard textbook on the subject. The Library's copy of *Conicorum* was edited by the astronomer Edmond Halley and published in Oxford in 1710. It is a folio-size book with both Greek and Latin text. Nearly every page has diagrams.

Few details of the life of Apollonius are known, but it is recorded that he was called the "Great Geometer." His greatness was reaffirmed eighteen centuries later when it was found that the orbits of planets were not always circular, as had been believed, but followed paths better described by conic sections. Since his conic sections included curves, ellipses, parabolas, and hyperbolas, the work of Apollonius became directly applicable to that of Kepler and Newton and from that point on was no longer regarded only as a work of ingenious mathematical play or diversion.

Greek contributions to mathematics did not end with this glorious explosion of thought in the third century B.C., although Salomon Bochner for one says that "after Apollonius darkness falls on the landscape of mathematics." The legacy of Greek mathematics is that it created the science as we know it today. Specifically, the Greek contributions to mathematics were in establishing the ideal of rigorously deductive proof as well as the method of developing a subject by a chain of theorems based on definitions, axioms, and postulates and the constant striving for complete abstraction and generality. The mechanics of Archimedes aside, the Greeks cared little for the applied or practical side of mathematics. According to Plato, each science is a science only insofar as it contains mathematics, for mathematics is the perfect mode of thought.

The decline of Greece and its influence had many repercussions, but none so antithetical to this Platonic way of life and thinking as the rise of pragmatic, imperial Rome. The Romans did not understand pure science and even scorned mathematics—regarding it as an art practiced by astrologers. Consequently, Rome contributed nothing to pure mathematics. With the fall of Rome and the much later burning of the Alexandrian library, the history of mathematics and of most sciences became a story of intellectual stagnancy and social disorder. In fact, some mathematical writers virtually discount these centuries and jump directly from the third century B.C. and Apollonius to the seventeenth century and Descartes with great ease.

But between Rome and the Renaissance there were certain mathematical milestones worthy of note that are represented in the Library's collections. The first and most ancient of these represents the work of a Roman born about the time of Rome's fall, usually given as A.D. 476. Boethius was the last Roman of any note to have studied the language and literature of Greece, and he prepared commentaries and translations of Aristotle and summaries of other subjects. These works of Boethius are significant more for their timing than for any original contributions they may have made, since they became the standard textbooks for the next six or seven centuries. As such, they served as a vital connecting link between the mathematics of classical and medieval times. The Library has one of Boethius's mathematical works, *De institutione arithmetica,* published in Augsburg in 1488. By the time it was published as a book, however, many Arabic works (from the original Greek) had been translated into Latin, and Boethius's work was quickly eclipsed. The Library's copy of this valuable incunabulum is a quarto of forty-eight leaves printed in two columns.

Isidore of Seville, who lived about a hundred years later than Boethius and was archbishop of that city in Spain, was also a conduit of sorts. His major work, called *Etymologiae,* first published in Augsburg in 1472, managed to salvage and to transmit some of the knowledge of the Greeks to the medievals. The Library has this work in first edition, Book 3 of which is titled *De vocabulo arithmetice disciplinae.* It is a large work with many eye-catching rubrics. A fairly early incunabulum, this volume is said to contain the first printed reference to arithmetic.

With the beginnings of a revival of learning in Europe at the start of the thirteenth century, a remarkable mathematician appeared on the scene to herald the commencement of the mathematical renaissance. Leonardo Fibonacci, also called Leonardo of Pisa, was the son of a merchant who traded with Islamic North Africa. Leonardo was tutored by an Arab and was able to travel widely. Because of this, he became aware of the great advantages of the Hindu-Arabic system of notation over the cumbersome alphabet system of the Romans. In 1202 he finished a work called *Liber abaci* or *Book of the Abacus* in which he explains the Arabic system, gives an account of algebra, and discusses some geometry. His work had a wide circulation and is credited with not only introducing the use of Arabic numerals into Christian Europe but giving the deathblow to the old system. The fact that no list of incunabula includes this popular and progressive work may seem surprising, but indeed, no early printer chose to publish it. Like printers of today, fifteenth-century printers were very

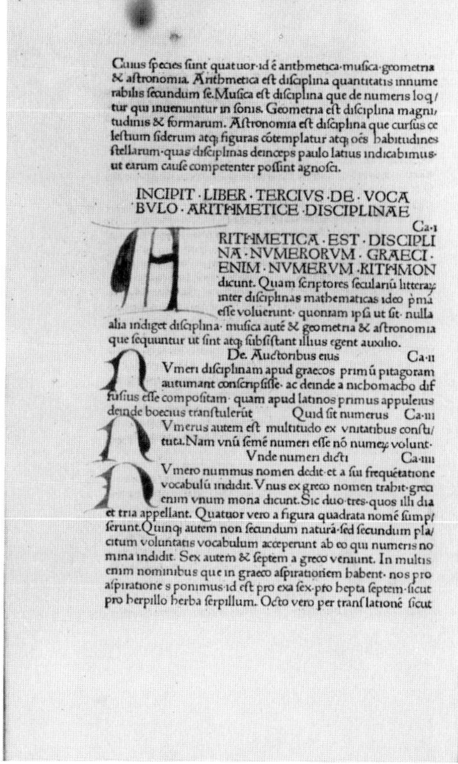

Cuius ſpecies ſunt quatuor·id ē arithmetica·muſica·geometria & aſtronomia. Arithmetica eſt diſciplina quantitatis innumerabilis ſecundum ſe. Muſica eſt diſciplina que de numeris loquitur qui inueniuntur in ſonis. Geometria eſt diſciplina magnitudinis & formarum. Aſtronomia eſt diſciplina que curſus celeſtium ſiderum atq; figuras cōtemplatur atq; oēs habitudines ſtellarum·quas diſciplinas deinceps paulo latius indicabimus· ut earum cauſe competenter poſſint agnoſci.

INCIPIT · LIBER · TERCIVS · DE · VOCABVLO · ARITHMETICE · DISCIPLINAE

Ca·i

ARITHMETICA · EST · DISCIPLINA · NVMERORVM · GRAECI · ENIM · NVMERVM · RITHMON dicunt. Quam ſcriptores ſeculariū litterarū inter diſciplinas mathematicas ideo p̄mā eſſe voluerunt· quoniam ipſa ut ſit· nulla alia indiget diſciplina· muſica autē & geometria & aſtronomia que ſequuntur ut ſint atq; ſubſiſtant illius egent auxilio.

De. Auctoribus eius Ca·ii

NVmeri diſciplinam apud graecos primū pitagoram autumant conſcripſiſſe· ac deinde a nichomacho diffuſius eſſe compoſitam· quam apud latinos primus appuleius deinde boecius tranſtulerūt

Quid ſit numerus Ca·iii

NVmerus autem eſt multitudo ex vnitatibus conſtituta. Nam vnū ſemē numeri eſſe nō numerū volunt·

Vnde numeri dicti Ca·iiii

NVmero nummus nomen dedit·et a ſui frequētatione vocabulū indidit. Vnus ex greco nomen trahit·greci enim vnum mona dicunt. Sic duo·tres·quos illi dia et tria appellant. Quatuor vero a figura quadrata nomē ſumpſerunt. Quinq; autem non ſecundum naturā·ſed ſecundum placitum voluntatis vocabulum acceperunt ab eo qui numeris nomina indidit. Sex autem & ſeptem a greco veniunt. In multis enim nominibus que in graeco aſpirationem habent· nos pro aſpiratione s ponimus·id eſt pro exa ſex·pro hepta ſeptem·ſicut pro herpillo herba ſerpillum. Octo vero per tranſlationē ſicut

As a versatile scholar, statesman, and bishop, Isidore of Seville was the most prominent man in seventh-century Europe. His encyclopedic work on the seven liberal arts, the trivium and quadrivium, consisted of twenty books, the third being on arithmetic (shown here). This book is basically a condensed version of Boethius which, in turn, was based on the arithmetic of Nicomachus and Euclid's geometry. The mathematics of such medievalists as Isidore was simple, rarely employing fractions and never irrational numbers. The Church encouraged the teaching of mathematics to keep its calendar in order, although good calculators in early medieval times were often regarded as practitioners of a black art. *De vocabulo arithmetice disciplinae*, 1472. Isidorus of Seville.

much businessmen who worked for profit, and often chose the encyclopedic or the sensational (astrological and other mantic treatises) over the truly worthy. Leonardo of Pisa is therefore represented in the Library of Congress collections by a two-volume work, *Scritti di Leonardo Pisano,* printed in Rome between 1857 and 1862. This contains both his *Liber abaci* and *Practica geometriae,* a compilation he produced in 1220, which also introduced some trigonometry.

One of the more significant books of early printing, however, is largely based on the works of Leonardo of Pisa. Printed in Venice in 1494, *Somma di aritmetica, geometria, proporzione e proporzionalità* was written by the Franciscan monk Luca Paccioli. Paccioli wrote in Italian and his

Ad illustrissimum Principem Gui. Ubaldum Urbini Ducē Montis feretri: ac Durantis Comitem. Grecis latinisq; litteris Ornatissimum: & Mathematice disciplinae cultorem ferventissimum: Fratris Luce de Burgo sancti Sepulchri: Ordinis minorum: & sacre Theologie Magistri. In artem arithmetice: & Geometrie. Prefatio.

A quantita Magnanimo Duca: e si nobile & excellēte cosa che molti phylosophi per questo lhano giudicata ala substantia para: ecōessa coeterna. Peroche hano cognosciuto per verū modo alcuna cosa in rerū natura sença lei nō potere existere. Per la qual cosa de lei itēdo (cō laiuto de colui che li nostri sensi reggi) tractarne: nonche per altri prischi e antichi phylosophi nonne sia copiosamente tractato: e in theorica e pratica. Ma per che lor dicti gia ali tempi nostri sonno molto obscuri: e damolti male apresi: e ale pratiche vulgari male applicati: diche i loro operationi molto variano: e con grandi elaboriosi affanni mettano in opera: si de numeri cōmo de misure: vnde di lei parlando non intendo se non quāto che ala pratica e operare sia mestiero: mescolandoci secōdo iluoghi oportuni ancora la theorica: ecausa de tale operare: si de numeri cōmo de geometria. Ma prima accio meglio qllo che sequita se habia apprehēdere: essa quantita diuid iremo secōdo el nostro proposito: ediuidendola aciascun suo membro assegnaremo sua propria e vera diffinitione e descriptione. E alora poi sequira quello che Arist. dici in secundo poster. Tūc enim maxime scitur aliquid cum habetur suum quid est &c.

Diffinitiones & diuisio discrete & continue quantitatis: articulus primus prime distinctionis.

Dico adōca. La quātita essere imediate bimembre: cioe continua e discreta. La continua e quella lechui parti sonno copulate e gionte a certo termine cōmune: cōme sōno legni: ferro: e saxa &c. La discreta oueramēte numero: e qlla lecui parti nō sōno giōte adalcuno termine cōe: cōmo e.1.2.3.&c. Diche prima dela discreta: cioe del numero: e poi dela continua cioe geometria: quāto alo intento aspecta chiaramente tractaremo.

Diffinitio numeri propriissima: articulus secundus.

Numero: e (secondo ciascuno phylosophāte) vna multitudine de vnita cōposta: et essa vnita nō e numero: ma ben principio de ciascun numero: ede qlla mediāte laqle ogni cosa e dicta essere vna. E secōdo elseuerin Boetio i sua musica: e la vnita ciascū nūero i potēria: & passiz i la sua arithmetica Regina e fondamento dogni numero lapella. Laqual piu magnificandola in le cose naturali disse in quello che fa de vnitate & vno. Omne quod est: ideo est: quia vnum numero est. E ne ancora el numero in infiniti membri diuiso: per quel che esso Aristo. dixe: cioe. Siquid infinitum est: numerus est. E per la terça petitione del septimo de Euclide: la sua serie in infinito potere procedere: et quocūq; numero dato: dari pōt maior: vnitatem addendo. Ma noi pigliaremo quelle parti a noi piu note e accomodate. E pero dico con glialtri alcuno essere primo: ede quello che solo dala vnita e numerato: e non ha altro numero: che integralmente aponto lo parta. Altro e ditto cōposto: ede quello che da altro numero e mesurato: ouero numerato. Exēplum primi Cōmo.3.7.11.13.e.17. &c. Exemplū secūdi. Cōmo.4. chel doi lo mesura e numera: e.8. chel.2.e.4. El.12.14.18. e simili: tutti sōno ditti numeri composti: nō solo che constino

a

The Franciscan monk portrayed within the initial *L* on this page is the work's author, Luca Paccioli. His *Somma* is a wide-ranging Renaissance work on mathematics which pulled together in Italian most of the existing knowledge on the subject and made no attempt to be original. However, its section on what he called the "method of Venice," or double-entry bookkeeping, presented this calculating art for the first time in printed form. As a young student, Paccioli had worked for a Venetian merchant and he always retained a practical and mercantile interest. Having taught at the universitites of Perugia, Naples, and Rome, he took a position teaching mathematics at the court of Ludovico Sforza, duke of Milan. It was there that he met Leonardo da Vinci (who made reference to Paccioli in his notebooks). Leonardo left Milan for Florence with his friend Paccioli once the French army entered their city. *Somma di aritmetica, geometria, proporzione e proporzionalità,* 1494. Luca Paccioli.

book received a wide circulation. The Library has this 1494 edition, which consists of two parts, the first dealing with arithmetic and algebra and the second with geometry. Apart from being the earliest printed book on arithmetic and algebra, it is best known for its introduction of double-entry bookkeeping. Paccioli's compendium linked mathematics with a variety of practical applications and presented methods of accounting which have long endured. Paccioli was an intimate friend of Leonardo da Vinci and was the first occupant of a chair of mathematics founded by the Sforza family in Milan. His book is also noteworthy because it contained much more than was taught in the universities and emphasized how mathematics could be used in the practical world of commerce. It should be noted also, however, that it did not add to what Leonardo of Pisa had contributed nearly three centuries before.

The ablest and the most influential mathematician of the fifteenth century was Regiomontanus, whose given name was Johannes Müller. Although he is better known today for his astronmical work analyzing Ptolemy's *Almagest,* he contributed also to the development of mathematics. His systematic work *De triangulis omnimodis,* written in 1464, is credited with the deliberate separation of trigonometry from astronomy. After translating the Greek works of Apollonius and Archimedes, the inspired Regiomontanus produced, in his *De triangulis,* a landmark work that proved to be the earliest Western treatise on plane and spherical trigonometry. It was a work of great synthesis, for it added to the work of the Greek pioneers the developments of the Hindus and Arabs—trigonometry being one of the branches of mathematics that both these Eastern cultures had developed. A generous patron had given Regiomontanus his own printing press as well as an observatory and no doubt he would have printed *De triangulis* himself had he not died suddenly in 1476. Consequently the work did not see print until 1533. The Library has a first edition copy of *De triangulis,* printed in Nuremburg. It is a book dense with geometrical diagrams, and the many tiny wormholes this copy now contains do not affect its usefulness or readability. The work is divided into five books, the first two devoted to plane trigonometry and the remaining three to spherical trigonometry. It also introduces the trigonometric functions sine and cosine. There are conflicting accounts as to the circumstances of the author's death—some say he died of the plague and others argue that he was poisoned. What is known for sure is that Regiomontanus died in Rome shortly after being invited there by Pope Sixtus IV to help reform the Julian calendar.

The time of Pope Sixtus was that of the Renaissance flowering, when artists began to discard the two-dimensional flatness of their religious paintings and exuberantly took to glorifying nature itself rather than its creator. But the realistic depiction of nature confronted the artist with an essentially mathematical problem: how to represent a three-dimensional world on a two-dimensional canvas. Although many artists studied and used mathematics during the early fifteenth century, it was not until the theoretical genius of Leone Battista Alberti that the mathematical laws of perspective were developed. In his *Della pittura,* written in 1435 and

DI LIONARDO DA VINCI. 89

Auuertimento circa l'ombre e lumi. CAP. CCCII.

AVVERTISCI che ſempre ne' confini dell'ombre ſi miſchia lume & ombra: e tanto più l'ombra deriuatiua ſi miſchia col lume, quanto ella è più diſtante dal corpo ombroſo. Mà il colore non ſi vedrà mai ſemplice: queſto ſi proua per la nona, che dice: La ſuperficie d'ogni corpo partecipa del colore del ſuo obbietto, ancora che ella ſia ſuperficie di corpo traſparente, come aria, acqua e ſimili; perche l'aria piglia la luce dal ſole, e le tenebre dalla priuatione d'eſſo ſole. Adunque ſi tinge in tanti varij colori quanti ſon quelli fra li quali ella s'inframette infra l'occhio e loro, perche l'aria in ſe non hà colore più che s'habbia l'acqua, mà l'humido che ſi miſchia con eſſa dalla mezza regione in giù è quello che l'ingroſſa, & ingroſſando, i raggi ſolari che vi percuotono, l'alluminano, e l'aria ch'è dalla mezza regione in sù reſta tenebroſa: e perche luce e tenebre compone colore azzurro, queſto è l'azzurro in che ſi tinge l'aria, con tanta maggior o minor oſcurità quanto l'aria è miſta con maggior o minor humidità.

Pittura, e lume vniuerſale. CAP. CCCIII.

VSA di far ſempre nella moltitudine d'huomini e d'animali le parti delle loro figure, ouero corpi, tanto più oſcure quanto eſſe ſono più baſſe, e quanto elle ſono più vicinè al mezzo della loro moltitudine, ancorche eſſi ſiano in ſe d'vniforme colore: e queſto è neceſſario, perche meno quantità di cielo, alluminatore de' corpi, vede ne' baſſi ſpatij interpoſti infra li detti

M

Leone Battista Alberti has been called one of the first scholar-artists of the Renaissance, and as a painter, sculptor, and architect he embraced the Renaissance philosophy of an appreciation of nature. It was the goal of Renaissance painters to faithfully depict the real world, just as Renaissance scientists sought to understand nature's secrets. To Alberti, mathematics was the essential tool for both science and art. It was he who conceived the principle that became the ultimate foundation of a mathematical system of perspective, and in his treatise on painting he stated that the first requirement of a painter is that he know geometry. Alberti offered the general rules of mathematical perspective and thus gave Renaissance painters an effective technique to depict nature in a realistic manner. This page is from a seventeenth-century version of his work which offers advice on shadow and light. *Della pittura e della statua*, 1651. Leone Battista Alberti.

published in 1511, Alberti furnished correct rules, theorems, and constructions that tell an artist how to paint three-dimensionally. The Library has an 1804 Milan edition of this masterpiece as well as a 1651 Paris edition edited by Raphael Trichet du Fresne. Alberti's book was thoroughly mathematical, and to him goes the credit for conceiving the principle that became the basis for the mathematical system today called projective geometry. His system of perspective was adopted and perfected by later artists and gave Renaissance art its distinctive naturalism.

mit einem anderen puncten aber also piß das du die gantzen lauten gar an die tafel punctirst / dann zeuch all puncten die auf der tafel von der lauten worden sind mit linien züsamē / so sichst du was dar-auß wirt / also magst du ander ding auch abzeychnen. Dise meynung hab jch hernach aufgerissen.

Vnd damit günstiger lieber Herr will jch meinem schreyben end geben / vnd so mir Got genad ver-leycht die bücher so jch von menschlicher proporcion vñ anderen darzü gehörend geschryben hab mit der zeyt in druck pringen / vnd darpey meniglich gewarnet haben / ob sich yemand vnder-steen wurd mir diß außgangen büchlein wider nach zü drucken / das jch das selb auch wider drucken will / vñ auß lassen geen mit meren vnd grösserem züsatz dañ ietz beschehen ist / darnach mag sich ein yetlicher richtē / Got dem Herren sey lob vnd eer ewigklich.

N iij

Gedruckt zü Nürenberg.
Im. 1525. Jar.

Having studied in Renaissance Italy, Albrecht Dürer returned to his native Germany fully informed about the theory and method of perspective and was determined to pass this knowledge on to his countrymen. Dürer was a natural-born geometrician and in his practical treatise on geometry he also taught the science of perspective. To Dürer, however, perspective was not solely a technique but rather was an important branch of mathematics, what we today call projective geometry. In this woodcut, Dürer shows one of several mechanical means (or "perspective apparatus") he invented to achieve an approximate image correctness. This device consists of a needle driven into the wall and a piece of string, the needle's eye replacing the artist's. Between the eye of the needle and the object to be drawn (a lute) is a wooden frame and a sheet of tracing paper hinged to the frame. Eventually a pattern of the lute is traced on the paper. *Underweysung der Messung*, 1525. Albrecht Dürer.

From the sixteenth century on, the theory of perspective was taught in all painting schools according to the principles of the masters. One of these great artists, who was certainly the best mathematician of them all, was Albrecht Dürer. Dürer had studied in Italy and, to pass his knowledge of perspective on to his fellow Germans, he wrote the work *Underweysung der Messung,* published in Nuremberg in 1525. The Library has this well-illustrated book in first edition. Bound with it is his *Bücher von menschlicher proportion* (Nuremberg, 1528). Dürer's work dealt primarily with geometry and, because its emphasis was so practical, it became a very influential text. This work, along with Dürer's artistic output, influenced the entire course of North European painting for generations.

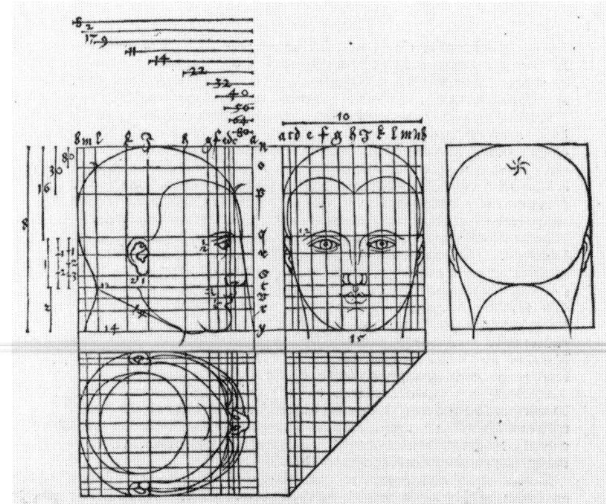

Bound with the Library's 1525 edition of Dürer's treatise on projective geometry is his *Treatise on Human Proportions.* In this work Dürer elaborates on the science of perspective as applied to the human figure. *Bücher von menschlichen Proportion*, 1528. Albrecht Dürer.

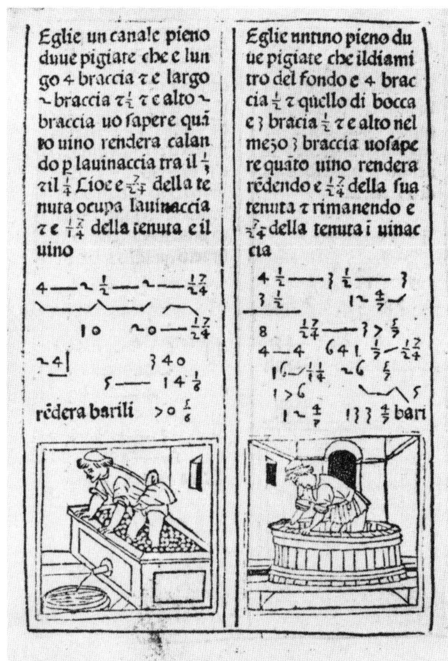

Eglie un canale pieno duue pigiare che e lungo 4 braccia e e largo 2 braccia 2¼ e e alto 2 braccia uo sapere quāto uino rendera calando p lauinaccia tra il ⅓ e il ¼ Cioe e 7/24 della tenuta ocupa lauinaccia e e 17/24 della tenuta e il uino

rēdera barili

Eglie uno tino pieno duue pigiate che il diamitro del fondo e 4 braccia ½ e quello di bocca e 3 braccia ½ e e alto nel mezo 3 braccia uo sapere quāto uino rendera rēdendo e 17/24 della sua tenuta e rimanendo e 7/24 della tenuta i uinaccia

Of the first generation of printed arithmetics, this work by Filippo Calandri is one of the more rare. Not only was it the first to contain examples of long division in the modern form but it was also the first printed Italian arithmetic to offer illustrated problems. Many of its woodcuts dealt with the familiar problems of everyday life, demonstrating the usefulness of arithmetic. *Aritmetica*, 1492. Filippo Calandri.

Mathematics can thus be said to be at the very heart of the Renaissance, since it provided artists with a dependable scientific method of achieving the realistic third dimension on canvas. This, in turn, was integral to both the rediscovery of nature and its accurate representation. Nothing so manifested and so symbolized the freshness of discovery and the feeling of liberation that was the Renaissance than did this new art; and mathematics—science, as it were—was responsible.

The vitality of the Renaissance permeated the everyday world as well as the special world of the artist. As interest in education increased and the fields of banking and commerce expanded, the significance of mathematics became more apparent. With this awareness came the early texts—the arithmetics of each country, written in Latin or, more often, in the vernacular. In addition to the Boethius and Pacioli texts already mentioned, the Library has two of the more significant arithmetics in its collections. Of these, the earliest is the 1492 edition of Filippo Calandri's *Aritmetica,* printed in Florence. Considered the first printed Italian arithmetic to contain illustrations, this very small volume also contains the first printed example of the modern process of long division. In keeping with its practical tone, it offers tables of money exchange and calculations of money and weight. The first separate arithmetic printed in England was written by Cuthbert Tunstall, bishop of Durham. Titled *De arte supputandi,* this book was based largely on Pacioli's *Somma* and was written in Latin. It was published in London in 1522. The engraved title page is signed HH, and is considered by some to be the work of Hans Holbein.

The most influential English textbook of the sixteenth century, however, was an arithmetic by Robert Recorde with the wonderful title *The Grounde of Artes.* First published in 1540, this work saw at least twenty-nine printings. The Library has the 1561 London edition, which is prefaced "To the Lovinge reader." Recorde served as physician to Edward VI and Queen Mary, but he died while confined in the King's Bench prison for debt. *The Ground of Artes* was written as a dialogue between master and student and used the signs + for plus and − for minus. In Recorde's words, "+ whyche betokeneth too muche, as this line −, plaine without a crosse line, betokeneth too little." The Library also has two of his other mathematical works, *Pathway to Knowledge* (London, 1551), an abridgment of Euclid's *Elements,* and the 1557 *Whetstone of Witte.* The latter is an algebra that contains the earliest introduction of the modern symbol = for equality. Recorde selected this particular sign, he said, "bicause noe 2 thynges can be moare equalle" than two parallel lines.

Improved algebraic symbolism was one of the more significant accomplishments of the sixteenth century, but one of its most spectacular achievements was the discovery of the algebraic solution of cubic and quartic equations. For centuries, the solution of equations involving the cube of the unknown quantity had defied mathematicians, and as late as 1494 Pacioli had announced its solution to be impossible. The story of that solution involves two of science's more fascinating personalities and tells a tale of ambition and treachery. One personality, Girolamo Cardano, is considered at best an unprincipled genius and the other, Niccolo Fon-

DE ARTE SVPPVTANDI
LIBRI QVATTVOR
CVTHEBERTI
TONSTALL I.

The first separate arithmetic printed in England was compiled by Cuthbert Tunstall, bishop of London and later of Durham. An outstanding classical scholar, Tunstall was a close friend of Sir Thomas More (to whom he dedicated this work) as well as Erasmus. As a compilation of many other works, Tunstall's arithmetic was more than serviceable, yet it never became popular in his own country nor was it ever translated into English. This bordered title page, which appears to be unfinished, was engraved by Hans Holbein (see the "HH" in the left, middle scroll). *De arte supputandi*, 1522. Cuthbert Tunstall.

tana, is sometimes called a plagiarist. Fontana came from extreme poverty and was mostly self-taught. He was known as "Tartaglia," or "the stammerer," for an affliction he received during the French massacre at Brescia in 1512. Taking refuge in a cathedral, the boy was attacked and left for dead, his skull, jaw, and palate being split. Remarkably, his mother was able to nurse him back to health, and he became a gifted mathematician. His reputation was made in 1535 when, having independently discovered the cubic solution, he demonstrated the results of his secret in a public contest, which he easily won. Tartaglia's reputation was later tarnished by a charge that he had presented a 1543 translation of Archimedes's work as

De subtilitate, 1554. Girolamo Cardano.

Quesiti et inventioni diverse, 1554. Niccolò Tartaglia.

That science can arouse passion is demonstrated by the two great enemies, Cardano and Tartaglia, pictured here. Sixteenth-century science had not yet developed a tradition of openness and it was not unusual for ideas to be held captive. Such was the case with the solution to the cubic equation which Niccolò Tartaglia discovered and later revealed to Girolamo Cardano, who pledged his secrecy. Cardano later found that the solution was known before Tartaglia by Scipione dal Ferro (who never published it) and Cardano, feeling no longer bound by his promise, published his version of the solution in 1545. The fact that history dubbed the solution "Cardan's formula" must certainly have contributed to the scientific practice of granting credit for a discovery to the one who first publishes it.

his own. (In fact, it was a thirteenth-century Latin translation by William of Moerbeke.) In the same year, however, Tartaglia did produce an Italian translation of Euclid's *Elements,* marking the first printed translation of this work into a modern language.

Cardano was a flamboyant contrast to the brilliant but stuttering Tartaglia. The illegitimate son of a lawyer of Milan, he traveled extensively and began his career as a physician. He is accused of many extraordinary acts, among them cutting off the ear of his son, being imprisoned for heresy for having published a horoscope of Christ's life, and even committing suicide to fulfill his own prediction of the date of his death. It is to such a man that Tartaglia revealed his algebraic discovery—only to have Cardano publish the method in 1545.

Had Cardano been merely a knave and opportunist rather than a genius, and had his 1545 book *Artis magnae,* commonly called *Ars magna,* the first great Latin treatise devoted to algebra, not had such a formidable influence on the rapid growth of algebra in Europe, neither man would be mentioned here. But Cardano's *Great Skill,* of which the Library has the second edition, published in Basel in 1570, did more than reveal a scientific secret told in confidence. It also contained novel ideas of negative roots and it even presented formal computations with imaginary numbers. Tartaglia protested long and hard, however, that what became known as "Cardan's formula" was in fact his by the primacy of discovery. In the Library's collections is his *Quesiti et inventioni diverse* (second edition), published in Venice in 1554, which presents his case and accuses Cardano of perjury. The Library also has the second edition of a more significant work of Tartaglia's, *La noua scientia,* published in Venice in 1550. More is said of this slim (it is only seventy-two pages) but important work in the chapter on physics. History now links inseparably the two great enemies, Tartaglia and Cardano.

During this time in France, algebra was being given a new name and an independent language. François Viète, known by his latinized name Vieta, called algebra "analysis" or "the analytic art" and introduced a symbolic language, based on an international shorthand, that used general symbols for quantities and operations in place of word abbreviations. His use of vowels for unknown quantities and consonants for known quantities is chiefly responsible for the rapid adoption of truly symbolic algebra. Because of this, he is known as the father of modern algebra. The Library has a 1646 Lyons edition of his major works, titled *Opera mathematica.* It is a large book and appears well-used. Vieta occupied a high position in the French court, and during the war with Spain he put his talents to work deciphering the code of Phillip II. Phillip considered his code to be unbreakable and appealed to the pope, contending that the French were using magic against Spain.

Vieta died in 1603, at the beginning of what many call the century of genius. For mathematics, it was the dawn of the modern age. The herald of this new age was a Scotsman named John Napier. Napier, who was born when his father was only sixteen years of age, was a violently anti-Catholic aristocrat and was regarded by the locals as either unbalanced or

MIRIFICI

Logarithmorum

Canonis deſcriptio,

Ejuſque uſus, in utraque *Trigonometria; ut etiam in* omni Logiſtica Mathematica, *Ampliſsimi, Facillimi, &* *expeditiſsimi explicatio.*

Authore ac Inventore, IOANNE NEPERO, Barone Merchiſtonii, *&c. Scoto.*

EDINBURGI, Ex officinâ ANDREÆ HART *Bibliopolæ*, CIↃ. DC. XIV.

This ornate title page belies the fact that between its covers this book is essentially a continuous list of tables. But its seemingly dry contents of logarithmic tables were recognized immediately as providing science with a new computational tool of immense potential. John Napier had been searching for a method of making astronomical calculations less tedious and time-consuming, and his analyses and comparisons of arithmetic and geometric progressions revealed that there was an alternate, briefer way to state certain numbers (for example 100 is 10 times 10 or 10^2) and that many complicated operations could be replaced by simple addition or subtraction. Napier spent twenty years on his logarithmic tables, which were embraced by a grateful scientific community. *Mirifici logarithmorum canonis descriptio*, 1614. John Napier.

a dabbler in black magic. He dabbled some in mathematics, too, and spent twenty years creating an essentially new method of calculation—a method with startling implications. Napier called his invention "logarithms," based on a word meaning ratio number. Having noted that all numbers could be expressed in exponential forms (the number 4 can be written as 2^2 and 8 as 2^3), Napier realized that multiplication could be done by adding exponents, and division by subtracting them. Napier made his invention known in a slim volume of ninety pages, *Mirifici logarithmorum canonis descriptio,* published in Edinburgh in 1614. The Library has a first edition of this ingenious work. Napier's invention was immediately and enthusiastically accepted throughout Europe and had an explosive impact on those areas most dependent on numerical calculations, such as astronomy, navigation, and engineering. Henceforth, the drudgery of calculation was relieved by the simplicity, quickness, and accuracy of Napier's new system. Napier's first book contained his logarithmic tables and rules for their use but offered no account of his calculations. This was given in a book published in 1619, two years after his death (although actually written before his *Descriptio*). It was titled *Mirifici logarithmorum canonis constructio*. The Library has an 1889 translation published in Edinburgh.

One of the most dominant names of this century of genius is René Descartes. With him began not only modern philosophy but modern mathematics as well. Philosophically, the precocious Descartes literally started over, calling into question the certitude of all he had been taught and eventually rejecting it in toto. His education had served only to allow him to recognize his ignorance, he said. The search for a basis of establishing truth in all fields led him to mathematics, which, he argued, provided the method both of achieving certainties and of demonstrating them. In this he sounds very much like Plato, saying that "all the sciences which have for their end investigations concerning order and measure are related to mathematics." Toward the end of securing exact knowledge in all fields by the use of mathematics, Descartes produced a classic of literature, philosophy, and science. In 1637 he published his *Discours de la methode pour bien conduire sa raison, & chercher la verité dans les sciences*. The Library has a first edition of this masterpiece, published in Leiden.

The *Discours* contains three appendixes, "La Dioptrique," "Les Météores," and "La Géométrie." The last of these, "La Géométrie," occupies about one hundred pages of the complete work and contains Descartes's contributions to analytic geometry. This breakthrough is usually described as the application of algebra to geometry or, in short, the creation of analytic geometry. Put simply, Descartes took all that was best in both and corrected the defects of one with the help of the other. In doing this, he broke with the static geometry of the Greeks and formulated a dynamic and powerful new geometry that used algebraic equations to represent and to study curves on a coordinate system. This advance did away with the dimensional limitation on algebra. Problems of motion then became amenable to solution, once a curve could be represented as an equation. With this new tool, problems dealing with the parabola that had defeated the great Archimedes became soluble by any good mathematician. Descartes departed even further from the Greeks in his conception of mathematics as a useful and constructive science. Mathematics for its own sake was idle play, he argued. Its real value and purpose was as a universal method with which to study nature.

But for a method which he believed to be universal, Descartes did little either to make it understandable or approachable. His *Géométrie* was written in a deliberately obscure and unsystematic manner, requiring the reader to reconstruct many of the details on his own. Descartes even boasted that few would understand his work. In 1649 a Latin translation that contained explanatory notes and commentary appeared in Lyons. The

Library has a copy of this work, titled *Geometria,* which had a wide circulation. Descartes was a complex man, with many seemingly contradictory habits and interests. He was both scholar and fop. A man of delicate health who spent all morning in bed, he nevertheless made a career of soldiering. A doubter of all knowledge, he posited his belief in God in mathematics. He died during the winter of 1650 in Stockholm, having accepted an invitation to instruct the young Queen Christina of Sweden. The frail Descartes was not up to giving lessons in an unheated library at five o'clock in the morning, nor to the rigors of a court that prized physical endurance and ignored hardship and discomfort. He died of pneumonia at fifty-four years of age.

The mathematical work of Descartes's brilliant contemporary Galileo is treated at length in the chapter on physics. There, the larger impact of Galileo's *Discorsi* is discussed. Pierre Fermat was also a contemporary of Descartes and a good-natured rival of the great man. Fermat's profession, like Vieta's, was the law, and he cultivated mathematics as a hobby. To his avocation he brought a genius that made him a mathematician of the first rank. He has been called the world's greatest amateur. Fermat led a quiet, orderly, leisured life and took satisfaction from his scientific work itself rather than from publication or publicity. He published very little and announced most of his discoveries in his voluminous letters to other French mathematicians. Because of this, Fermat's independent discovery of analytic geometry ten years before that of Descartes went unrecognized. Fermat's genius applied to all branches of mathematics, and he contributed basic concepts to the theory of numbers as well as to the theory of probability (as did Pascal). He also anticipated Newton, discovering some features of differential calculus. Fermat was an accomplished linguist and classicist who knew the Greek mathematic masterpieces firsthand. He did not discover his genius for mathematics until he was nearly thirty and even then seemed not to fully realize the magnitude of his ability. His work and much of his correspondence were published by his son five years after his death. The Library has the large, slim first edition of this compilation, *Varia opera mathematica,* published in Toulouse in 1679.

At this time, the names of Newton and Leibniz were becoming inextricably linked. Born within a few years of each other, the two men—each a scientific giant in his own nation—were destined to share the claim to the discovery of the differential calculus. The controversy as to who should get the credit for the calculus has occupied a place disproportionate to its significance in the history of mathematics. By now, most agree that each discovered the calculus independently of the other. Priority seems to rest with Newton, who used his method in 1666 but gave no printed account until 1693. Leibniz was in possession of the calculus in 1675 and published in 1684. While the term *calculus* implies any system of mathematics, what is known as *the calculus*, or more properly differential or infinitesimal calculus, refers to the method of Newton and Leibniz. Its discovery completed what the Greeks had left undone—squaring the circle. Geometrically, it is a way of reducing curve-sided figures to straight-sided ones by regarding the curves as a series of very small but straight lines. This

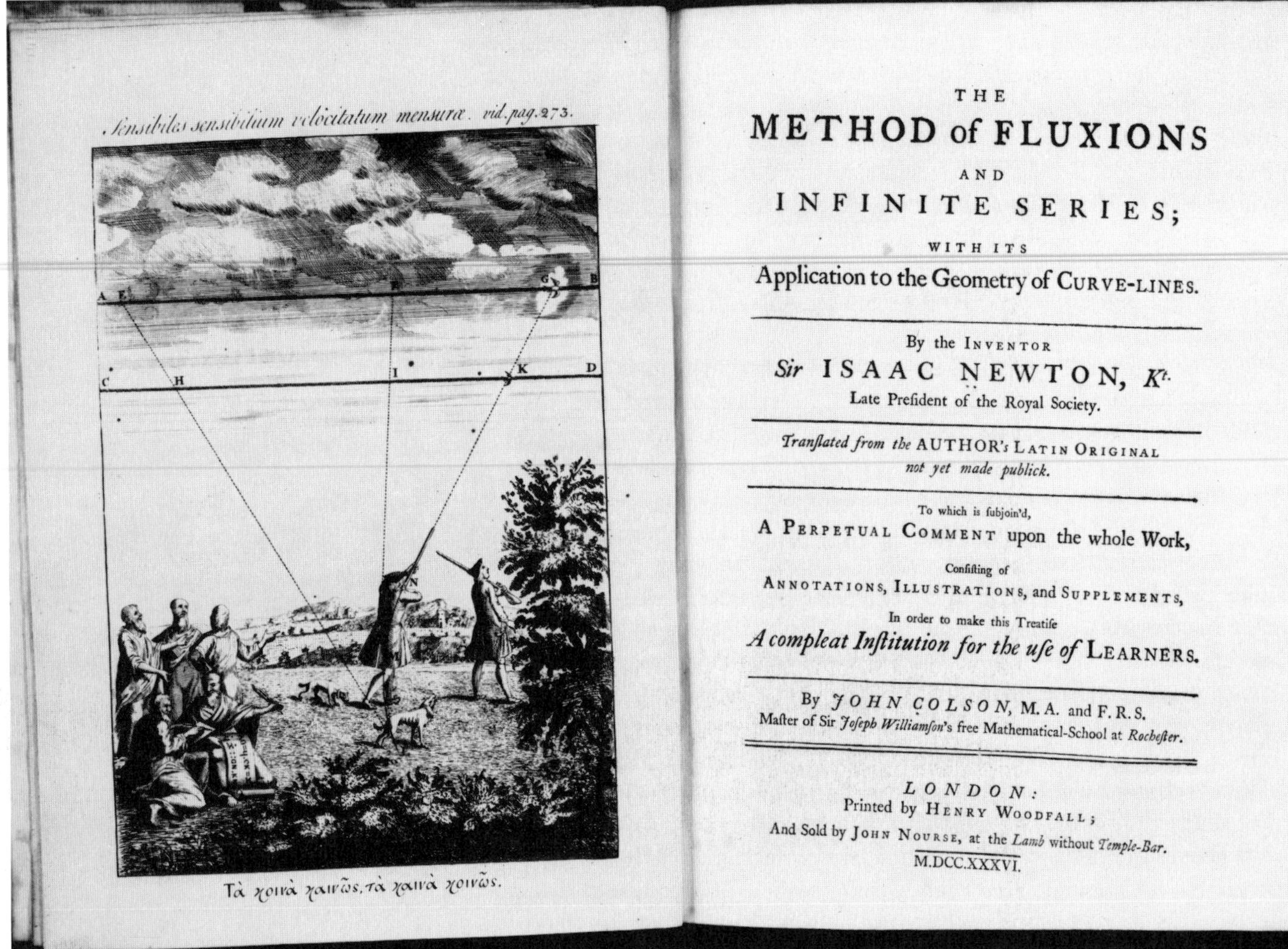
THE
METHOD of FLUXIONS
AND
INFINITE SERIES;
WITH ITS
Application to the Geometry of CURVE-LINES.

By the INVENTOR
Sir ISAAC NEWTON, *K^t.*
Late Prefident of the Royal Society.

Tranflated from the AUTHOR's LATIN ORIGINAL
not yet made publick.

To which is fubjoin'd,
A PERPETUAL COMMENT upon the whole Work,
Confifting of
ANNOTATIONS, ILLUSTRATIONS, and SUPPLEMENTS,
In order to make this Treatife
A compleat Inftitution for the ufe of LEARNERS.

By *JOHN COLSON*, M. A. and F. R. S.
Mafter of Sir *Jofeph Williamfon*'s free Mathematical-School at *Rochefter*.

LONDON:
Printed by HENRY WOODFALL;
And Sold by JOHN NOURSE, at the *Lamb* without *Temple-Bar.*
M.DCC.XXXVI.

This is one of three mathematical works by Newton that are the basis for the historical claim of his priority over Leibniz as inventor of the calculus. First written in 1671 in Latin, this work was so severely mathematical that no printer would publish it. It was not until nine years after Newton's death that his treatise was published as translated and edited by John Colson. *The Method of Fluxions and Infinite Series*, 1736. Isaac Newton.

method had immense implications for science—especially for physics—in that it made possible the analysis of movement. Henceforth, such changing phenomena as accelerated motion could be represented geometrically by curves. The changing and the changeable no longer were impenetrable, but could be calculated and studied by the method which Newton named "fluxiones."

The two men were strikingly dissimilar. Leibniz was a remarkably precocious child whose talents grew to almost universal proportions. His eminence was as great in philosophy as in science. He was a man of the world and a practicing diplomat involved in the high-level political matters of his day. Sadly, he died neglected and forgotten, discarded by the king he had served. Newton, in contrast, was regarded with awe by his countrymen, who thought of him as a true colossus. After a desultory youth, his transcendental genius emerged and he discovered the universal law of gravitation, the composite nature of white light, and the calculus—all during his annus mirabilis of 1665–66. Socially and personally he was the opposite of Leibniz, remaining a somewhat solitary and sober individual all of his life. Whereas Leibniz in his diplomatic missions had the ear of Louis XIV and Peter the Great, Newton labored for nearly thirty years as Master of the Mint. The old age of these two very different but brilliant men was made unhappy by their bitter feud over the invention of the calculus.

The Library has first editions of both Newton's and Leibniz's work on the calculus. Although Newton's first publication involving his calculus is the great *Principia mathematica* (London, 1687), he first circulated a paper titled "De analysi per aequationes numero terminorum infinitas" in 1669 in which he gave the bare essentials of his new discoveries. This paper was not published until forty-two years later. However, Newton amplified this work and produced in 1671 his larger "Methodus fluxionum et serierum infinitarum," which also was not published. After a long hiatus, Newton turned to the calculus again in 1692 and focused specifically on the quadrature of curves. By the end of 1693, he had produced a tract that eventually was published, in 1704, as "Tractatus de quadratura curvarum."

The Library has each of these works as they were first published. The first to appear in print, "Tractatus de quadratura curvarum," was appended to Newton's *Opticks* (London, 1704). In 1711, "De analysi per aequationes numero terminorum infinitas" appeared in a work edited by William Jones titled *Analysis per quantitatum series, fluxiones, ac differentias.* This work also contained the "de quadratura curvarum." Finally, Newton's "Methodus fluxionum" was translated by John Colson and published as *The Method of Fluxions and Infinite Series* (London, 1736).

Nearly all the mathematical papers Leibniz wrote were produced during the years 1682 to 1692, and most of them were printed in a journal, *Acta eruditorum.* Founded by Leibniz and Otto Mencke in 1682, the journal had a wide circulation on the Continent. It was in this journal that Leibniz's work on the calculus, "Nova methodus pro maximus et minimis," first appeared, in 1684. The calculus controversy continued even after the deaths of both men and unfortunately assumed nationalistic overtones. Although both men had invented the calculus, their methods of approach and discovery were decidedly different, as were their systems of notation. The British rigidly adhered to Newton's notation, ignoring the broader and more useful method of Leibniz, and consequently failed to keep up with the Continental advance in mathematics. Because of this, British mathematics became moribund for over a century and the country failed to produce a single first-rate mathematician during that time.

If the seventeenth century saw the birth of the calculus, the eighteenth century was the era of its elaboration and extension. In this, the key figure and the greatest mathematician of his time was Leonhard Euler. Euler was a Swiss who studied mathematics under his famous countryman Jean Bernoulli. Bernoulli and his brother Jacques were among the first mathematicians to recognize the power of the calculus as a widely applicable tool. It fell to Euler then to harness that power and direct it toward the new sciences of the emerging modern era—mechanics and astronomy. Euler was, without a doubt, the most prolific mathematician ever. His

METHODUS
INVENIENDI CURVAS
MAXIMI MINIMIVE PROPRIETATE
GAUDENTES.

CAPUT PRIMUM.

De Methodo maximorum & minimorum ad lineas curvas inveniendas applicata in genere.

DEFINITIO I.

1. 

ETHODUS *maximorum & minimorum ad lineas curvas applicata*, eſt methodus inveniendi lineas curvas, quæ maximi minimive proprietate quapiam propoſita gaudeant.

COROLLARIUM I.

2. Reperiuntur igitur per hanc methodum lineæ curvæ, in quibus propoſita quæpiam quantitas maximum vel minimum obtineat valorem.

Euler *De Max. & Min.* A Co-

The name Euler is found in all branches of mathematics as well as in physics, astronomy, chemistry, geography, philosophy, religion, and even *belles lettres*. To a man whose memory was so prodigious that he knew the entire *Aeneid* by heart, memorizing the formulas of trigonometry and analysis of the first six powers of the first 100 prime numbers was short work. Euler did not let his talents lie fallow and deservedly earned his reputation as the most prolific mathematician ever. During his lifetime alone, about 560 of his books and articles appeared, and in 1911, the editing of his *Opera omnia* began. To date, 65 of the projected 72 volumes have been published. Euler was an innovator and a creator of new mathematical ideas and methods. This page is from one of his most significant contributions to mathematics, in which he systematically treats the calculus of variations and creates the new field of analytical mechanics. *Methodus inveniendi lineas curvas maximi minimive proprietate gaudentes*, 1744. Leonhard Euler.

name is linked with every branch of mathematics, and some of his papers are still unpublished. He spent his career serving two royal courts, holding the chair of mathematics at the St. Petersburg Academy of Peter the Great and later leading the Berlin Academy of Frederick the Great. A masterful writer of textbooks, Euler produced books which instantly became classics. These texts accomplished the essential tasks of systematizing, unifying, and clarifying partial results and isolated ideas. But it was his *Methodus inveniendi lineas curvas,* published in Lausanne in 1744, which the Library has in first edition, that revealed him to be not only a great instructor but a great discoverer. It was in this work on the calculus of variations that Euler created analytical mechanics (as opposed to the older geometrical methods). Because of this, he has been called the founder of the science of pure mathematics. Euler's unmatched productivity was not gained at the expense of his family or by his ignoring the world. As the father of thirteen children, he would often work with a child in his lap, easily doing the most difficult problems while surrounded by family members. When he lost his sight seventeen years before he died, his mathematical productivity actually increased. Blessed with a phenomenal memory, Euler knew Virgil's *Aeneid* by heart and could perform lengthy calculations mentally. He died suddenly in Russia at age seventy-six while playing with his grandson, having returned to St. Petersburg in 1766.

The only other eighteenth-century mathematician to rival the greatness of Euler was Joseph Louis Lagrange. Ironically, it was Euler himself who, after recognizing the brilliance of the nineteen-year-old Lagrange's insights into the calculus of variations, encouraged the boy to continue. Euler went so far as to hold back his own work so as to allow the young Lagrange to publish first and make a name for himself. Soon, his mathematical reputation was such that many considered him the greatest mathematician of his century. Based on his masterpiece, *Mécanique analytique,* that reputation is well deserved. Although Lagrange always said he had thought the work out when he was a nineteen-year-old professor at Turin, he wrote it many years later in Berlin during his stay at the court of Frederick the Great. When Euler left Berlin to return to St. Petersburg in 1766, Lagrange succeeded him in Prussia as head of the Berlin Academy. There, for twenty years, Lagrange applied his mathematical genius toward a systematization of mechanics and discovered the general formula from which all problems of motion could be solved. In method, he was a thoroughgoing analyst (in the sense Vieta used the word) and not a geometrician. "No diagrams will be found in this work," he boasted in the preface to his masterpiece.

The *Mécanique analytique* was first published in 1788, after Lagrange had left Berlin and returned to Paris. Despite its worth and eventual significance, a willing publisher was not easy to find. Only after a friend of Lagrange's guaranteed to purchase all unsold copies did the work see print. Lagrange remained in France during the turbulent years of the Revolution—an experience which deepened his world-weariness. By the time he was fifty, his penchant for overwork had so exhausted him that when a copy of his book was finally delivered to him, he left it unopened on his desk for two years. Many years later, a renewed Lagrange poured his last scientific efforts into a revision and extension of the *Mécanique analytique*. The Library has this copy, published in two volumes in Paris between 1811 and 1815. Unlike the intuitive and profusely detailed work of Euler, that of Lagrange is rigorous and concise. His *Analytical Mechanics* crowned Newton's work on mechanics, establishing it as branch of mathematical analysis. It has been described as "a sequel to Newton's *Principia.*"

The genius Carl Friedrich Gauss not only towered above his nineteenth-century contemporaries, but ranked with Archimedes and Newton in the history of mathematics. Called the "Prince of Mathematicians," Gauss exhibited an extraordinary intellectual precocity as a child—something not easily accomplished in a field full of prodigies. The little Gauss showed his ability before he was three years old. The "wonder child" lived up to his promise and while still in his teens made a number of stunning mathematical discoveries. At eighteen years old, he began thinking on the theory of numbers while a student at the University of Göttingen. Three years later his *Disquisitiones arithmeticae,* which many consider his greatest masterpiece, was completed. Although not published until he was twenty-four, the book was immediately recognized as a classic. It elevated the theory of numbers to the level of algebra, analysis, and geometry. It also contained Gauss's discovery of the law of quadratic reciprocity and the algorithm of congruences. The initial demand for his *Disquisitiones* far surpassed the supply, owing to the bankruptcy of a bookseller. In the Library's collections is an 1801 first edition, published in Leipzig.

After this publication, Gauss's career had many other stages and facets. His interests broadened to include astronomy, geodosy, and electricity, and he made essential and significant contributions to those fields. Although a genius of the first rank, Gauss published relatively little of his work—holding back many of his discoveries. He was a man whose striving for intellectual perfection never relaxed—a man who preferred to perfect one idea rather than offer the broad outlines of many. His seal was a tree with few fruit, bearing the inscription *"Pauca sed matura"* (few, but ripe). Nearly a half century after his death, the first serious study of his diary revealed scientific nuggets sufficient for a dozen reputations. It also demonstrated how much mathematics Gauss had anticipated. This diary, along with his letters and unpublished papers, contains sufficient innovative raw material to establish his unclaimed priority to many mathematical discoveries. One discovery which most now concede to be his was that of non-Euclidean geometry. At the age of twelve he began to view the foundations of Euclidean geometry with a critical eye, and by sixteen, he realized there might be another geometry besides Euclid's. From about 1813 on he became convinced that a new geometry was both logically consistent and applicable.

But Gauss published no definitive account of his work, and the honor of discovery went to others. Gauss was a man of almost universal interests and ability, who applied himself successfully to many disparate scientific fields. To describe all his outstanding contributions to mathematics alone would take a very long time. Interestingly, the name Gauss does not evoke anything like the popular recognition that many lesser men can claim. But Gauss determined this himself—remaining throughout his life a self-sufficient, austere, and extremely isolated individual.

The honor of discovering non-Euclidean geometry is shared by two men, Lobachevsky and Bolyai, in yet another case of independent scientific discovery. The development of a self-consistent geometry other than Euclid's was a profoundly significant mathematical event in the nineteenth century and led to the mathematics of curved surfaces. Traditional Euclidean geometry dealt mainly with plane surfaces and broke down when applied to curves. The immediate consequences of the discovery of non-Euclidean geometry was the final settlement of the age-old problem of Euclid's parallel postulate. This postulate had long plagued mathematicians, since it was the only one of Euclid's axioms that was not actually verifiable by experience—because it dealt with a line being drawn to infinity. The postulate stated that through any given point, one and only one line can be drawn, infinitely and in both directions, that is parallel to a given straight line. Surprisingly, when the postulate was denied—by assuming that more than one parallel line could be drawn through a point—the results showed no contradictions but rather were as self-consistent as Euclid's. The now obvious conclusion—that there could be other geometries as valid as Euclid's—went against two millenia of mathematical thinking.

Gauss learned this heresy first, but for some reason he never published his finding. Some say he deliberately suppressed it. Yet the revolutionary idea that there could be many different systems of geometry liberated geometry from 2,200 years of tradition. The enlightening realization that pure mathematics need not be true or false in the same sense as physics—but need only be self-consistent—paved the way for some mind-stretching hypotheses. It is safe to say that without non-Euclidean geometry, Einstein would have been unable to develop his general theory of relativity.

The Russian Nicolai I. Lobachevsky is credited with first publication of this particular non-Euclidean geometry. His findings were published in a series of five papers that appeared in the *Kazan University Courier,* published between February 1829 and August 1830. The first of these was titled "On the Principles of Geometry." The Library of Congress does not have this ephemeral journal in its collections, but Lobachevsky's important paper is represented in its collections by a translated version, *Études géométriques sur la théorie des parallèles,* published in Paris in 1866, as well as in a 1946–51 Moscow edition of Lobachevsky's collected works. Also included in the Library's collections is an 1856 Kazan edition of his *Pangéométrie,* in which the then-blind Lobachevsky dictated a completely new exposition of his geometry. The earliest Lobachevsky in the Library is an 1834 work, published in Kazan, on trigonometry, *Ob izchezanii trigonometricheskikh strok,* a book containing more equations than text.

Lobachevsky was the son of a peasant and entered the new University of Kazan on merit at the age of fourteen. At twenty-one he was assistant professor there, and by age thirty-four he had become its rector or president. Through his efforts, Kazan achieved a considerable eminence, yet the government saw fit to relieve him as rector and professor in 1846, giving no explanation.

The work of the brilliant Russian did not become known in Europe for some years, owing to the language barrier and the generally awkward and haphazard methods of information dissemination. In Hungary another mathematician was making a similarly bold and even heretical leap almost concurrently with Lobachevsky. Janos Bolyai, the son of a mathematician, is said to have worked out the ideas of non-Euclidean—or what he called "absolute"—geometry by 1825. On his way to the realization that this new geometry was self-consistent, the young man wrote to his father on November 23, 1823, "I have made such wonderful discoveries that I am myself lost in astonishment." His father urged him to continue his research, and when the elder Bolyai published his own *Tentamen iuventutem studiosam in elementa matheseos purae elementaris . . . introducendi,* a large two-volume semi-philosophical work on elementary mathematics, in

1832–33, it contained the landmark work of his son. His father had arranged that his son's seminal work be included as a twenty-six page appendix under the young man's name. The son's few pages, entitled *Scientiam spatii absolute veram exhibens,* outshone the rest of the book several times over. Indeed, it has been described as "the most extraordinary two dozen pages in the history of thought."

The Library has a two-volume Budapest printing, accomplished between 1897 and 1904, of the original father-son work, as well as a separate Leipzig treatment of *Appendix scientiam spatii absolute veram exhibens* published in 1903. The dashing Bolyai, who in the best Hungarian tradition could wield skillfully both sword and violin, never published anything further on the matter, but he left a great mass of associated manuscripts. When Bolyai first saw Lobachevsky's work (which had been published before his own), he naturally thought it had been copied from his own work.

At this time in England, a young man was struggling to overcome the privations of his family's shopkeeper status by attempting to educate himself. So determined and obviously brilliant was the young George Boole that by the age of twenty he had mastered the enigmatic and profound *Mécanique céleste* of Laplace as well as the highly abstract *Mécanique analytique* of Lagrange. It was at this time that British mathematics was beginning to recover from what has been called its "Newtonian sleep"—that inward-turning disregard for Continental developments in mathematics. None so characterized this insular attitude as did George Boole, who read the Continental masters and then proceeded to his own work as if they had never existed. In 1854, Boole's main work, *An Investigation of the Laws of Thought* was published in London. The Library has a first edition of this landmark work. In it, Boole did what others had dreamed of but never done—he added logic to the domain of algebra. More generally, he treated logic as a branch of mathematics—assuming that logical and mathematical operations are, to a certain extent, interchangeable. Boole's splendidly original invention, now called symbolic logic, reduced logic to an easy and simple type of algebra. By discovering the deep analogy between the symbols of algebra and those which can be made to represent logical forms, Boole was able to reduce logical propositions to the form of equations.

Boole's mathematization of logic was regarded for many years as a mere novelty and a philosophical curiosity, and it was not until the work of Whitehead and Russell that it received any serious attention. Today, Boolean algebra has a wide field of application and is generally considered to have provided the basis for modern computer systems and languages. Boole died in his fiftieth year—having achieved the chair of mathematics at Queen's College, Dublin, with no university training himself—his fame only just beginning. He was posthumously accorded the rare scientific honor of having his name linked with an intimate and real part of the scientific process.

The monumental work of Alfred North Whitehead and Bertrand Russell used Boolean algebra in an attempt to place mathematics on a rigidly

SECTION B] ON THE ARITHMETICAL PRODUCT OF TWO CLASSES 107

$[*113 \cdot 143] \supset . P=(R‘\alpha) \downarrow (R‘\beta) . Q=(R‘\alpha) \downarrow (R‘\beta) .$

$[*13 \cdot 172] \quad \supset . P=Q \qquad (1)$

$\vdash . *21 \cdot 33 . \supset \vdash :. \text{Hp} . \supset : PTQ . PTR . \supset .$

$(\exists x, y, z, w) . x, z \in \alpha . y, w \in \beta . P=x \downarrow y = w \downarrow z . Q = x \downarrow \alpha \cup y \downarrow \beta . R = z \downarrow \alpha \cup w \downarrow \beta .$

$[*113 \cdot 143] \supset . Q = \text{D}‘P \upharpoonright \iota‘\alpha \cup \text{Ɑ}‘P \upharpoonright \iota‘\beta . R = \text{D}‘P \upharpoonright \iota‘\alpha \cup \text{Ɑ}‘P \upharpoonright \iota‘\beta .$

$[*13 \cdot 172] \quad \supset . \breve{Q} = R \qquad (2)$

$\vdash . *33 \cdot 13 . \supset \vdash : \text{Hp} . \supset .$

$$\text{D}‘T = \hat{P} \{(\exists R, x, y) . x \in \alpha . y \in \beta . P = x \downarrow y . R = x \downarrow \alpha \cup y \downarrow \beta\}$$

$[*11 \cdot 55 . *13 \cdot 19] \quad = \hat{P} \{(\exists x, y) . x \in \alpha . y \in \beta . P = x \downarrow y\}$

$[*113 \cdot 101] \quad = \beta \times \alpha \qquad (3)$

$\vdash . *33 \cdot 131 . \supset \vdash : \text{Hp} . \supset .$

$$\text{Ɑ}‘T = \hat{R} \{(\exists P, x, y) . x \in \alpha . y \in \beta . P = x \downarrow y . R = x \downarrow \alpha \cup y \downarrow \beta\}$$

$[*11 \cdot 55 . *13 \cdot 19] \quad = \hat{R} \{(\exists x, y) . x \in \alpha . y \in \beta . R = x \downarrow \alpha \cup y \downarrow \beta\}$

$[*80 \cdot 9] \quad = \epsilon_\Delta‘(\iota‘\alpha \cup \iota‘\beta) \qquad (4)$

$\vdash . (1) . (2) . (3) . (4) . \supset \vdash . \text{Prop}$

Note to $*113 \cdot 144$. In virtue of $*113 \cdot 143$ and $*55 \cdot 61$ we have

$$\vdash :. \text{Hp} *113 \cdot 144 . \supset : PTR . \equiv . R \in \epsilon_\Delta‘(\iota‘\alpha \cup \iota‘\beta) . P = (R \parallel \breve{R})‘(\alpha \downarrow \beta).$$

At a later stage (in $*150$) we shall put

$$R \dagger S = (R \parallel \breve{R})‘S \quad \text{Df.}$$

Thus we shall have, anticipating this notation,

$$\vdash : \text{Hp} *113 \cdot 144 . \supset . T = \{\dagger(\alpha \downarrow \beta)\} \upharpoonright \epsilon_\Delta‘(\iota‘\alpha \cup \iota‘\beta).$$

Hence we have

$$\vdash : \alpha \neq \beta . \supset . \{\dagger(\alpha \downarrow \beta)\} \upharpoonright \epsilon_\Delta‘(\iota‘\alpha \cup \iota‘\beta) \in (\beta \times \alpha) \,\overline{\text{sm}}\, \epsilon_\Delta‘(\iota‘\alpha \cup \iota‘\beta).$$

$*113 \cdot 145$. $\vdash : \alpha \neq \beta . \supset . \beta \times \alpha \,\text{sm}\, \epsilon_\Delta‘(\iota‘\alpha \cup \iota‘\beta) \quad [*113 \cdot 144]$

$*113 \cdot 146$. $\vdash : \alpha \neq \beta . \supset . \alpha \times \beta \,\text{sm}\, \epsilon_\Delta‘(\iota‘\alpha \cup \iota‘\beta) \quad [*113 \cdot 141 \cdot 145]$

$*113 \cdot 147$. $\vdash : \text{Hp} *113 \cdot 144 . \beta \times \alpha = \mu . \supset .$

$$T = \hat{P}\hat{R} \{P \in \mu . R = \text{D}‘P \upharpoonright \iota‘s‘\text{D}‘‘\mu \cup \text{Ɑ}‘P \upharpoonright \iota‘s‘\text{Ɑ}‘‘\mu\}$$

Dem.

$\vdash . *113 \cdot 114 . \text{Transp} . \supset \vdash : \text{Hp} . P \in \mu . \supset . \exists ! \alpha . \exists ! \beta .$

$[*113 \cdot 142 . *53 \cdot 22] \quad \supset . \alpha = s‘\text{D}‘‘\mu . \beta = s‘\text{Ɑ}‘‘\mu \qquad (1)$

$\vdash . *113 \cdot 101 \cdot 143 . \supset \vdash :. \text{Hp} . P \in \mu . \supset : PTR . \equiv . R = \text{D}‘P \upharpoonright \iota‘\alpha \cup \text{Ɑ}‘P \upharpoonright \iota‘\beta \qquad (2)$

$\vdash . *113 \cdot 144 . \quad \supset \vdash : \text{Hp} . PTR . \supset . P \in \mu \qquad (3)$

$\vdash . (1) . (2) . (3) . *113 \cdot 101 . \supset \vdash . \text{Prop}$

The advantage of this proposition is that it exhibits the correlator of $\beta \times \alpha$ and $\epsilon_\Delta‘(\iota‘\alpha \cup \iota‘\beta)$ as a function of $\beta \times \alpha$.

In this typical page from the collaborative work of pupil and teacher (Russell and Whitehead), the notion that mathematics is both derived from and an extension of logic is most startlingly illustrated. The entire work resembles the symbolic form of this page, resulting in a book that is extremely difficult, if not impossible, to read. The authors attempted to start at the very beginning of mathematics by vigorously analyzing all of its fundamental concepts. Thus, after 347 pages of symbolic argument and demonstration, they are able to arrive at a logically acceptable definition of the number one. *Principia mathematica*, 1925–27. Alfred North Whitehead and Bertrand Russell.

logical basis. First published in Cambridge between 1910 and 1913, the three-volume *Principia mathematica* echoed Newton's famed treatise in both its title and ambitiousness. This collaborative work of teacher and pupil (Whitehead taught Russell at Cambridge) was the basis for what is called the logistic school of mathematics. Essentially, it argued as Boole had, that mathematics is both derived from and an extension of logic—the two being almost synonymous. Russell and Whitehead then went a step further and said that only through mathematics-logic can any certain knowledge be attained. The basic ideas of logicism were first outlined by Russell in 1903 in his *Principles of Mathematics*. The Library has this first edition. The *Principia mathematica,* which contains the definitive theory, is available in the Library in a 1925–27 Cambridge edition.

The work of Whitehead and Russell reflects the concern of twentieth-century mathematicians with the foundations of mathematics. Their logistic view of the nature of mathematics is countered by two other major schools of thought—the formalist school and the intuitionist school. Formalism, as founded by David Hilbert, regards the axioms of Russell and Whitehead as entirely man-made, claiming that mathematics is a meaningless game with meaningless symbols. Intuitionism, systematically founded by L. E. J. Brouwer, takes exception to both schools, especially formalism, and argues that mathematical ideas in the human mind precede language, logic, and experience. All three schools are basically concerned with the same question: What is mathematics? Such a query is not a fruitless exercise in self-examination but rather indicates how the study of mathematics has come almost full circle.

Any final conclusions as to the ultimate nature of mathematics were shown to be more of a mystery than ever with the publication in 1931 of what is known as Gödel's proof. In that year, the twenty-five-year-old Austrian published a paper in the Vienna journal *Monatschefte für Mathematik und Physik* that formally examined all systems of mathematical definition. The Library has this twenty-six page tour-de-force entitled "Über formal unentscheidbare Sätze der Principia mathematica und verwandter Systeme I." Here, Kurt Gödel dealt a blow to every mathematical system, but an especially shattering one to both Russell's and Hilbert's systems. Essentially, Gödel showed that within any rigidly logical mathematical system there are propositions that cannot be proved or disproved on the basis of the axioms within that system. Stated another way, "If the game of mathematics actually is consistent, the fact of this consistency cannot be proved within the rules of the game itself." But Gödel's proof did not call the game off. The outcome of discovering that certainty in mathematics does not exist has, surprisingly, been a stimulus to further research rather than a cause for abandoning this line of scientific inquiry.

Mathematical tradition is essentially different from that of most other disciplines, for it is basically a tradition of abstraction. Despite its many and obvious practical applications, mathematics operates in the realm of the ideal and distinguishes itself as a separate discipline by its "purity"—that is, its intrinsic logic and consistency. It is in this pure state that mathematics becomes most unlike other traditional observational disciplines in that it need not really take into account the exigencies of the physical world. Yet it is in discovering the nature and the laws of the physical world that mathematics is probably most useful, for it gives scientific expression to mankind's natural search for order and symmetry. Essentially abstract, mathematics is also essentially universal. Its language transcends time and place, as the conciseness and elegance of Euclid's geometry still inspires and enlightens mathematicians today. The abstract tradition of mathematics was given its ultimate expression twenty-five hundred years ago, when Pythagoras stated paradoxically, "Things are numbers."

8. *Physics: The Elemental Why*

The study of physics may be the most ambitious of all of mankind's scientific undertakings, focusing as it does on that simply stated but most elusive "nature of things." To the Greeks, physics meant "knowledge of nature," and despite the oceans of time and culture that separate us from that ancient period, the meaning and purpose of physics remain essentially unchanged. Then, as now, physics sought to know the physical world at its most elemental, concerning itself with the fundamental realities of nature—matter, motion, forces, and energy. Representing perhaps the height of scientific effrontery, physics best exemplifies the intellectual strivings of mankind. What could be more audacious—or natural—than to repeatedly ask "why" after receiving an answer. In a sense, physics plays this childlike game—sometimes with results.

Physics may also be our most influential scientific pursuit, for it not only affects all other scientific disciplines but insinuates itself into our culture as well, almost automatically and with great ease. So it has been, seemingly forever, that the concepts of physics have come to shape the epistemological and cosmological ideas of the nonscientist as well the scientist, molding our perceptions of ourselves and of our world. We have only to compare the assumptions and implications of the teleological approach to nature that dominated most thinking before Newton with the dynamic and mechanistic notions of the post-Newton world to appreciate the significance of our "world-picture." And to compare the assured and orderly universe of Newton's successors with that of today is almost to make a point too well.

Pervasive in influence and bold in its goals, physics over the centuries has attracted the passionate concern of the best minds and has always been regarded as an estimable pursuit. Whether it was the physics of the wheel, lever, and pulley that facilitated the placement of stone upon stone or the physics of ideas that offered glimmerings of the infinite, the search for what the Greeks called "the single-root of the multiplicity of things in the physical world" has been constant. In sum, the history of physics could be described as a journey from the concrete and the practical to the increasingly abstract and theoretical, notwithstanding the Greeks who began with theory and deduced from its assumptions. Following the Greeks, physics became a scientific pursuit concerned mainly with simple, individual natural forces, until the genius of Galileo sought to understand the complex and dynamic movements in nature. His intellectual stamp on physics is as great as that of Newton. Since his time, physics has experienced several great "unifications" or amalgamations—great leaps in the

Opposite page:
This busy woodcut illustrates several discoveries in the field of optics—itself only one of the many provinces in the complex realm of physics. The battle scene shows Archimedes's mirrors using the sun to defend the city of Syracuse. The rainbow at the top left indicates that the role of light in its formation is already known. The elephants in the foreground begin to shrink in size and the bridge becomes narrow at a distance, illustrating the mystery of perspective. The naked man standing in a pool shows how water distorts an image by refraction. And the man with the mirror may simply be demonstrating its ability to duplicate an image. *Opticae thesaurus*, 1572. Alhazen.

nature and content of physical thought that served to consolidate and to explain apparently diverse phenomena. These great surges of physical knowledge, such as Newton's laws and Maxwell's theories, have usually been both innovative and synthetic, personal and insightful. Such intellectual outbursts, although preceded by diligent experimentation, often emerge or are only fully realized because of an ineffable "something" which that particular scientist brings to the problem being studied. This process of creative speculation is a unique and wonderful intellectual phenomenon, usually recognized only after the fact. The history of physics can highlight several such instances, and the "truths" that they establish often become a generation's dogma.

So the truths of classical physics, as given by Isaac Newton, remained valid and unassailable until they too were revised by the "new truths" of modern physics. Although this certainly is the case, and classical physics has indeed been eclipsed by the revolutionary offerings of Einstein and Planck, both the old and the new do admit of a sort of coexistence. Providing that classical physics remains in the world as we know it—that is, the reality accessible to our senses—its premises and laws remain valid and useful. But since 1900, in the world of modern physics—that is, in the invisible microworld of the very fast and the very small, and the world of elementary particles and bent light—a new set of rules has taken precedence. Here the common sense-defying principles of quantum mechanics rule—here truth is found in contradiction, matter has a dual nature, and absolute certainty is unattainable. Here space contracts, time dilates, and simultaneity is relative. To do modern physics is "to deny the intuitively obvious," something akin to a literary suspension of disbelief. It is an irony of the twentieth century that its central physical concept of relativity is the direct result of the discovery of the only known absolute in the universe—the speed of light. All other physical quantities are relative to an observer, but the velocity of light remains constant.

In discussing modern and classical physics, it has been said that classical physics built the pyramids and the cathedrals, whereas modern physics built the atomic bomb. The connotations of both are obvious and they are especially unfair to modern physics. For what physics achieved in the middle of this century, despite its hideous application, was the transformation of the elements—the dream and the goal of alchemists and natural philosophers since the beginning of science. The new physics has shown us that it is this transmutation of matter that powers the sun. It has revealed one of nature's hidden keys—the idea that matter itself evolves. It has been in the use of this knowledge and not in its attainment that mankind has shown itself to be morally juvenile, despite our apparent intellectual maturing.

The following will trace the course of physics as represented in the collections of the Library of Congress. Central to this story is the notion of physics as a pursuit to understand "the nature of things." Little mention is therefore made of the control of natural forces or their useful application. Neither is mention made of the central role played by astronomy in the development of physical thought—astronomy having been treated here separately.

The legacy the Babylonians and the Egyptians left to physics, although not devoid of theoretical concepts, was characterized by its utilitarian aspects. The measurement of time (a seven-day week, twenty-four-hour day, sixty-minute hour, and sixty-second minute) is Babylonian in origin, as is the measurement of angles (the 360° circle, each degree having 60 minutes of arc, each minute 60 seconds). From the Egyptians come such basic tools as the drill, the lever, and the wedge. The level of physical achievement reached by these pre-Greek ancients is well illustrated by the great Cheops Pyramid, whose base differs from square by only half an inch in a side length of 756 feet. Such a spectacular accomplishment does not, however, indicate that the ancients had any special insight into the laws of the physical world. It is not necessary to know the principles of combustion to be able to use fire to one's advantage.

For the most part, speculation as to the essential nature of the physical world—seeking to know "why" as well as "how"—began in earnest with the Greeks. Their speculative intensity regarding the phenomena of nature was so great that it has been said that they "jumped one stage ahead of induction" and posited a systematic theory of physics that was logically deduced from certain assumptions. These assumptions, different as they might be, all sought to answer one question: "What is the single root of the multiplicity of things in the physical world?" From the pre-Socratic period to well past the Hellenistic Age, this concern for the essential structure of matter best characterized Greek physics. Over the centuries, Greek answers to this fundamental question had both a material and an immaterial foundation. There were those who believed in a single, primordial substance (for Thales it was water, for Anaximenes, air, and for Heraclitus, fire) which was transformed into other substances. Others argued that a limited number of elements combined to form nature's substances (Empedocles's famous earth, air, fire, and water). The atomic theory of Democritus was a type of variation on the single element theme and stated that all atoms were the same but that the difference in their number or arrangement accounted for the diversity of nature. At the opposite pole were Anaximander and Anaxagoras, who believed in an infinite multiplicity of substances from the beginning of time—what the former called "the unlimited." These four schools of thought all agreed, however, that the basic principle of nature was in essence material, or something tangible. A very different answer was offered by the Pythagorean school, which held that the basis of the physical world was not found in matter but in the immaterial—specifically in numbers and their relationship. The real world was not "real" then, but only an illusion.

It was from this rich mix of theories offered by his predecessors, as well as from his teacher, Plato, that Aristotle constructed his physics. Of his immense body of writings, those that can be classified as his physical writings are the *Physica,* in eight books, the *De generatione et corruptione,*

Vōniã qđē ītelligerē et ſcire ɔtingit circa oēs
ſciās q̃ꝝ ſūt pricipia ā cāe ā elemēta ex hoꝝ cog
nitōe tūt ē cognoſcere arbitramur unūqđq̃
cū cauſas primas et prima pricipia cognoſcim⁹
et vſq̃ ad elemēta māifeſtū qđē ꝙ et q̄ ſūt circa prīcipia ſcīē q̄
de nā ē pri⁹ det'mare tēptādū Innata āt ē ex notiorib⁹ nobis
via et certiorib⁹ ī notiora nē et certiora nō ē eadē nobis no
ta et ſiplr ūde necē ɓm modū hūc ꝓducere ex īcertiorib⁹ nē
nobis āt certiorib⁹ ī certiora nē et notiora Sūt āt primū no
bis māifeſta et certa ɔfuſa magis p⁹teri⁹ āt ex hiis fiūt nō
elemēta et prīcipia diuidētib⁹ hec vn̄ ex vniūſalib⁹ ī ſingula
ria oꝫ ꝓcedere Totū ē ɓm ſēſū noti⁹ ē vniūſale āt totū qđ/
dā ē multa ē ɔphēdit vt pꝛes vniūſale ſuſtinēt āt idē hoc
qđāmō et nōia ad rōꝫ totū ē qđdā et īdiſtincte ſignificāt vt
puta circul⁹ diffitō āt ipi⁹ diuidit ī ſigularia et pueri primū
appellāt oēs uiros patres et mrēs feiās p⁹tei⁹ āt detīnāt hoꝝ

Necē aūt ā vnū ēē prīm ā plura et ſi vnū ā ¶ vnūqđq̃
īmobile ſicut dicūt p menides et melliſſ⁹ ā mobile mo/
tū ſicut phſici hii qđē aerē dicentes ēē alii vero aq̄ primū
prīm ſi āt plura ā ſita ā īfinita etſi ſita plura āt vno ā duo
ā tria ā q̃tuor ā ɓm aliū aliquē numeꝝ etſi īfinita ā ſic ſicut
democrit⁹ gen⁹ vnū figura ā ſpē dr̃a ā et ɔria ſilr āt q̃rūt
et q̄ ſūt q̃rētes q̄t ſūt ex q̄bus ē ſūt q̄ ſūt q̃rūt primū hoc
vtrū vnū ā plura et ſi mlta ā finita ā īfinita q̃re prinūm et
elemētū q̃rūt vtꝝ vnū ā mlta Id qđē igit' ſi vnū et īmobile
ſit qđ ē intēdere nō de nā ē intendere ſicut ē geometre nō
āplius rō ē ad deſtruētē principia ſed ā alteri⁹ ſciē ā omibus
cōis ſic neq̃ huic de principiis nō ē āpli⁹ principiū ē ſi vnū
ſolū et ſic vnū ē prīcipiū ē cuiuſdā ā q̃rūdā ē ſilr igit' intē
dere ē ſi ſic vnū ē et ad aliā pōnē q̃libꝫ diſputare verbigrā
dictā vt eracleanā ā ſi aliq̄s dicat hōiem vnū ... ā ſolue
rōnē litigioſā qđ ſane vtreq̃ qđē hn̄t rōnes meliſſi et pme/

This stately page from an early printed edition of Aristotle's *Physica* conveys the admiration and respect in which his work was held. *Physica*, ca. 1475. Aristoteles.

in two books, the *De caelo et mundo,* in four books, and the *Meteorologica,* in three books. The Library's copies of these works are all in Latin and are all incunabula. The *Physica,* which deals primarily with motion, was first published in Padua between 1472 and 1474. The Library's copy is one of an edition issued separately in Louvain in 1475. The *De generatione et corruptione,* first published in Padua in 1474, presents Aristotle's theories of the four elemental qualities. The Library's copy has commentary by Paul of Venice and was published in Venice in 1498 and with it is bound the astronomical text *Sphaera mundi* of Johannes de Sacrobosco. The *De caelo et mundo* was first published by the same Paduan press in 1473 and had commentary by Averroës. Books I and II treated Aristotle's astronomical theories, and Books III and IV his elements. The Library's copy has commentary by Thomas Aquinas and Petrus de Alvernia and was published in Venice in 1495. Finally, the *Meteorologica,* which deals with what might today be called the phenomena of aerospace, was also first printed in Padua in 1474, but is not found separately in the Library's collections. The Library's earliest collected works of Aristotle is a Venice edition in Greek, *Opera,* published by Aldus Manutius between 1495 and 1498. It contains all four of the above-mentioned works.

The works of Aristotle are cited first in this chapter not only because of their merit (although some argue that Aristotle was a much better biologist) but also because for nearly two thousand years Aristotle was the supreme authority on all physical matters. His work was greatly admired by the Arabs, who translated and thus preserved his intellectual legacy. Christian Europe recovered Aristotle from the Arabs, translating his work into Latin manuscripts during the twelfth and thirteenth centuries and then printing his books in the post-Gutenberg era. Until Galileo, the dominating world view was that of Aristotle, whose words were regarded as dogma. As noted, Aristotle inherited several varied theories concerning the structure of matter, and pick and choose from among them he did. Foremost was his modified acceptance of Empedocles's four elements. For Aristotle, however, the elements of earth, air, fire, and water were only different aspects of one substance, which he called "primary matter." It was this

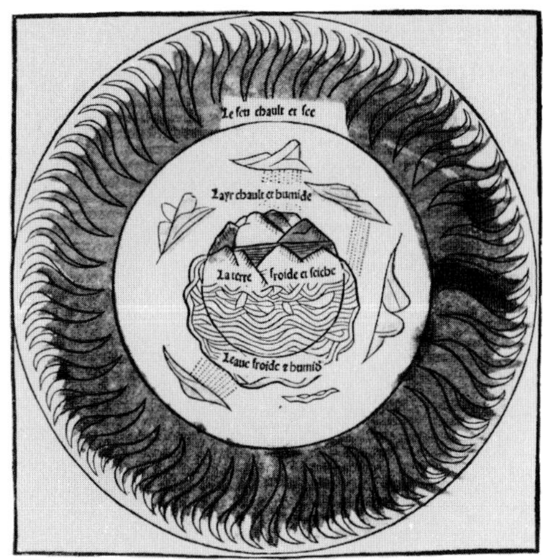

Variations on Aristotle's theory of the structure of matter abounded, but all fixed on his notion that the elements of earth, air, fire, and water were different aspects of a single substance called "primary matter." It was this substance, he said, that exhibited the qualities of hot, cold, wet, and dry. Here, in a late fifteenth-century encyclopedic work, the elements and their qualities are illustrated. *De proprietatibus rerum*, 1486. Bartholomaeus Anglicus.

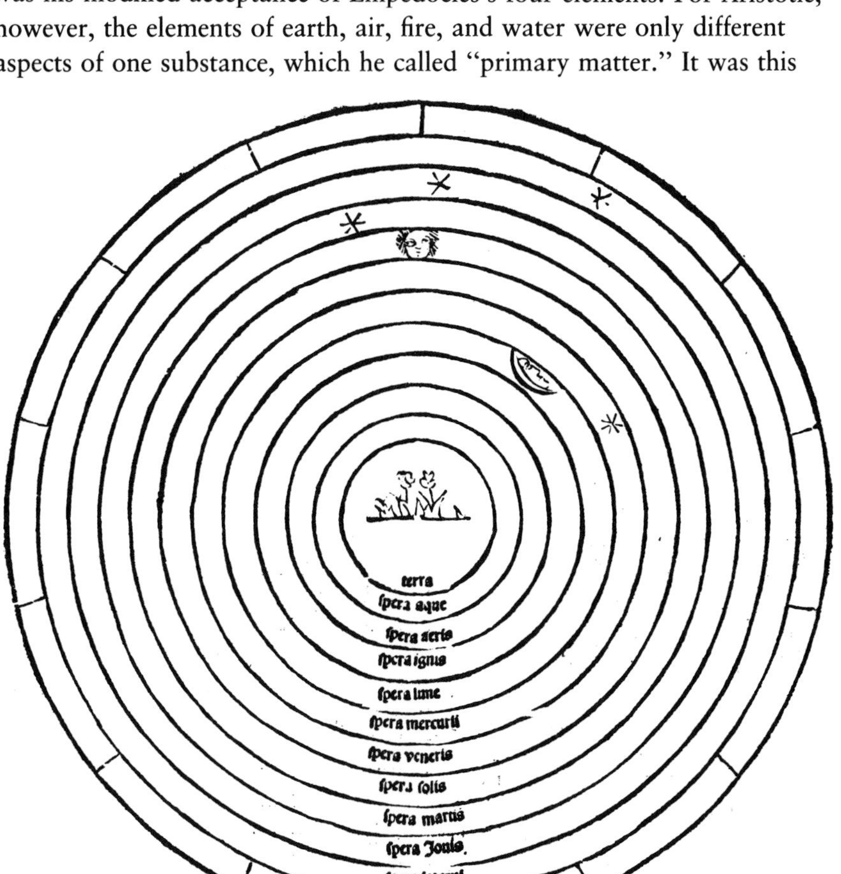

All of the major cosmological notions that dominated Western physical thinking for so long are represented here: an earth-centered universe, the perfectly circular paths of planets, and a finite universe. As long as this orderly but erroneous world view remained entrenched, no real progress was made in physics. *De caelo*, 1495. Aristoteles.

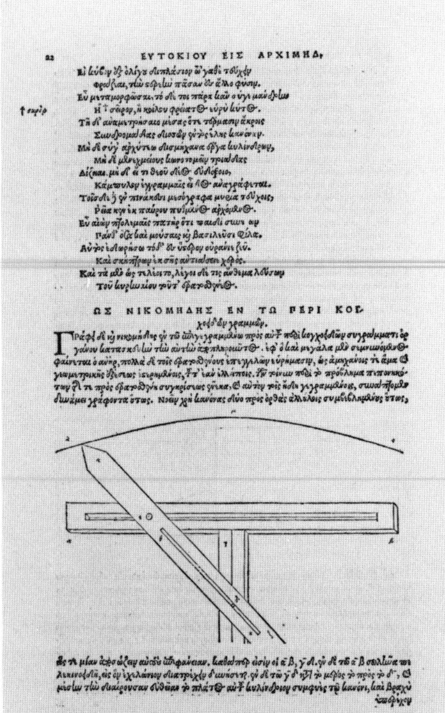

ΕΥΤΟΚΙΟΥ ΕΙΣ ΑΡΧΙΜΗΔ.

ΩΣ ΝΙΚΟΜΗΔΗΣ ΕΝ ΤΩ ΠΕΡΙ ΚΟΧ-

The complete works of Archimedes was published for the first time in 1544, twenty years before Galileo's birth. It is difficult to overestimate the importance of this work in shaping the scientific outlook and approach of the man who was to lay the foundations of modern mechanics. To Archimedes, nature's laws were mathematical laws, and it was he who first linked mathematics to physics. Galileo did the same nearly two thousand years later. *Opera, quae quidem extant, omnia*, 1544. Archimedes.

new substance, he said, that exhibited the qualities of hot, cold, wet, and dry. Variations on this theme would come to dominate all aspects of Western science. Aristotle also introduced the idea of a fifth element, one derived from Plato and often called the aether.

Aristotle's rejection of the enlightened atomism of Democritus may be said to have effectively cut off those ideas from consideration for centuries. His influence on cosmology was no less great. One of the main features of Aristotle's physics was his separation of heaven and earth. This dichotomy made for a finite universe—a perspective from which any progress in science was made difficult. Another dominant and, to the Scholastics very friendly, idea of his was that nature was purposive and goal-oriented. From this teleological point of view, happenings in the world were perceived as events occurring to fulfill a plan. Aristotle himself likened nature to an architect who builds a house according to a well-designed plan. The implications for physics of this seemingly innocuous and orderly world view were disastrous, leading its followers down scientific dead ends until the seventeenth century. Some examples of where Aristotelian assumptions could lead are the following beliefs: the natural motion of celestial bodies is that of an eternal circle; the acceleration of a falling body depends upon its mass; and nature abhors a vacuum. Such totally false assumptions about the mechanics of the physical world went unchallenged for nearly two milennia and led to a sterile science of physics. Aristotle the scientist is faultless, however, in the perpetuation of such errors. Rather it is the Scholastic values and mind-set with their blind obedience to Aristotle that should be faulted.

The Scholastic reverence for Aristotle became in later centuries a point of ridicule. A joke is told of a medieval dispute in which the participants are hotly arguing the correct number of teeth in a horse's mouth. The matter cannot be resolved because no copy of Aristotle's work is available. When a youth suggests checking the teeth of a horse nearby he is ridiculed, scolded, and driven away.

Surprisingly, the teachings of Aristotle that were taken as dogma by the medievalists had little impact on his own immediate successors. In fact, two of the most brilliant, Archimedes and Hero of Alexandria, completely ignored his work. Archimedes is one of those names linked to the golden age of Greek science and he is ranked as one of the greatest mathematicians of all time. But most people associate Archimedes with physics or mechanics, because his famous inventions made such a dramatic mark on the history of science. His name is linked to the hydraulic screw, the compound pulley, and the explanation of the principle of the lever. The most celebrated of his discoveries is the principle of the buoyancy of liquids. Underlying all of these accomplishments was his greatest achievement—the linkage of physics and mathematics. He is represented in the Library's collection by his *Opera, quae quidem extant, omnia,* which came out of Basel in 1544. None of his work was printed until the sixteenth century, and this editio princeps in Greek and Latin is the first complete edition of his work. The rare combination of brilliant theorist and practical genius that was Archimedes would not be seen again until the dazzling Leonardo da Vinci.

The greatest technician of antiquity may have been Hero of Alexandria, who flourished a century or so after Archimedes. He has been called the greatest experimentalist of antiquity as well as its first engineer. Hero used his mathematical skills to fabricate many ingenious mechanical devices—notably in the fields of pneumatic and automatic machines. He produced a treasurehouse of working toys, artistic pieces in themselves, each of which demonstrated a principle of physics and a manipulative and daring approach to natural forces. It was Hero who first established the motive power of steam—and demonstrated it vividly with his famous aeolipile or aeolian wheel. His knowledge of the properties of air—he recognized that it was compressible—was far ahead of his time. His beautiful and amusing mechanical toys would whistle, pivot, siphon water, move gears, and flap wings. But Hero was no mere talented tinkerer. He founded a school at Alexandria that specialized in technical instruction and research. His many devices led to his formal explanation of many physical phenomena and the elaboration of some basic principles of physics. "That which is gained in force is lost in time" is a familiar and useful example. He was a serious writer who took great pains to explain his experiments. His most famous works deal with air and mechanics. The Library has the earliest Latin translation of his *Pneumatica,* titled *Spiritalium liber* and published in Urbino in 1575. His *Automata* is represented in the Library by the 1589 Ferrara edition entitled *Gli artifitiosi et curiosi moti spiritali di Herrone,* translated into Italian by Giovanni Battista Aleotti. Also significant is the Italian translation by Bernardino Baldi, published in 1589 in Venice,

HERONIS
ALEXANDRINI
SPIRITALIVM
LIBER.

VM ſpiritalis tractatio maximo ſtudio digna ab antiquis tum philoſophis tũ mechanicis exiſtimata ſit, illis quidem ratione eius vim, ac facultatem tractantibus; his vero etiam per ipſam ſenſilium actionem: neceſſarium fore iudicauimus, quæ ab antiquis tradita ſunt in ordinem redigere; & quæ nos ipſi inuenimus exponere. ita enim fiet ut qui poſthac in mathematicis verſari volent, ex his maxime adiuuentur. Præterea cum arbitraremur conſentaneũ eſſe tractationem hãc cohærere ei, quæ de aquaticis horoſcopijs in quattuor libris eſt tradita, de ipſa conſcribe-

A re

Hero has been called the greatest technician of antiquity as well as its first engineer. His daring and manipulative approach to natural forces, combined with a real understanding of some of the basic principles of physics, produced amazing results. This page is from his *Pneumatics*, a popular work which consists mostly of practical descriptions of his pneumatic apparatus and instruments. *Spiritalium liber*, 1575. Hero of Alexandria.

88

FAR CHE CON VN DRACONE, CHE STIA ALLA GVARDA DE I POMI D'ORO COMbatta vn' Hercole, con vna mazza, & mentre ch'egli l'alza ſibili il Dracone, & nel punto che Hercole lo percuoterà in capo: far che eſſo le ſpruzzi l'acqua nella faccia.

Theorema Primo.

IA la baſe A.B.C.D. vna parte della quale C.D.E.F.G.H.K. ſia eccellentiſsimamente chiuſa: sì che non poſſa di eſſa vſcirne l'aria. Sopra di queſta ſotto il canale S. ſia poſto lo infundibulo T. la coda del quale riſtretta uerſo il fine: come dimoſtra la parte di eſſo nottata V. tanto ſtia diſſopra dal fondo della baſe G.H.K. quanto per il fluſſo dell'acqua parrà che baſti: in queſto fondo ſiano aſſaldate le due infleſſe ſiffoni X. & Y. ma la Y. ſia ſottile molto piu della X. indi ſia poſto oue è la P. lo Hercole, doue è la N. il pomo d'oro, & ſotto di eſſo oue è la L. ſiaui poſto il Dracone. Fatto queſto pongaſi nel lato della baſe E.F.G.H. la canna M. che in O. ſi uolga, & arriui alla bocca del Dracone in maniera accommodata, che mandi ſibilo, mentre l'aria (dall'acqua del canale S. che per lo infundibulo entra nella baſe) ſcacciata conuerrà (non hauendo altro exito) vſcire per eſſa canna; & ſia anco accommodata in maniera (che mentre per la ſiffone Y. ſi uoterà la baſe non potendo eſſa d'altronde, che per la bocca di detta canna riceuer aria, che in queſto anco mandi ſibilo maggiore, come non è difficile à niuno il ciò fare per mio auiſo. Sia doppò queſto dal perno. OO. ſoſtenuto il Regolo DD.CC. ſotto l'un capo, del quale CC. ſia accommodato il conno uoto RR. Nela parte di dentro con circoli perfetti, & linee rettiſsime giuſtiſſimamente con il torno lauorato. Et dentro di eſſo ſiaui accommodato il conno ſodo BB. che in eſſo giuſtiſsimamente ſtia; queſto nella parte ſuperiore habbia un'anello à cui legata una fune ſia in CC. attaccata ſtando il Regolo DD.CC. in perfetto piano. Et ſotto DD. vi ſia appeſo il Vaſo Z. che (uoto) ſia aſſai piu leggieri del cõno BB.

& queſto

AGGIVNTI. 89

& queſto, & il Regolo, & il conno ſiano in maniera diſpoſti che il uaſo Z. ſtia ſotto la gamba eſteriore della infleſſa ſiffone X. & al ſuo manico ſia legata una fune che per la gamba deſtra d'Hercole paſſi, & per il corpo aggiunga nella ſnodatura delle braccia di eſſo, le quali da una chiaue in figura d'una T. ſiano in bilico ſoſtenute lo eſſempio è la ⊣. 3. è la ſpalla deſtra 2. la ſpalla ſiniſtra, & 4. la ſcitala ſtando dunque 2. 3. in bilico ſia la fune allegata in 4. capo della ſcitala.

Et ſia dopò queſto poſto nelle mani d'Hercole la mazza Q. indi ſia ſottopoſto alla gamba della infleſſa ſiffone X. il uaſo AA. & queſta cãna nel coperto di detto uaſo ſia beniſsimo ſaldata, & eſſo coperto al uaſo: ſuori del quale eſca la canuccia TT.R. la quale ponga capo nel uoto conno RR. che con lei ſia bucato & habbia in queſta bocca un'aſſario, ò cartella che nella parte di dentro di eſſo ſi apra. Scontro di queſto buco ue ne ſia fatto un'altro, & in eſſo ſia aſsaldata la canna uota QQ. la quale anco lei arriui alla bocca del Dracone: queſto eſequito corra l'acqua per il canale S. nell'infundibulo T. ch'ella ſcenderà nella baſe fuor della quale conuien che ſe ne fuga l'aria per la canna M.O. la quale farà ciuffollare il Dracone & ripiena d'acqua la baſe ella ſi uoterà, & l'aria ritornãdo in dietro per la canna M.O. dara maggior ſibilo, & ſtridore. Si uoterà dico per la infleſſa ſiffo

M ne X.

This late sixteenth-century Italian version of Hero's automatic theater describes a sort of puppet show whose figures move by themselves. In this diagram, Hercules clubs a dragon on the head and water streams from the mouth of another figure. The driving mechanism appears to be a system of water and siphons, and no springs or cogwheels are used. *Gli artifitiosi et curiosi moti spiritali di Herrone*, 1589. Hero of Alexandria.

entitled *Di gli automati.* The Library's copy of this small book is heavily annotated. All three of these works contain beautiful engravings of Hero's mechanical devices. Hero must be recognized as a genius of mechanics whose insights and abilities presaged and provoked the later renaissance of physics. After him, there was no real progress in physics in the West until the rational mechanics of Stevin and Galileo.

Physics was no exception in the long hiatus of productive theoretical research that descended on the West after the Greek decline and persisted through the rise and fall of an empire. During this time of Arab scientific eminence, real strides were made in only one branch of physics, that of optics. Responsible for these advances was Alhazen, the greatest of all Arab physicists. Alhazen was the Latin name of Ibn al-Haitham, who was a mathematician and physician as well. Born at Basra (a city that is now part of modern Iraq) around A.D. 965, Alhazen went far beyond the optics of his time, which essentially involved only the construction of mirrors, to study not only the eye but the nature of light as well. His writings did more than preserve the ancient knowledge, for in his main work he broke with Ptolemy's theory that the eye emitted light rays which were then reflected back and offered the correct view that the object and not the eye was the source of those rays. This theory was such a fundamental break with traditional Greek assumptions that it later established Alhazen as the most original scientific mind that the Arab culture produced. Alhazen's most famous and most significant work, the *Opticae thesaurus,* was first published in the West in 1572 in Basel, and the Library has this first edition folio as part of its collection. The second of *Opticae thesaurus* contains a treatise on optics by the Polish monk Vitello and also Alhazen's smaller treatise on atmospheric refraction and twilight, the *De crepusculis et nubium ascensionibus,* which had been published separately in Lisbon thirty years earlier.

This work of Alhazen's was translated by the most famous and probably the most brilliant of the twelfth-century translators, Gherardo da Cremona. He left Italy and went to Spain in search of a copy of Ptolemy's *Almagest,* which he did translate from the Arabic. Gerard stayed in Spain and eventually translated many of the Greek classics—Archimedes, Hippocrates, Galen, and Euclid, to name a few. Between his arrival in Toledo in 1160 and his death in 1187, he translated more than seventy works from Arabic. He seemed to select for translation the best from each field, and it was mainly through Gerard that the West received its Greco-Arabic inheritance.

Once the Latin translation of *Opticae thesaurus* became available, Alhazen became the standard authority on optics. Alhazen has been described as an eccentric who feigned madness to escape a death sentence imposed by his crazed caliph, al-Hakim. He was obviously both an original thinker and a resourceful man.

The optical achievements made by Alhazen around A.D. 1000 stand in even greater relief when compared to the centuries that preceded and followed them. Roman emphasis on the expedient and the useful was as stifling to the scientific spirit as was the later dominance of theology. Not until well after the twelfth-century translators had done their work did anything notable occur in the field of physics. By this time, disagreement with Aristotelian concepts was becoming fashionable (as well as fruitful, scientifically) and later, real progress almost seemed linked to the degree of the departure. Traditionally, the final break with Aristotelian physics was made by Galileo. He was preceded, however, not only by what one writer has called "the inspired guesses of Renaissance science," but by the real contributions of a few individuals. The rival mathematicians Niccolo Tartaglia and Girolamo Cardano both wrote on motion and, to a degree, offered some modifications on Aristotle. Tartaglia wrote first, in 1537, of the "new science" of ballistics. The Library has a 1550 edition of his *Noua scientia* published in Venice which, despite its false notions, stands as the first real theoretical discussion of what had been a purely practical matter. In 1554, Cardano's *De subtilitate* was published in Nuremberg. In this large volume, which the Library has in its collections, Cardano departed a bit further from Aristotle than did Tartaglia. Like Tartaglia, he discussed ballistics, but he focused more on velocities than on trajectories. He also distinguished between the electrical attraction generated by friction and the magnetic attraction of a lodestone. The book's varied subjects reflect the breadth of Cardano's interests—ranging from cosmology to cryptology—and it contains as much information about the occult as it does on the natural sciences. This volume is also one of the few books known to have been printed by Robert Granjon, the famous type-cutter and type designer.

As Tartaglia and Cardano studied motion and mechanics using mathematical methods, so did an older contemporary of Galileo, Simon Stevin. Stevin was a Dutch military engineer and mathematician, an eminently practical man who came to be known as the "second Archimedes." Through his highly original approach and unique experimental methods, Stevin may be said to have founded the science of hydrostatics and to have given statics its first major advance since Archimedes. In hydrostatics, his elegant arguments demonstrated the "hydrostatic paradox," in which the pressure of a liquid on a vessel's surface is shown to be independent of the vessel's shape and size and dependent only upon the depth of the liquid and the area of the surface. To statics he gave the powerful new tool of the parallelogram of forces, a representational figure for physics that enabled him to determine the strength and direction of the force produced when forces from two different directions act upon the same body. Furthermore, his intuitive assumption that perpetual motion is impossible enabled him to solve the historic problem of the equilibrium of heavy bodies on an inclined plane. These several accomplishments not only were important in themselves but indicated that physics was maturing beyond the study of simple, individual forces toward a more serious consideration of complex and dynamic realities. Stevin was a remarkably original thinker who is credited also with the introduction of the regular use of decimal fractions to mathematics. The Library does not have in first edition his major works in physics, which were published at Leiden and written in Dutch. It does have, however, the best and most complete edition of his

nomise diuide in.90.parti eguale & cadauna di quelle chiamano grado. Pero la mita di quello (cioe.g f.) uerria a esser gradi.45. Ma per acordarse con quello che se ha da dire lo hauemo diuiso in.12. parti eguali & accioche uostra Illustrissima.D.S.ueda in figura quello che di sopra hauemo con parole depinto hauemo qua disotto designato il pezzo con la squara in bocca assettato secondo il proposito da noi conchiuso al detto nostro amico. La qual conclusion a esso parse hauer qualche consonantia pur circa cio dubitaua alquanto parendo a lui che tal pezzo guardasse troppo alto. Il che procedeua per non esser capace delle nostre ragioni, ne in le Mathematice ben corroborato, niente di meno con alcuni isperimenti particolari in fine se uerifico totalmente cosi essere.

Pezzo elleuato alli. 45. gradi sopra al'orizonte.

Ma piu nel anno MDXXXII. essendo per Prefetto in Verona il Magnifico misser Leonardo Iustiniano. Vn capo de bombardieri amicissimo di quel nostro amico. Vene in concorrentia con un altro (al presente capo de bombardieri in Padoa) & un giorno accadete che fra loro fu proposto il medemo che a noi propose quel nostro amico, cioe a che segno si douesse assetare un pezzo de artegliaria che facesse

Renaissance physics was to become obsessed with the long-ignored subject of dynamics or motion. Before Galileo, the first to discuss theoretically the "new science" of ballistics (the notion of a projectile in flight) was Niccolò Tartaglia. In this first work of his on ballistics he offered a totally wrong idea of the path of a fired cannonball, saying that its trajectory must consist of a straight line out from the cannon mouth followed by a small circular arc and then a straight downward line. Tartaglia corrected his theory somewhat in a later work. This illustration shows Tartaglia's gun quadrant, a device that measures the angle of elevation. *La noua scientia*, 1550. Niccolò Tartaglia.

12 **Liber Primus**

deſcendat primo, deinde aſcendat, ut in figura ſequente ex A in B, inde in E, & poſtmodum in C, & in D, tunc peruenire poterit, ſi D minus diſtet à linea B C, quàm A locus, ex quo deſcendit. Sed oportet in ſingulis ſpacijs certam eſſe differentiam altitudinis A & D. Quanto enim longior uia fuerit, eo maiorem eſſe differentiam A & D, iuxta altitudinis menſuram oportet. Hinc errores quorundam, qui ad libramentum cũ conati eſſent aquas deducere, maximas iacturas impenſarum ſuſceperunt. In ſingulis igitur millibus paſſuum A altius palmo eſſe debet quàm D, ut in decem millibus paſſuum decem palmis. Cauſa huius eſt aquæ rotunditas euidens, quæ etiam in urceorum ſuperficie apparet. Vnde ad libramentum licet A ſit altius quàm D, non tamen erit altius quandoq; loco medio inter A & D, indiget etiam impetu quodam: ſed hæc nunc præter intentum quaſi ſunt. Volui tamen ob magnitudinem periculi, & erroris frequentiam hæc ſubieciſſe.

Ratio ducẽdæ aquæ.

Sed iam ad elementorum motum ſimplicem ueniamus exemplis explicandum. Igitur grauis motus exemplum præbent ponderum horologia, quæ ſenſim trahendo rotas uertunt. Huiuſce autem generis infinita facilè eſſet inuenire exempla. At motus leuis hoc unum ſubijciatur exemplum. Cum naues freto merguntur, quas eruere conſilium eſt: cimbæ onuſtæ ſaxis per funes alligantur nauigio ab urinatoribus, ſicut funes quantum fieri poteſt tendantur, inde totidem cimbis uacuis lapides ex prioribus detracti excipiũtur: quo fit ut alleuatæ cimbæ nauigium paululum ex profundo ſecum trahant. Nam aer cimbas, quæ pondere lapidũ fermè mergebantur, cum aquæ ſubeſſe nolit, in ſuperficiem aquæ attollit: unde nauigium fermè pro cimbæ altitudine ſuperius trahitur. Trahatur igitur ex A in B, tunc cimbæ quæ plenæ ſunt lapidibus illi annectãtur funibus, transfuſisq; lapidibus nauigium trahetur in C. Rurſus priores cimbæ, in quas lapides transfudiſti, nectũtur tenſis funibus nauigio in C exiſtenti, trahentq; deductis lapidibus ipſum in D, atq; perpetua tranſmutatione ad aquæ ſuperficiem tandem deducetur. Verum dices plurimis cimbis opus erit ad triremem educendam, uerum eſt, ſed ratio ſic cõſtat: quælibet nauis aut cimba tantum ferre poteſt ponderis, quantum eſt pondus aquæ quam continere poteſt. Velut ſi triremis capiat in flumine mille amphoras aquæ, quarum pondus

Modus quo naues demerſæ gurgitibus recuperantur.

The problem of motion was also considered by another Renaissance savant, Girolamo Cardano. Like Tartaglia, his dynamics was almost as traditional or Aristotelian in many of its assumptions and conclusions as it was premodern, but his moderate defense of the impetus theory makes his contribution noteworthy. This theory was in complete opposition to the incorrect Aristotelian theory of motion, which said that a body can move only if at every moment a mover continues to act upon it. Cardano's book seemed to contain a bit of all of the physical knowledge of his time, and in this illustration he offers a method for raising sunken ships. *De subtilitate*, 1554. Girolamo Cardano.

works, *Les oeuvres mathematiques,* edited by Albert Girard and published in Leiden in 1634. This very thick folio volume contains his classic work on statics. It was not his statics that earned him renown in his lifetime, however, but rather a wind-driven chariot that he built in 1600, which could carry twenty-eight passengers.

DE LA SPARTOSTATIQUE. 505

COROLLAIRE III.

Or pour venir à la declaration de la qualité des peſanteurs ſuſpenduës par cordages, ſoit AB une colomne, de laquelle C ſoit le centre, ſuſpenduë à deux lignes CD, CE (venans dudit centre C) és poincts fermes D, E, leſquels ſeront diametres de gravité par la 5 definition: parquoy menant HI entre DC, CF, parallele à CE, alors par la 13 definition, CI ſera elevation droite, CH oblique; tellement que comme CI à CH, ainſi ceſt elevant direct à l'elevant oblique: mais l'elevant direct de CI eſt egal au poids de la colomne: Donc comme CI à CH, ainſi le poids de la colomne entiere, au poids qui avient en D; & de meſme maniere trouvera-on le poids qui advient en E, en menant de I juſques à CE, la ligue IK, parallele à DC; & diſant, comme l'elevation droite CI à l'elevation oblique CK, ainſi le poids de la colomne, au poids qui advient ſur E.

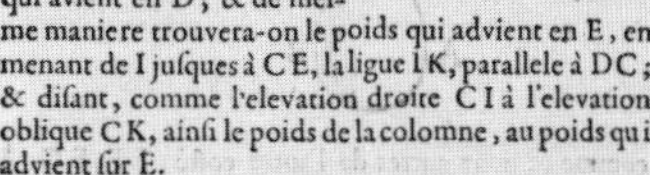

Mais CK eſt touſiours egale à HI; parquoy il n'eſt pas beſoing de mener ceſte ligne derniere IK, car ſans cela les termes neceſſaires ſont cognus au triangle HIC, avec lequel on dira: comme CI à CH, aſnſi le poids de la colomne, au poids qui advient ſur D. D'avantage CI à IH, ainſi le poids de la colomne, au poids qui advient ſur E. Derechef comme CH à HI, ainſi le poids qui advient ſur D, au poids qui advient ſur E.

COROLLAIRE IV.

Et pour proceder plus avant, ſoit AB la colomne abaiſſée, comme cy joignant, & par la troiſieſme peti-

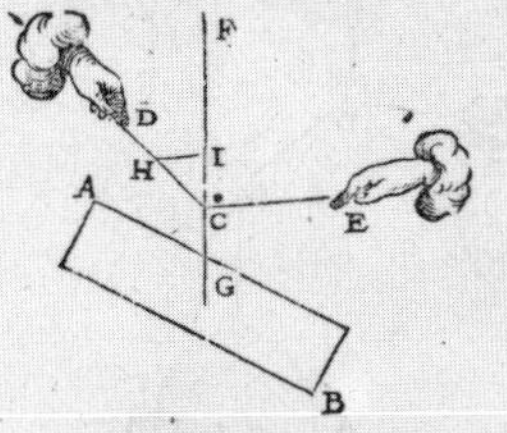

tion, il n'y a en cela aucune alteration, & partant la meſme proportion que deſſus y ſera encore.

COROLLAIRE V.

Soit maintenant au lieu de la colomne precedente, un corps d'egale peſanteur à icelle, de figure & matiere

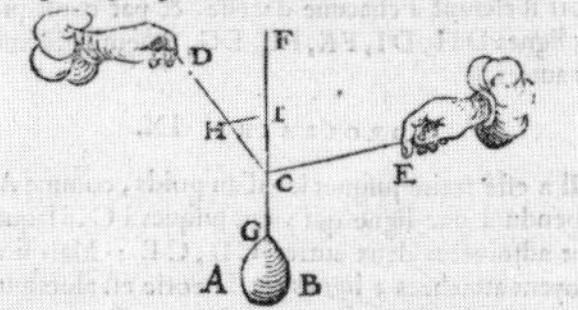

quelconque, alors la meſme proportion demeurera; aſſavoir, comme CI à IH, ainſi AB au poids qui advient en E. Derechef comme CH à HI, ainſi le poids que D ſouſtient, à celuy que E ſouſtient.

D'icy eſt manifeſte, que s'il y avoit à la ligne DCE comme corde, un poids AB cognu, & les angles FCD, FCE, auſſi cognus, qu'on pourra dire quel poids advient ſur chaque partie, comme DC, CE.

COROLLAIRE VI.

Mais s'il y avoit pluſieurs poids ſuſpendus en une meſme ligne, comme icy la ligne ABCDEF, ſes poincts fermes extremes A, F, à laquelle ſont ſuſpendus 4 poids cognus, G, H, I, K; il eſt manifeſte qu'on peut dire quel effort ils font à la corde, à chacune de ſes parties AB, BC, CD, DE, EF: Car par exemple, produiſant GB enhaut vers L, & MN parallele à BC: Je dis BN donne BM, combien le poids G? viendra l'effort qui eſt fait à AB.

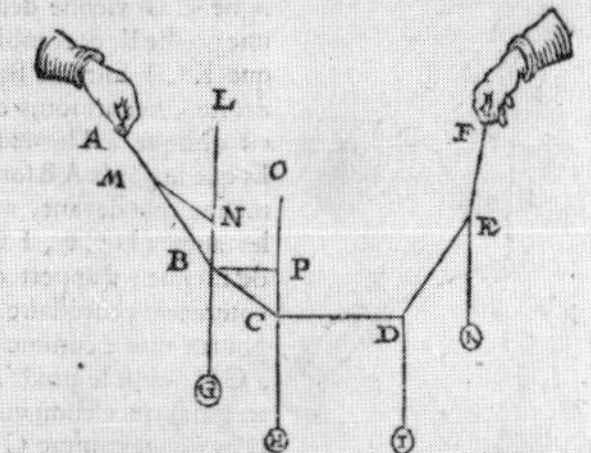

Derechef BN donne MN, combien le poids G? ce qui viendra ſera l'effort qui eſt fait à BC.

Soit encore HC produite juſques en O, & BP parallele à GD: Je dis alors, CP donne CB, combien le poids H? Ce qui en ſortira ſera pour la force qui eſchoit ſur BC, d'où s'enſuit, qu'il faudra trouver le meſme qu'à BC cy-deſſus. De ces choſes, & de pluſieurs autres S. EXCEL. a trouvé que la practique s'accorde du tout avec la Theorie.

La proportion de la 27 propoſition peut encor eſtre autrement expoſée, que cy-deſſus, d'où s'enſuit une operation plus facile. Soit par exemple la figure de la 27 propoſition, où eſt dit, que comme l'elevant oblique au direct, ainſi chaque elevant oblique à ſon elevant direct. Mais pour dire cecy d'une autre maniere, d'où

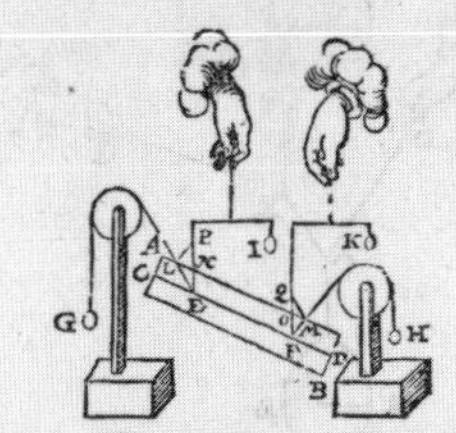

reſulte une operation plus facile: Soit menée entre les elevations droite, & oblique une ligne, comme LP, parallele à FM: Ce qu'eſtant ainſi, je dis maintenant, que comme l'elevation directe, à l'elevation oblique, ainſi la peſanteur de la colomne entiere, à ſon elevant

VV oblique;

As a branch of physics dealing with the relations of forces producing equilibrium among bodies, sixteenth-century statics was a very practical science used in such simple machine operations as those of the lever, pulley, wedge, wheel and axle, and inclined plane. Not until the Dutch engineer and mathematician Simon Stevin, however, did statics progress much beyond where Archimedes had left it. Stevin gave to statics the laws which determine the behavior of bodies on an inclined plane as well as the principle of the parallelogram of forces, demonstrated here by this diagram for corollary 3 from his collected works. Stevin's resolution of the parallelogram of forces—a complex problem in which two forces act upon one body from different directions—was not expressed mathematically but rather in this form of geometric figures. *Les oeuvres mathematiques*, 1634. Simon Stevin.

The intellectual daring exhibited by Stevin's foray into the unknown areas of complex movement was met and far surpassed by the boldness of Galileo's attempt to formulate laws governing the dynamic movements found in nature. Galileo not only succeeded in this bold venture by founding modern mechanics, but in a broader context his lifework served as the model for the scientific revolution. It was Galileo's greatest work, the *Discorsi e dimostrazioni matematiche, intorno à due nuove scienze*, which founded and demonstrated the principles of what has become the

DISCORSI
E
DIMOSTRAZIONI
MATEMATICHE,
intorno à due nuoue ſcienze
Attenenti alla
MECANICA & i MOVIMENTI LOCALI,
del Signor
GALILEO GALILEI LINCEO,
Filoſofo e Matematico primario del Sereniſſimo
Grand Duca di Toſcana.
Con una Appendice del centro di grauità d'alcuni Solidi.

IN LEIDA,
Appreſſo gli Elſevirii. M. D. C. XXXVIII.

Published when he was seventy-four, this was Galileo's final work. With it, the great old man not only saved his best scientific work for last but justified his prudent capitulation of 1633. Although humiliated and abused by the institutional terror of the Inquisition and forced to renounce his work and deny his conscience, Galileo had the resilience and farsightedness to know he would eventually defeat his tormentors. Written while he was under house arrest and smuggled out of Italy to be published, this work is a triumph of both the scientific and the human spirit. In it, Galileo's brilliant, beautiful, and lucid language conclusively defeated the Aristotelian physics of the Scholastics and offered to physics, and indeed to all the physical sciences, the essence of the experimental method. *Discorsi e dimostrazioni matematiche, intorno a due nuoue scienze,* 1638. Galileo Galilei.

dominant form of the scientific method. Of his published work, it is this treatise that is most responsible for dealing the fatal blow to Aristotelian physics and for beginning modern experimental science. Some details of Galileo's life and work are given in the chapter on astronomy, where the significance of his earlier work, the *Dialogo,* is discussed. The *Dialogo* was published in 1632 and was the immediate cause of his celebrated brush with the Inquisition. In addition to championing the heliocentric system of Copernicus, the *Dialogo* is related to the 1638 *Discorsi* in that the latter analyzes mathematically what was considered philosophically in the former. In that respect, the *Discorsi* is the first modern textbook in physics.

The Library's collections contain the first edition of the *Discorsi,* published in Leiden in 1638. Galileo prudently offered his book to the Netherlands Elzevier family for publication, knowing he could never obtain an imprimatur in Venice. The book's text is written defiantly in the Italian vernacular, although its many mathematical proofs are in Latin. This was Galileo's last publication, and it was issued when he was seventy-four years old. Among several other first editions of his works in the Library's collection is one considered the most scarce of Galileo's works, *Le operazioni del compasso geometrico, et militare.* Printed in Padua in 1606, this book describes an invention, the "geometric and military compass," which functioned as a combination slide rule and proportional divider. It was Galileo's first published work and he indicates in its preface that only sixty copies were printed. The book and the instrument were to be presented to the Grand Duke of Tuscany and his friends.

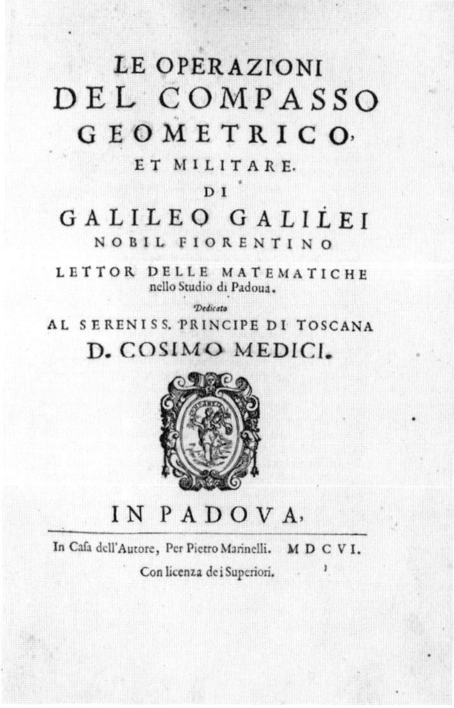
LE OPERAZIONI
DEL COMPASSO
GEOMETRICO,
ET MILITARE.
DI
GALILEO GALILEI
NOBIL FIORENTINO
LETTOR DELLE MATEMATICHE
nello Studio di Padoua.
Dedicato
AL SERENISS. PRINCIPE DI TOSCANA
D. COSIMO MEDICI.

IN PADOVA,
In Casa dell'Autore, Per Pietro Marinelli. M D C V I.
Con licenza dei Superiori.

This book was Galileo's first published work and describes his invention of an improved military compass—a computational device that served the same purpose as a slide rule. This work is also one of the rarest of Galileo's publications, since its preface indicates that only sixty copies were printed. *Le operazioni del compasso geometrico, et militare,* 1606. Galileo Galilei.

For physics and the history of science, the *Discorsi* was a turning point. In its clear and at times beautiful language, it offered a rational conception of a knowable universe. The world was not unpredictable and chaotic, it said, but functioned according to knowable laws and contained calculable forces. It was in keeping with this new mode of thought that Galileo offered a method whereby the essential pattern of natural forces might begin to be discovered. This method had three stages. First came intense observation of the phenomenon, which would reduce it to its essentials. Second followed abstract reflection and the setting up of a hypothetical assumption from which the consequences were deduced. Third came what he called "resolution," the confirmatory experiments that would test the hypotheses. Galileo was trained as a mathematician and it was as such that he used this approach to search for the mathematical laws of the phenomena he so astutely observed. Toward that end, his *Discorsi* laid the foundations of modern mechanics, formulating what has come to be known as the first law of motion (or the law of inertia), as well as the laws of cohesion and of the pendulum. It also provided a definition of momentum. Furthermore, the *Discorsi* marked the departure from the classical Aristotelian method of seeking the "why" of a phenomenon and turned instead to the more limited and soluble "how." In this reorientation, Galileo was Greek in his method, but more of the Pythagorean or Platonic school than the Aristotelian. As with Pythagoras, nature to Galileo meant mathematics, and to his credit, he realized that the formulation of mathematical laws that indicated *how* bodies move under the action of natural forces must precede the more ambitious search for *why* they behave as they do.

In summary, then, Galileo's *Discorsi* was a pioneering work of specialization, most characteristic of modern scientific research. Galileo did not so much disprove Aristotle as turn future science away from Aristotle's predominant concern with purposes and reasons—the why—to offer instead a new conceptual structure that made the era of Newtonian physics possible. Comparing the decidedly less ambitious approach of Galileo with the grandiose teleological schemes of Aristotle is unfair to both. Each approach emanates from a different world view and each therefore asks very different questions. Both suffer from comparison. It was Galileo's key insight, however, that investigation of his limited field of inquiry was as valid and as worthwhile as the Aristotelian search for the ultimate explanation. It could be said that Galileo showed science its future by demonstrating the essential worth of a particular, albeit incomplete, law of nature. His was a patient, positive, and ultimately productive incrementalism that was to bring about an intellectual revolution.

The seventeenth century became the locus of this scientific revolution. For physics it was an especially heady time, when the physical world seemed to reveal itself as eminently understandable and explainable. The work of René Descartes had much to do with such an optimistic outlook. The impact of his mechanistic view of nature, although affecting all the sciences, was particularly strong in physics. The most fundamental general conclusion of the Cartesian philosophy was that all natural phenomena operated like a machine and could hence be understood by mechanical example and analogy. This view was expressed in Descartes's famous *Discours de la méthode,* published in Leiden in 1637, and in his *Principia philosophiae,* which came out of Amsterdam in 1644, both of which the Library has in first edition. It was this mechanistic approach to nature, so widely followed in the seventeenth century, that to a great degree determined what questions were asked and what answers would be accepted. Descartes was a great scientist—his analytical geometry and work in optics attest to that—but his contributions to physics are found more in the intellectual stimulation his ideas provided to other scientists, whether they articulated or opposed them.

Under the intellectual stimulus of the two great contemporaries Descartes and Galileo, the progress of physics began to change from a steady, linear growth to a more specialized, almost radial evolution. The specialized development of single fields or specific areas of investigation came to characterize post-Newtonian physics, and certainly it best describes modern physics. Anticipating this later branching-off were the seventeenth-century scientists who investigated a particular physical phenomenon such as heat, magnetism, or light. Those who studied the physical properties of air and the existence of a vacuum best exemplify the beginnings of specialization in the seventeenth century. The names Torricelli, Pascal, Guericke, and Boyle are linked to the study of a phenomenon, the investigation of which has become a standard feature of most elementary physics courses. First among them was a pupil of Galileo's, Evangelista Torricelli. Serving

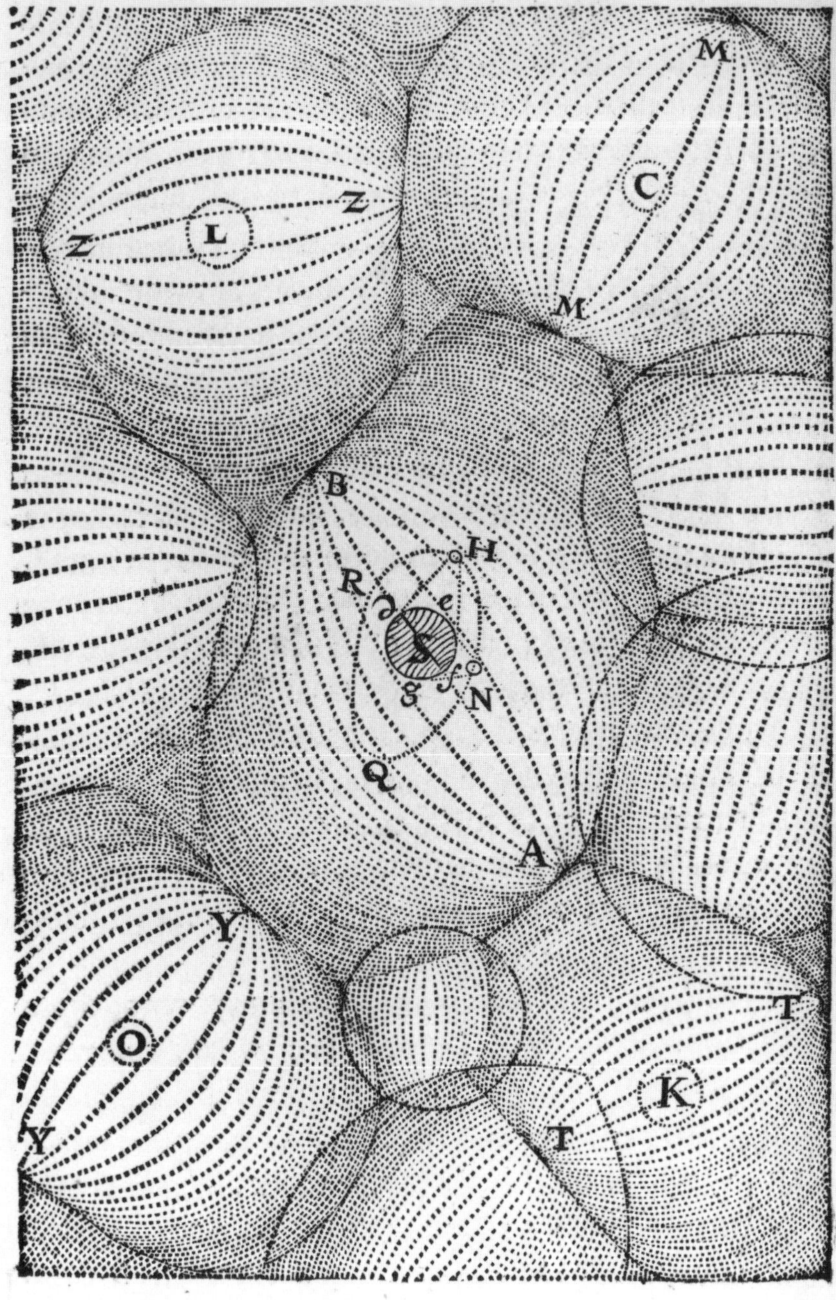

According to Descartes's mechanical ideas, the universe operated as a continuously running machine which God had set in motion. Descartes rejected the notion of a void in space and dismissed Newton's action at a distance (gravitation). Instead, he argued that space was full of a subtle matter he called "plenum" which swirled around in vortices. Planets thus moved by actual mechanical contact with the plenum. Here, Descartes's system of vortexes carries the planets around the sun (S). *Principia philosophiae,* 1644. René Descartes.

xxviii

ſi ſente nel voler fare il vacuo, indarno mi pare ſi cercherebbe d' attribuire al vacuo quella operazione, che deriva apertamente da altra cagione, anzi che facendo certi calculi faciliſſimi; io trovo, che la cauſa da me adattata, cioè il peſo dell' aria, dovrebbe per ſe ſola far maggior contraſto, che ella non fà nel tentarſi il vacuo. Dico ciò, perche qualche Filoſofo vedendo di non poter fuggir queſta confeſſione, che la gravità dell' aria cagioni la repugnanza, che ſi ſente nel fare il vacuo, non diceſſe di conceder l' operazione del peſo aereo, ma perſiſteſſe nell' aſſeverare, che anche la Natura concorre a repugnare al vacuo. Noi viviamo ſommerſi nel fondo d' un pelago d' aria elementare, la quale per eſperienze indubitate ſi ſa, che peſa, e tanto, che queſta groſſiſſima vicino alla ſuperficie terrena, peſa circa una quattrocenteſima parte del peſo dell' acqua. Gli Autori poi de crepuſcoli hanno oſſervato, che l' aria vaporoſa, e viſibile ſi alza ſopra di noi intorno a cinquanta, o cinquanta quattro miglia, ma io non credo tanto, perche moſtrerei, che il vacuo doverebbe far molto maggior reſiſtenza, che non fa, ſebbene vi è per loro il ripiego, che quel peſo ſcritto dal Galileo, s' intenda dell' aria baſſiſſima dove praticano gli uomini, e gli animali, ma che ſopra le cime degli alti monti, l' aria cominci ad eſſer puriſſima, e di molto minor peſo, che la quattrocenteſima parte del peſo dell' acqua. Noi abbiamo fatti molti vaſi di vetro come i ſeguenti ſegnati A, e B groſſi,

e di

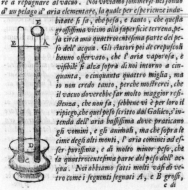

Torricelli's writings on mechanics so impressed Galileo that he invited the young man to visit him at Arceti, near Florence, where he was under house arrest. Torricelli accepted and moved into Galileo's home in October 1641. Galileo died in January 1642, and Torricelli's presence considerably helped the blind, ill old man in his last days. Torricelli was led to his major discovery in physics while investigating a problem suggested by Galileo—that of why water could not be pumped higher than about thirty-two feet. Torricelli sought a mechanical solution, positing not only that air had weight but that it pressed its weight down upon the water outside the pump. To demonstrate this, he created the first barometer—a four-foot-tall glass tube filled with mercury and inverted into a bowl of mercury. The daily fluctuations in height of the mercury column demonstrated that the liquid was responding to the daily changes in atmospheric pressure. *Lezioni accademiche,* 1715. Evangelista Torricelli.

the blind old man during the last three months of his life, Torricelli was guided by Galileo to an investigation of the Aristotelian maxim that "nature abhors a vacuum." The "horror vacui" principle provided the physical explanation as to why water rose in a suction pump—the water was impelled to rise to prevent a vacuum from being formed. But the well-known limit of any suction pump to raise water beyond eighteen "braccia" (about thirty-two feet) seemed to limit seriously the applicability of this traditional explanation. When led to consider the problem of Galileo, Torricelli arrived at the startling hypothesis that since the air itself had weight (something Galileo knew) it therefore must exert a force upon the water outside the pump. Torricelli's famous mercury experiment of 1643, which confirmed this hypothesis, signified a radical departure from traditional thinking in that it posited both the existence of an external force produced by the air (atmospheric pressure) and the reality of a vacuum. Torricelli died at thirty-nine and lived to see only one of his works reach print, the *Opera geometrica,* published in Florence in 1644. The Library has this volume, nearly every page of which has a diagram, in its collections as well as the more significant and complete *Lezioni accademiche* published in Florence in 1715, some sixty-eight years after Torricelli's death. His now-legendary mercury experiment, which created the first barometer, demonstrated that liquid is pushed up inside a tube by the pressure of the atmosphere on the surface of the liquid outside the tube, and that the height of the liquid varies slightly from day to day. Torricelli also created the first man-made vacuum in the tube space above the liquid, called to this day a "Torricellian vacuum."

News of his experimental findings went from Florence to Rome in a letter from Torricelli to his friend Michelangelo Ricci, who then sent it to a Parisian priest, Marin Mersenne. Père Mersenne, called the postbox of Europe because of his extensive correspondence, acted as an intermediary among Europe's scientific elite. From him, the information was passed to the brilliant young mathematician Blaise Pascal, who repeated the experiment with water at full scale (Torricelli used the much denser mercury because his tube then needed to be only four feet in length). With the mental clarity and intuition that had made him a child prodigy, Pascal reasoned that if air had a finite weight, it was then logical that it should also have a finite height. The weight of the atmosphere should decrease, he supposed, with an increase in altitude, and this decrease should be demonstrable on the new barometer. Pascal lived in Rouen near no mountains and, because of his own frail health, he asked his brother-in-law, who lived in Clermont-Ferrand, to climb the nearby Puy de Dôme with a barometer in 1648. The barometer was upended and read at various times up and down the 4,800-foot mountain. These readings proved that the air pressure decreased as one ascended. With what he called his "great experiment," Pascal had offered a factual demonstration that the atmosphere did not extend upward infinitely and that well above earth there existed what had been traditionally considered impossible—a vacuum. In the same year as the experiment, Pascal underwent a religious experience that made him forsake any further sciences. His work on the weight of air is represented in the Library's collections by a small, posthumously published volume which almost fits in one hand, called *Traitez de l' équilibre des liqueurs,* published in Paris in 1663.

The long-held "horror vacui" doctrine was overthrown not only by this work in Italy and France but by the independent experimental research of Otto von Guericke in Germany. In 1650, Guericke produced a vacuum by constructing the first pneumatic machine or air pump. He had begun his experiments using wine and beer casks and eventually progressed to a spherical copper vessel from which most of the air could be removed. Once able to produce a vacuum, using his new air pump, Guericke set to work documenting its properties. He found that in a vacuum, a flame is extinguished, an animal dies, and a clock cannot be heard to strike. His most famous experiment—and probably one of the most elaborate and dramatic technical demonstrations in the history of science—was a spectacle he produced in 1654 before Emperor Ferdinand III. Guericke produced a vacuum in his "Magdeburg hemispheres," two hollow hemispheres tightly sealed, to which teams of horses were attached on each side. The straining horses could not pull the sphere apart until the stopcock was opened, air entered, and the two halves parted easily. The great power of a vacuum was made apparent by Guericke's natural ability as a showman. Ferdinand was so impressed that he ordered Gaspar Schott, a Würzburg University professor, to document the experiment. So the first accounts of Guericke's air pump and vacuum experiments were seen in Schott's *Mechanica hydraulico-pneumatica,* published in Frankfurt in 1657. The Library has Schott's book in first edition. It contains scores of engravings depicting all manner of devices similar to Hero's. Not until 1672 did Guericke himself publish some of the results of his long career in mechanical experimentation. Published that year in Amsterdam, his *Experimenta nova magdeburgica* documents his many and varied experiments and contains some fascinating engravings. The Library has this treasure-house of experimental science in first edition.

Guericke is said to have first heard of Torricelli and his experiments on the occasion of his demonstrations before Ferdinand III in 1654—eleven years after Torricelli's discovery. But news of his own work spread quickly after the publication of Schott's book, and so it came that Robert Boyle learned of Guericke's air pump in 1657. Intrigued with the new device, he set out to devise an air pump of his own and with the help of his brilliant twenty-two-year-old assistant, Robert Hooke, produced an improved version. Further experiments with air and a vacuum resulted in what has become known as "Boyle's law" (called "Mariotte's law" by the French, in recognition of Edme Mariotte's independent discovery). This law first established the relationship between volume, density, and pressure of gases by stating that the volume of air in a confined space varies inversely to the pressure. Boyle was the boy genius who had entered Eton at eight years old and was in Italy studying the works of Galileo at fourteen. His seminal contributions to chemistry are cited in chapter 5. To physics, his new law was more than the accretion of greater knowledge and understanding. Its method of discovery and elaboration, which wedded the abstract to the

DE PROPRIIS EXPERIMENTIS. 109

CAPUT XXVII.

De Vase Vitreo, plures quàm viginti vel 50 & plures homines robustos attrahere valente.

Quædam Experimenta, quæ author exhibuit Cæsar. Majestati in Comitiis Ratisbonæ Anno 1654.

CUm in Comitiis Ratisponæ anno 1654 habitis, Reipubl. causâ degerem, & *Cæsar. Majestati* FERDINANDO TERTIO, ut & *Rom. Regiæ Majestati* FERDINANDO QUARTO, utriusque memoria sit beatissima, quorundam Experimentorum meorum, inter quæ ceu nova antecedens numerabatur, specimen jussus præberem, atque de illis prius edissererem, quod si quis hominum halitum oris sui in hoc vitrum *Recipiens nempe dictum* efflaret, ille simul animam esset exspiraturus; Princeps *de Auersberg, dictæ Regiæ Majestatis Aulæ Præfectus*, ante adduci non potuit, quò hisce verbis meis fidem haberet, quam ipse, quæ in lagenâ vitreâ facta essent, oculis suis vidisset. Sed cum etiam tunc temporis

O 3 alia

Although Torricelli had artificially created a vacuum in the top space vacated by the mercury in his glass barometer, the impressive reality of a vacuum was most clearly demonstrated by Otto von Guericke, burgomaster of Magdeburg. Independent of Torricelli, Guericke had long sought to refute Aristotle's dictum that nature abhors a vacuum. In the course of his experiments he invented the air pump and demonstrated the elasticity of air. This illustration shows the properties of a vacuum created by his pump. When a vacuum is made in the cylinder (to which a piston and a rope are attached) by connecting it to the glass sphere, Guericke says that not even a large group like this can prevent the piston from moving downward, lifting the men off the ground. *Experimenta nova (ut vocantur) magdeburgica de vacuo spatio,* 1672. Otto von Guericke.

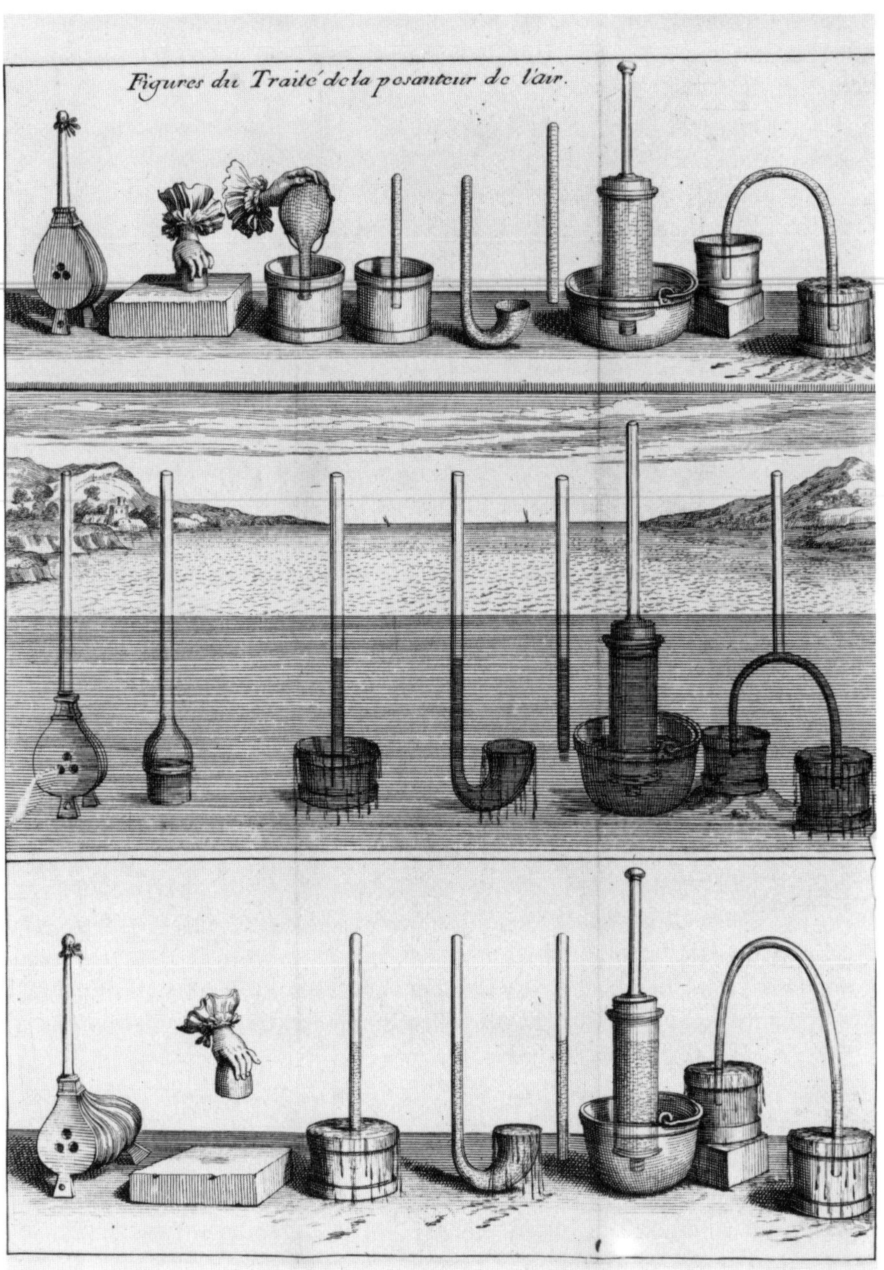

When Blaise Pascal heard of Torricelli's experiments, he first repeated them for himself and then went beyond them. Pascal reasoned that if the height of the mercury column was caused by the pressure or weight of the air, then the liquid ought to rise when brought to a higher elevation since logically there would appear to be less air above it pushing down. This experiment was conducted on the Puy de Dôme, and the difference of as much as three inches of mercury proved Pascal correct. Pascal wrote passionately of his great experiment, "which ravished us with admiration and astonishment." Here Pascal shows some of the experiments he conducted to demonstrate not only that air has weight but that it produces all the effects which had for so long been attributed to nature's supposed abhorrence of a vacuum. Both Pascal and Torricelli died at age thirty-nine. *Traitez de l'equilibre des liqueurs, et de la pesanteur de la masse de l'air,* 1663. Blaise Pascal.

Guericke was a politician and diplomat well before he became interested in science, so perhaps that accounts for the theatrical flair of this spectacular demonstration staged by him for the Emperor Ferdinand III. With his new air pump, Guericke evacuated all the air from the attached copper hemispheres in the center of the picture. Not even two teams of eight straining horses could pull them apart, vividly demonstrating the powerful force of air pressure. When the stopcock was opened allowing air into the sphere, it separated easily. Guericke's experiments helped demolish the long-held notion of "horror vacui." *Experimenta nova (ut vocantur) magdeburgica de vacuo spatio,* 1672. Otto von Guericke.

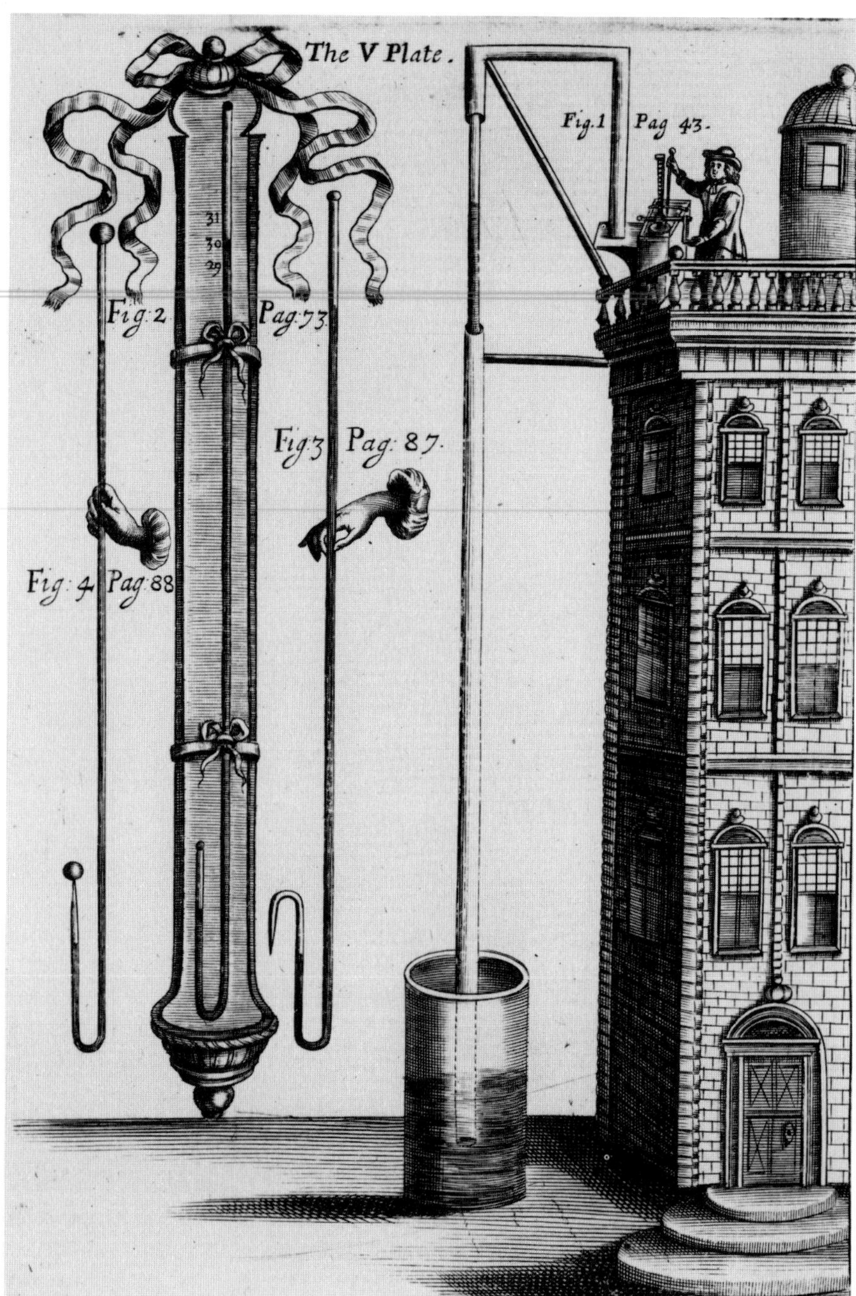

Working with the discoveries of Torricelli and Pascal and an improved version of Guericke's air pump, Robert Boyle performed experiments that enabled him to go beyond the limited working hypotheses of his predecessors and to forge broad, new conceptual schemes. This illustration shows two experiments he performed to prove that the height to which a fluid could be raised by suction or pressure varied inversely to its specific gravity. On the right he demonstrates how he stood on top of a four-story house and pumped water from a cistern through a glass tube to a height of thirty-three feet, beyond which it could not be raised. The illustration on the left shows his eight-foot mercury barometer, which obeyed the same laws as his water apparatus. *A Continuation of New Experiments, Physico-Mechanical, Touching the Spring and Weight of the Air, and Their Effects,* 1669-82. Robert Boyle.

experimental so fruitfully, was to become the mark of future physics and future science in general. Boyle's landmark experiments on the physical nature of air were models of rigor and completeness. The property of air that most impressed him was what he characterized as its "spring," that is, its ability to compress or expand like a coiled spring in response to a force exerted upon it. Boyle first reported his findings in 1660 in his *New Experiments Physico-Mechanicall, Touching the Spring of Air, and Its Effects,* published in Oxford. Boyle's views were immediately attacked by Thomas Hobbes and Franciscus Linus, and Boyle responded with a second edition in defense of his work. It was in this 1662 second edition of *New Experiments* that Boyle first made public his law. This marks one of the few cases in the history of science in which the second edition of a work became more important than the first. Although the Library has neither edition in its collections, it does have the 1680 *Nova experimente physico-mechanica* published in Latin in two small volumes in Geneva. It was through this edition that Boyle's work became universally known. A still-later edition in English (1669–82) contains this and some related works. The Library also has three separate collections of his works: a four-volume edition published in London just after his death, *The Works of the Honourable Robert Boyle* (1699–1700), a six-volume 1772 edition of the same title, and a 1680 *Opera varia* published in Geneva. The connections between Boyle, who died in the last decade of the seventeenth century, and Torricelli, born in its first decade, are of the sort that characterized that century's scientific revolution. Both men, as well as Pascal, Guericke, and many others, challenged the accepted—and unproven—values and notions of traditional science and achieved revolutionary scientific breakthroughs through rigorous experiment focusing on a single problem or phenomenon. The old qualitative physics of Aristotle was in fact and in deed coming to be replaced by the new quantitative physics of such heralds as these.

In a century that produced an astonishingly large number of great scientists, two of the greatest have not yet been discussed. Christiaan Huygens and Isaac Newton were these two great contemporaries. Huygens was a prodigiously gifted child mathematician who first made a name for himself by improving the telescope and then making several new astronomical discoveries. His greatest achievement, however, rested on his highly original studies of the dynamics of bodies in motion—studies that have their origin and elucidation in the seemingly pedestrian notion of the measurement of time. Up to his day, even the best clocks were not accurate enough to allow good astronomical observations, since no devices for keeping a constant periodic motion were known. Huygens therefore constructed the first pendulum clock, an achievement that required both great mathematical theorizing and superior mechanical ability. Huygens used Galileo's discovery of isochronicity (that a pendulum swings in constant time, irrespective of the width of its swing) and adapted it most ingeniously to the inner workings of a clock. Huygens patented his clock in 1657, but it was not until 1673 that he published his *Horologium oscillatorium* in Paris. This work not only gave a full description of his "grandfather clock" but offered its theoretical underpinnings as well. As a general work on dynamics it stands as a fundamental and highly original treatise. It concludes with thirteen theorems that deal with centrifugal force in circular motion—ideas that preceded Newton's work on universal gravitation. The Library has Huygens's 1673 *Horologium* in first edition. Also in its collections are the 1703 *Opuscula postuma,* published in Lyons eight years after his death, a 1728 *Opuscula* published in Amsterdam, and the only complete edition of Huygens's work, *Oeuvres complètes,* published in The Hague in twenty-two volumes (1888–1950). Because the *Horologium oscillatorium* not only offered physics a way to measure very

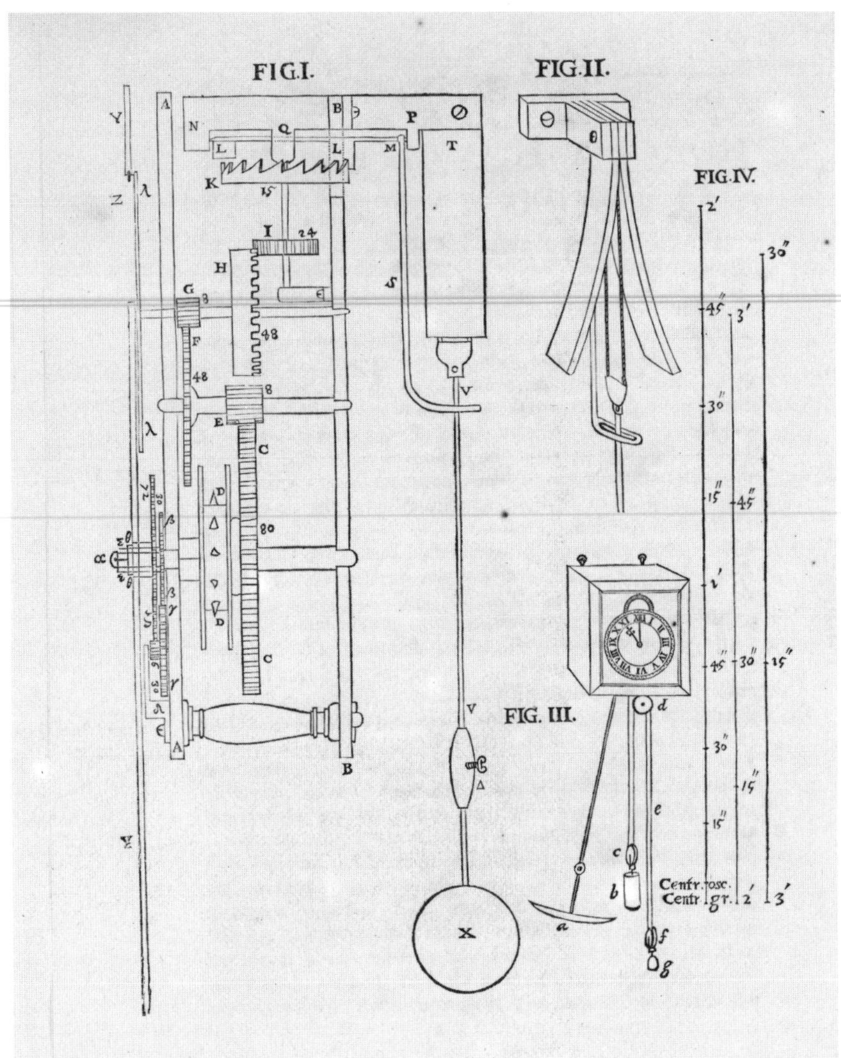

The realization of the first modern clock required the wedding of both superior theoretical and mechanical abilities—a description of the talents of Christiaan Huygens. This illustration of his pendulum or oscillating clock displays the two unique features that make it an accurate and revolutionary design. Figure I (K) shows the critical escapement mechanism that translates the regular motion of the pendulum to the wheelwork, and figure II shows the two curved metal strips that check the pendulum's swing. Not only did this treatise in which Huygens fully described his "grandfather clock" give science a method of accurately measuring very small units of time but it was also a highly theoretical work of the first order that made several fundamental contributions to the science of mechanics. *Horologium oscillatorium,* 1673. Christiaan Huygens.

small units of time but also was the first work to consider mathematically the action of bodies along curved paths, Huygens's work is ranked by some on a par with Newton's *Principia.*

But Newton is preeminent. For his brilliant and penetrating intellect and the scope and depth of his work, Isaac Newton stands unquestionably above the rest. It was his genius that united knowledge of the heavens and of the earth in the mathematics of what became known as classical physics. So comprehensive and fundamental was his work that it guided and formed the thought of thinkers for the next two centuries. Scientist and layman alike bore the intellectual stamp of Newton's world view. His legacy is stunning, prodigious, and fundamental—it includes the calculus, the composition of light, and the law of gravity. Any one of these discoveries would place an individual among science's all-time elite, but when seen as the work of one person they assume an even greater significance. Newton was a shy, complex man, sometimes arrogant and sometimes modest, who was unequivocally the greatest mathematical physicist in history as well as one of the best experimental practitioners. In his various discoveries he was at times pragmatic, at times synthetic, and at times most singular and daring. His invention of the calculus is treated in chapter 7. In physics, he is seen as both a radically speculative thinker and "a winnowing genius." In his decisive *Philosophiae naturalis principia mathematica* published in London in 1687, he is shown at his synthetic best, winnowing from the piecemeal work of his predecessors—Kepler's planetary laws, Descartes's mechanism, and Galileo's kinematics—and fabricating an essentially Newtonian universe. The *Principia* provided the great synthesis of the universe, proving that all aspects of the natural world, near and far, were subject to the same law of gravitation and that this universal law could be demonstrated in mathematical terms within a single physical theory. Furthermore, it achieved Newton's goal of uniting mathematics and mechanics, so that the latter could be considered mathematically as a systematic scientific discipline. The *Principia* thus proceeded systematically from definitions and axioms to theorems and finally to conclusions. It was those conclusions, his three laws of motion, which for the first time established the relationship of mass, force, and direction and offered the world a new, unified cosmology. Newton's great work linked the motions of planets with that of motion on earth, showing that the same laws rule everywhere. His *Principia,* written entirely in Latin, is difficult to the point of being intractable. It may be the most influential and significant book in the history of science and yet it is also probably the least read. The Library has the *Principia* in its 1687 first edition.

The Library also has Newton's first scientific publication, his famous letter on light, published in the Royal Society's *Philosophical Transactions* in 1672. This letter recounted his prism experiments of 1666, which led to his startling conclusions on the nature of light and color. Here in twelve pages, Newton presented his experimental findings on the relationship between light and color—findings that revealed for the first time the true nature of light. In this letter he recounted the classic experiment in which he let a ray of sunlight enter a darkened room through a small hole in a window shutter and then passed the ray through a prism onto a screen. The light was refracted and a band of consecutive colors in the order of the rainbow appeared on the screen. Many before Newton had gotten this far, but no one had gone further. Newton then passed each separate color through another prism and noted that although the light was refracted its color did not change. To this now complete experiment, Newton applied his genius and deduced from its results the essence of the phenomenon—that sunlight (or white light) consists of a combination of these colors. Here Newton is both the rigorous experimentalist and the speculative genius, defying the intuitively obvious and adhering instead to that which seems to contradict common sense. Newton's letter resulted in an ongoing debate with Hooke, Linus, and others whose arguments are found along with Newton's in the *Philosophical Transactions* for several years following 1672. Thirty-two years later, Newton gave a more comprehensive account of his work on the nature of light in his *Opticks,* published in London in 1704. In the first edition, which the Library has in its collec-

(3086)

about three foot radius (ſuppoſe a broad Object-glaſs of a three foot Teleſcope,) at the diſtance of about four or five foot from thence, through which all thoſe colours may at once be tranſmitted, and made by its Refraction to convene at a further diſtance of about ten or twelve feet. If at that diſtance you intercept this light with a ſheet of white paper, you will ſee the colours converted into whiteneſs again by being mingled. But it is requiſite, that the *Priſme* and *Lens* be placed ſteddy, and that the paper, on which the colours are caſt, be moved to and fro; for, by ſuch motion, you will not only find, at what diſtance the whiteneſs is moſt perfect, but alſo ſee, how the colours gradually convene, and vaniſh into whiteneſs, and afterwards having croſſed one another in that place where they compound Whiteneſs, are again diſſipated, and ſevered, and in an inverted order retain the ſame colours, which they had before they entered the compoſition. You may alſo ſee, that, if any of the Colours at the *Lens* be intercepted, the Whiteneſs will be changed into the other colours. And therefore, that the compoſition of whiteneſs be perfect, care muſt be taken, that none of the colours fall beſides the *Lens*.

In the annexed deſign of this Experiment, A B C expreſſeth the Priſm ſet endwiſe to ſight, cloſe by the hole F of the window

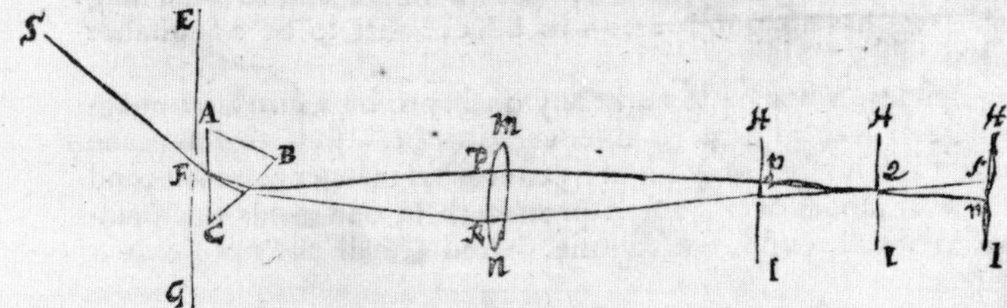

E G. Its vertical Angle A C B may conveniently be about 60 degrees: *M N* deſigneth the *Lens*. Its breadth $2\frac{1}{2}$ or 3 inches. S F one of the ſtreight lines, in which difform Rays may be conceived to flow ſucceſſively from the Sun. F P, and F R two of thoſe Rays unequally refracted, which the *Lens* makes to converge towards Q, and after decuſſation to diverge again. And H I the paper, at divers diſtances, on which the colours are projected: which in Q conſtitute *Whiteneſs*, but are *Red* and *Yellow* in R, r, and ϱ, and *Blew* and *Purple* in P, p, and ϖ.

If

In 1668, the twenty-six-year-old Newton invented the reflecting telescope. Three years later he offered an improved model to the Royal Society, which responded by electing him a fellow and publishing his account of the optical discovery that led him to his technical invention. In this simple diagram, Newton depicted his prism experiments of 1666, which led to his startling conclusion that "Light in itself is a Heterogeneous mixture of differently refrangible Rays." The triangle *ABC* is the first prism that separated the light into a spectrum of colors. The circle *MN* is the second prism, which refracted a single color beam but did not change its color. "A Letter . . . Containing his New Theory about Light and Colors," *Philosophical Transactions,* 1672. Isaac Newton.

tions, Newton begins by stating, "My Design in this Book is not to explain the Properties of Light by Hypotheses, but to propose and prove them by Reason and Experiments." He then proceeds to describe more than a dozen varied experiments that establish his theory beyond question. The *Opticks* appeared in three editions during his lifetime, and a fourth edition appeared shortly after his death. The book also offers his explanations of such optical phenomena as the rainbow, the double refraction of the Icelandic spar, and "Newton's rings." Unlike the *Principia,* it was written in English and is much more approachable, although still quite formal. Newton died in 1727 at the age of eighty-five. His many monumental accomplishments belie the fact that he actually spent much of his life

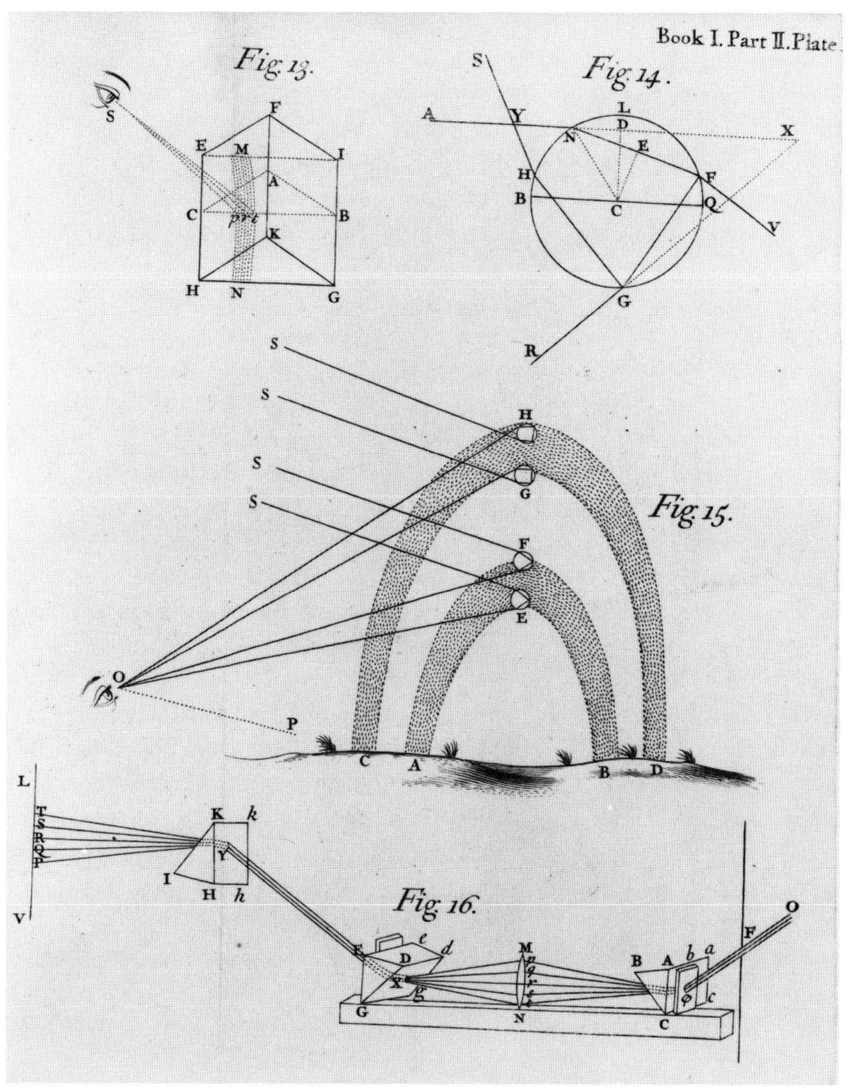

In 1704, Newton published a comprehensive work that presented his main discoveries and theories concerning the nature of light and color. As a revision of his optical lectures, the work reviews his earlier discoveries in a logical manner and explains such optical phenomena as the rainbow and the double Icelandic spar, as well as offering his corpuscular or emission theory of light. *Optics,* 1704. Isaac Newton.

dabbling in alchemy, speculating on theology, and functioning as a royal administrator. Nonetheless, his great impact was felt on both science and philosophy, and it was Newton who shaped the world picture of the next two centuries. With him, physics became a science that would strive for comprehensive explanations of natural phenomena within a single system of thought. The obligatory couplet by Alexander Pope captures the uniqueness and power of the Newtonian achievement:

> *Nature and Nature's laws lay hid in night:*
> *God said, Let Newton be! and all was light.*

"There could only be one Newton," Joseph Lagrange supposedly said to Napoleon, since "there was only one world to discover." In very much the same spirit is Alfred North Whitehead's well-known remark that all the mechanics of the eighteenth century was but a marginal note to Newton's *Principia.* Generally speaking, both eighteenth- and nineteenth-century

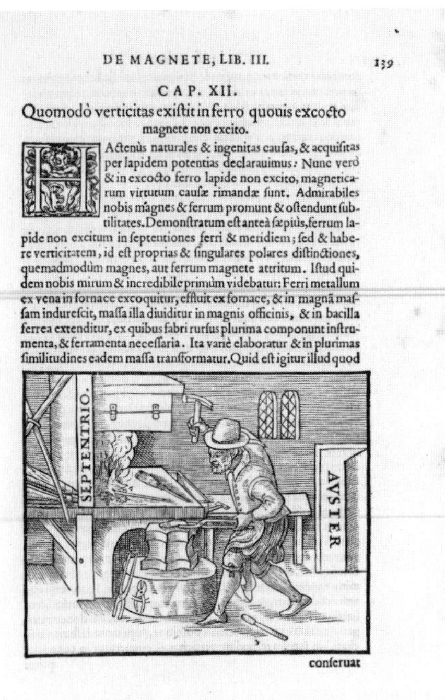

DE MAGNETE, LIB. III. 139

CAP. XII.

Quomodò verticitas exiſtit in ferro quouis excocto magnete non excito.

HActenùs naturales & ingenitas cauſas, & acquiſitas per lapidem potentias declarauimus: Nunc verò & in excocto ferro lapide non excito, magneticarum virtutum cauſæ rimandæ ſunt. Admirabiles nobis magnes & ferrum promunt & oſtendunt ſubtilitates. Demonſtratum eſt anteà ſæpiùs, ferrum lapide non excitum in ſeptentiones ferri & meridiem; ſed & habere verticitatem, id eſt proprias & ſingulares polares diſtinctiones, quemadmodùm magnes, aut ferrum magnete attritum. Iſtud quidem nobis mirum & incredibile primùm videbatur: Ferri metallum ex vena in fornace excoquitur, effluit ex fornace, & in magnã maſſam indureſcit, maſſa illa diuiditur in magnis officinis, & in bacilla ferrea extenditur, ex quibus fabri rurſus plurima componunt inſtrumenta, & ferramenta neceſſaria. Ita variè elaboratur & in plurimas ſimilitudines eadem maſſa transformatur. Quid eſt igitur illud quod

conſeruat

One of the most influential early works of the scientific revolution is William Gilbert's *De Magnete*. Studied by Bacon, Galileo, and Kepler alike, this work was not only the first serious study of electricity and magnetism but also a model of systematic investigation, with its emphasis on experimentation and its aversion to the easy generalization. Here, Gilbert shows how a heated iron bar is magnetized if hammered while aligned in an exact north-south (*septentrio-auster*) orientation. *De magnete*, 1600. William Gilbert.

physics were characterized more by the organization and elaboration of Newtonian physics than by any major conceptual innovations. The field that best exemplifies this application and elaboration of his basic laws and methods is that of electricity and magnetism. Historically, the two subjects had usually been considered together, although even in Newton's time no basic connection between the two had been established. But during the two centuries following Newton's famous prism experiment in 1666, an astounding story was evolving, whose conclusion would link not only the phenomena of electricity and magnetism but that of light as well, all within the scope of a single comprehensive theory.

Although the heyday of discovery for electricity and magnetism was certainly the eighteenth and nineteenth centuries, the scientific investigation of these phenomena actually was begun by a contemporary of Shakespeare's named William Gilbert. After Gilbert, there was little real development or discovery, aside from the peripheral researches of Otto von Guericke and Robert Boyle, until the electrical work of Benjamin Franklin in the mid-eighteenth century. William Gilbert was a physician by profession and served the English Crown in that capacity, having been appointed court physician to Queen Elizabeth I. Earlier, Gilbert had conducted a series of thoroughly detailed experiments on magnets, attempting to discover the real nature of magnetism, which was surrounded by legends. Conventional notions as to how a lodestone worked usually involved the magical and mystical. The Greeks believed magnetism to be a spirit trapped in iron, and during medieval times it was suspected to be a devil's trick. The prevailing idea during Gilbert's time was that a magnet stopped working when rubbed with garlic or when a diamond was near. A washing with goat's blood was believed to restore its power.

The results of Gilbert's work were published in 1600 in *De magnete,* a historic folio that was the first major scientific work based on systematic experimentation. The Library has a handsome first edition in its collections. Gilbert's great work was a harbinger of the "new science" that was eventually to lead to that century's scientific revolution. It was published in London, during the first year of the new century, and stood as a beacon to seventeenth-century scientists. All the greats of that century were influenced by Gilbert's methods, and to Galileo he was a particular hero—being the one person who, wrote Galileo, led the fight against that "pusillanimity of the mind which is content with repetition and spurns all innovations." Apart from his admirable methodology, Gilbert also contributed another critical dimension to the emerging scientific method—that of theoretical insight. At the core of his work was his theory that the earth itself was one great magnet—a landmark idea that exemplifies the extent of Gilbert's original genius. His creative experiments on his "terrella" (little earth), a lodestone ground into the spherical shape of the earth, explained why compass needles pointed not to the heavens, to the pole star, but roughly north and south to the "magnetic poles" of the earth. He noted that this magnetic influence was distributed throughout his terrella but was weakest at its equator and strongest at the poles. It was through careful and repeated experiment and observation that Gilbert was confident enough to translate his findings directly from his model to the earth itself. The epochal nature of both Gilbert's approach and his findings cannot be overemphasized—coming from an essentially sixteenth-century man, born but a year after the publication of the *De revolutionibus* of Copernicus. The courage, indeed the audacity, it took to regard the whole earth and its mysterious forces as the subject of his experiments and theories is most remarkable. Gilbert was well aware of how innovative his approach was, describing it as a "new style of philosophizing." He was openly contemptuous of those whose learning "comes only from books" and dedicated his work to the true philosophers who "in things themselves look for knowledge." No better description of the modern scientific method could be given.

Although Gilbert's work was concerned primarily with magnetism, he dealt with electricity in Book II of *De magnete* and coined the terms *electricity, electric force,* and *electric attraction.* His invention of the "versorium," a tiny rotating needle on a pivot, provided the first device designed to study electrical phenomena. For nearly two centuries after the publication of *De magnete,* the study of magnetism focused mainly on its practical aspects, such as mapping the earth's magnetic field. Electricity received somewhat more attention earlier, with devices such as Guericke's "frictional machine" being able to "make electricity." Eighteenth-century parlor tricks involving electricity offered divertissements similar to those supplied by the laughing-gas parties of the next century. But serious work was being done in both fields, and after each great discovery of the eighteenth and nineteenth centuries, there followed the mathematical physicists who gave to each field an intellectual order and unity through mathematical laws.

The first to investigate the phenomenon of electricity in the true spirit of Gilbert was Benjamin Franklin. Well before he became involved in politics and statecraft, Franklin had spent twenty years, from 1738, studying and experimenting with static electricity. This research led him to guess that atmospheric lightning was an electrical phenomenon similar to the spark produced by electric machines of his day. (These mechanical contrivances were similar to Guericke's frictional machines and generated small electrical charges when rotated quickly.) To test his hypothesis, in 1752 Franklin conducted his famous experiment in which he flew a kite with a wire connected in a thunderstorm and attached a key to its string. The key soon sparked and Franklin was able to charge a Leyden jar with electricity from the sky. This simple and very dangerous experiment provided formal proof that lightning was a form of electricity. Franklin, however, was lucky not to have been killed, as others who repeated his experiment sometimes were. In 1753, an experimenter in St. Petersburg achieved posthumous fame when, after he was struck dead by lightning attracted to his kite, the state of his organs was described in the publications of many a scientific society.

Throughout his career Franklin had kept in contact with the scholarly circles of Europe, and it was to Peter Collinson in London that Franklin first communicated his experimental work on electricity. The publication

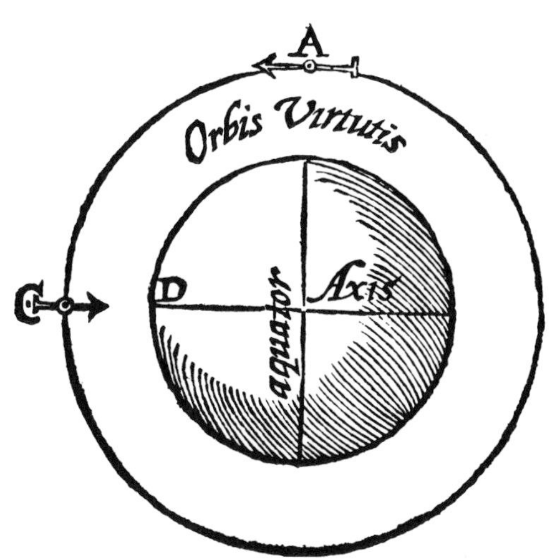

While repeating the thirteenth-century experiments of Petrus Peregrinus on spherical magnets, William Gilbert inferred the existence of the earth's magnetic properties. This led him to devise his "terrella," a lodestone ground into a spherical shape, upon which he could experiment. This "little earth," shown here, functioned as Gilbert's model from which he was able to transfer his findings directly to the earth itself. Gilbert's conclusion that the earth is one huge magnet served to explain many of its peculiarities of attraction and polarity. *De magnete*, 1600. William Gilbert.

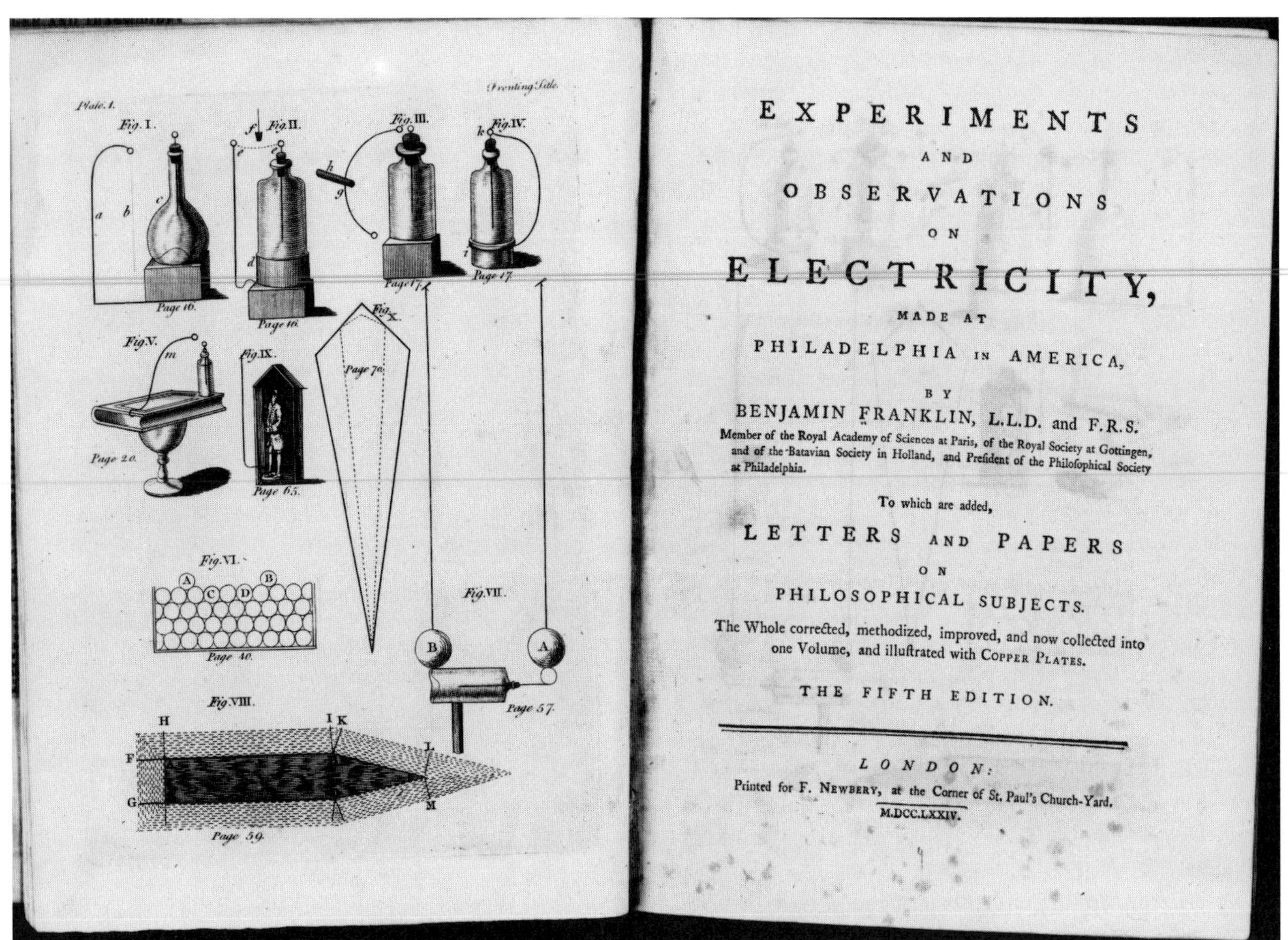

EXPERIMENTS AND OBSERVATIONS ON ELECTRICITY, MADE AT PHILADELPHIA IN AMERICA, BY BENJAMIN FRANKLIN, L.L.D. and F.R.S. Member of the Royal Academy of Sciences at Paris, of the Royal Society at Gottingen, and of the Batavian Society in Holland, and Prefident of the Philofophical Society at Philadelphia.

To which are added, LETTERS AND PAPERS ON PHILOSOPHICAL SUBJECTS.

The Whole corrected, methodized, improved, and now collected into one Volume, and illuftrated with COPPER PLATES.

THE FIFTH EDITION.

LONDON: Printed for F. NEWBERY, at the Corner of St. Paul's Church-Yard. M.DCC.LXXIV.

It was through the publication of this work, first communicated to Peter Collinson of London in a series of letters, that Benjamin Franklin became the first American to gain an international reputation through his scientific work. The book was translated into French the following year and within two years the main components of Franklin's electrical ideas were known to all of Europe. Although Franklin may be best remembered for his invention of the lightning rod and his experimental discovery that lightning is an electrical phenomenon, his foremost achievement was his theory of general electrical "action." *Experiments and Observations on Electricity,* 1774. Benjamin Franklin.

in London from 1751 to 1753 of these letters in *Experiments and Observations on Electricity Made at Philadelphia in America* established Franklin's international scientific reputation. The Library has this volume in first edition. Unquestionably it was the most significant scientific book to come out of eighteenth-century America. By 1769, five editions had been printed, as well as later translations into French, German, and Italian.

The ever-practical Franklin applied his discovery to the invention of the lightning rod—the first of which was erected in 1760 on the house of the Philadelphia banker Benjamin West. Although Franklin is popularly thought of today primarily as an experimentalist, his foremost achievement was the formulation of a theory of general electrical "action"—a theory that became widely accepted and earned him an international reputation in pure science.

Research in electricity soon became as popular internationally as Franklin's lightning rods were ubiquitous in Philadelphia. So it was that a professor of anatomy and obstetrics at Bologna who noticed the convulsive movement made by a frog's leg he was skinning when it was touched by a scalpel in certain ways chose to pursue the electrical aspects of this

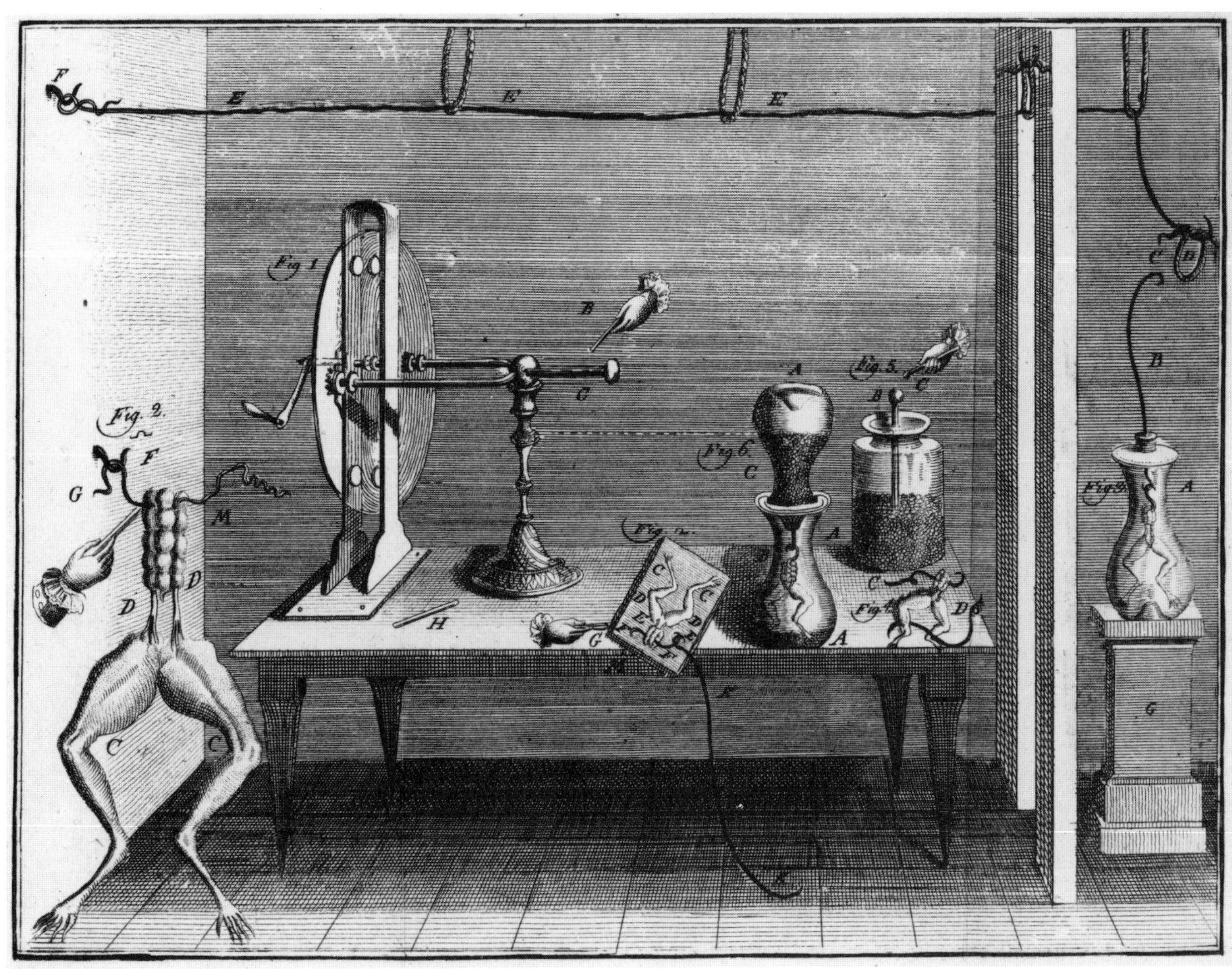

phenomenon. Luigi Galvani was enough of a scientist to know he had discovered something, but perhaps too much of a physiologist to go beyond the frog. Indeed, he claimed to have uncovered what he called "animal electricity"—stating that "in the animal itself there was an indwelling of electricity." His findings were first made public in 1791 in the transactions of the Bologna Academy of Science and in a separate private printing in Bologna the same year, entitled *De viribus electricitatis in motu musculari.* The Library has a facsimile copy of the 1791 edition and a 1793 German translation, *Abhandlung über die Krafte der thierischen Elektrizität auf die Bewegung der Muskeln* published in Prague.

Galvani's observations were startling, but it was not until his paper came into the hands of a countryman in Pavia that their real significance was appreciated. Alessandro Volta had occupied the chair of physics at the University of Pavia for twenty-five years and was a diligent electrical experimenter during that time. Intrigued by Galvani's assertion of animal electricity, Volta proceeded in the best tradition of scientific inquiry, at-

The claim of an Italian anatomist that he had proof of electricity flowing in animals generated enormous excitement throughout Europe. In his dissections of frogs, Luigi Galvani had noted how a frog's leg would twitch if touched with a scalpel. Although Galvani had discovered a phenomenon of enormous potential—that an electric current can be produced from the contact of two different metals in a moist environment—he wholly misinterpreted its real meaning. *Abhandlung über die Krafte der thierischen Elektrizität auf die Bewegung der Muskeln*, 1793. Luigi Galvani.

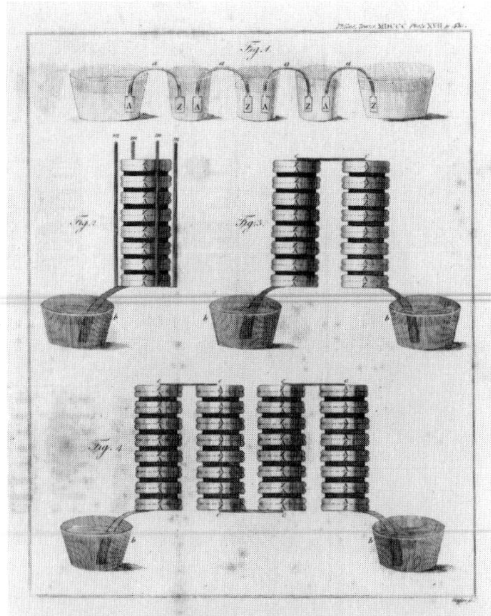

Alessandro Volta initially supported his countryman's new theory of animal electricity but became skeptical upon investigation. Within two years, Volta had discarded Galvani's conclusions altogether. In attempting to prove Galvani wrong, Volta performed the experiment without the animal and was able to demonstrate that the twitch of the frog's leg was caused by the contact of dissimilar metals that Galvani used to tie down the frog's leg. Volta then was able to produce and sustain an electric current by means of a "crown of cups" (figure 1). Figures 2, 3, and 4 show his piles—alternating layers of metals (zinc and copper) and saline-soaked leather pads—which also generated a constant current of electricity. The first battery had been invented and the electrical age had begun. "On the Electricity Excited by the Mere Contact of Conducting Substances of Different Kinds," *Philosophical Transactions*, 1800. Alessandro Volta.

tempting to discover the essentials of Galvani's phenomenon. Scores of experiments indicated that the factor common to all was the contact of different metals—and Volta was able to produce similar charges without the frog. At this point, the physicist took charge and Volta realized what Galvani had not—that a new source of continuous electric flow had been discovered in chemical action. The animal, he argued, had nothing to do with the production of electricity. Volta then used his condensing electrometer on different metals to measure the relative positive and negative charges of each. Once he knew the charge intensity of each metal, he arranged suitable pairs of metallic plates in order, separated by pieces of leather soaked in brine. This "pile" was to become the first battery—a device that for the first time could produce a continuous and controllable electric current. Volta had also grasped the idea of intensity, realizing that the larger the "pile," the stronger the current.

It was to the Royal Society that Volta first communicated his results, writing in French to Sir Joseph Banks, the society's president. The letter was dated March 20, 1800. It was translated by Tiberius Cavallo and was published in the society's *Transactions* the same year under the title "On the Electricity Excited by the Mere Contact of Conducting Substances of Different Kinds." The Library has a complete set of the society's *Philosophical Transactions* in its collections. With Volta's discovery, a source of natural energy had been summoned on command as it were, and the electrical age had begun.

Volta owed much of his success to the availability and the precision of his electrical measuring instrument, the condensing electrometer. In the best tradition of Huygens's oscillating clock and Volta's electrometer was the torsion balance of Charles Augustin de Coulomb. Without such a delicate and sensitive instrument, a complete understanding of electromagnetism would have been greatly delayed. Coulomb was a military engineer steeped in Newtonian thought who established his reputation with his invention of the torsion balance in 1777. His apparatus gave remarkably precise measurements of force. In 1784 Coulomb used it to measure the forces between electrically charged bodies and those between magnetic poles, and he discovered that for both the reaction is inversely proportional to the square of the distances between the reacting centers. And this relationship was identical to the one Newton described in his law of gravitation. Coulomb first published the results of his experiments in two *Mémoires de l'Académie royale des sciences* for the year 1785, published in 1788. The Library's collection of this journal lacks the volume for 1785, but Coulomb's two articles can be found in the Library in the *Collection de mémoires relatifs a la physique,* volume 1 of which is titled *Mémoires de Coulomb.* In addition to his two 1785 articles, this collection reproduces eleven of his more significant papers. It was Coulomb's accuracy that enabled him to establish mathematically the fundamental laws of electrostatics and magnetism and to demonstrate that they obeyed the same rules as Newton's gravitational forces. Nevertheless, neither Coulomb nor any of his contemporaries were able to define the real nature or relationship of such forces.

The nineteenth century could be described as the age of electricity, for it was during this brilliant period of discovery and achievement that the physical research of electricity and magnetism came not only to influence other disciplines (notably chemistry) but to lay the groundwork for the next century's revolutionary advances. In the relatively short span of forty-five years, the work of Oersted, Ampère, Faraday, and Maxwell transformed knowledge of the twin subjects of electricity and magnetism from a belief that no fundamental link existed between the two to a grandiose theory that embraced and unified not only electricity and magnetism but light as well.

This journey toward unification began at Copenhagen in the spring of 1820, with the classroom demonstration of Hans Christian Oersted, professor of natural philosophy. Oersted placed a compass needle near a wire carrying an electrical current and saw the needle move until it was at right angles to the wire. With this simple experiment he had demonstrated—and therefore finally established—an essential connection between electricity and magnetism. Well aware of the fundamental importance of his discovery, Oersted wrote a four-page paper for the *Journal für Chemie und Physik* in Latin, titled "Experimenta circa Effectum Conflictus Electrici in Acum Magneticam," and sent copies to the major scientific journals of Europe. Dated July 1820, this paper generated a frenzy of scientific activity, led directly to Ampère's great achievements, and heralded a new era in the history of physics. The Library does not have the *Journal für Chemie und Physik* in its collections but it does have Oersted's paper as it appeared translated in German, Italian, French, and English scientific journals of the same year.

Within four months of the publication of Oersted's paper, a Frenchman, André Marie Ampère, had taken Oersted's conclusions, conducted his own experiments, and published the first of a series of papers in the French Academy's *Annales de chimie et de physique,* Series 2, which would eventually lead to Ampère's being called "the Newton of electricity." In his 1820 paper, "De l'action mutuelle de deux courans électriques," Ampère argued that it was the current that produced the magnetic field—revealing that he had discovered the action of one current upon another current and realized that parallel currents in the same direction attract, while those in opposite directions repel. From this paper, which is in the Library's collections, Ampère proceeded to found the subject of electrodynamics, basing this study on the simple notion that electricity in motion generates a magnetic force. The genius of this unhappy scientist (Ampère lost his father to the Revolution's guillotine and his young wife to illness) and his accomplishments are apparent in his brilliant deduction of the quantitative physical laws of electrodynamics. In his *Théorie des phénomènes electrodynamiques,* published in Paris in 1826, Ampère offered the mathematical laws that governed the new field of electricity in motion or electrodynamics. The Library has a first edition copy of this most significant work, a large but thin book.

Ampère's work in electrodynamics led Michael Faraday, who became the greatest experimental physicist of all time, to focus on the nature and potential of electricity. Faraday's achievements are especially remarkable given his humble origins and his rudimentary education. Although assisted greatly by the famous chemist Humphrey Davy, Faraday nonetheless was a self-made scientist whose natural genius flourished when cultivated. As with many of his contemporaries, Faraday was impressed by Oersted's classical conversion of electricity into a magnetic force but, unlike them, he was driven to try the reverse—converting magnetism into electricity, and achieving what came to be known as electromagnetic induction. For ten years Faraday labored on this project and in 1831, for the first time, he created an induction current in a metal object by using a magnet. Further work led to the generation of a continuous current by rotating a round copper disk between two poles of a horseshoe magnet. Faraday had created the forerunners of both the dynamo and the transformer—devices that would come to power and interconnect the modern world. Faraday furthermore stated not only the laws governing the generation of these currents from magnets but the laws of electrolysis as well. But his interests went beyond these practical applications, to the very nature of electricity itself.

From his discovery of electromagnetic induction in 1831 through the late 1850s Faraday labored toward an essential understanding of electricity. In this pioneering work his lack of any real background in mathematics actually became a positive asset. As a mathematical innocent, Faraday had to think and to express himself in concrete terms, lending a sort of physical reality to his notions and concepts. Pursuing this fertile and intuitive mode of scientific investigation, Faraday demonstrated the unity and identity of all forms of electricity however produced—static, dynamic, or chemical. Furthermore, he came to regard electricity as a force or a field that was an essential part of physical reality. This personal manner of concrete visualization reflected Faraday's natural affinity for the physical aspects of the natural world. His intuitive imagination led one scientist to say Faraday could simply "smell the truth." Both Maxwell and Helmholtz later commented in wonder about how Faraday could discover, as Helmholtz said, "a large number of general theorems . . . by a kind of intuition, with the security of instinct, without the help of a single mathematical formula."

Faraday offered physics a new way of thinking about electricity and magnetism, and his fields or "lines of force" stimulated another great scientist, James Clerk Maxwell, to investigate Faraday's ideas mathematically. Faraday was a quiet, gentle, unassuming man whose entire life was devoted to science. In his last years he stopped work completely, realizing he was not at his best. At times his memory was so bad he forgot how to spell many words. Some have attributed this decline to low-grade poisoning contracted after years of electro-chemical experimentation. The lifework of this great physicist is well represented in the Library's collections. All of his papers to the Royal Society from 1831 to 1854 are available as they appeared in the society's *Philosophical Transactions.* These were subsequently collected and published in three volumes in London under the title *Experimental Researches in Electricity.* The Library has all three

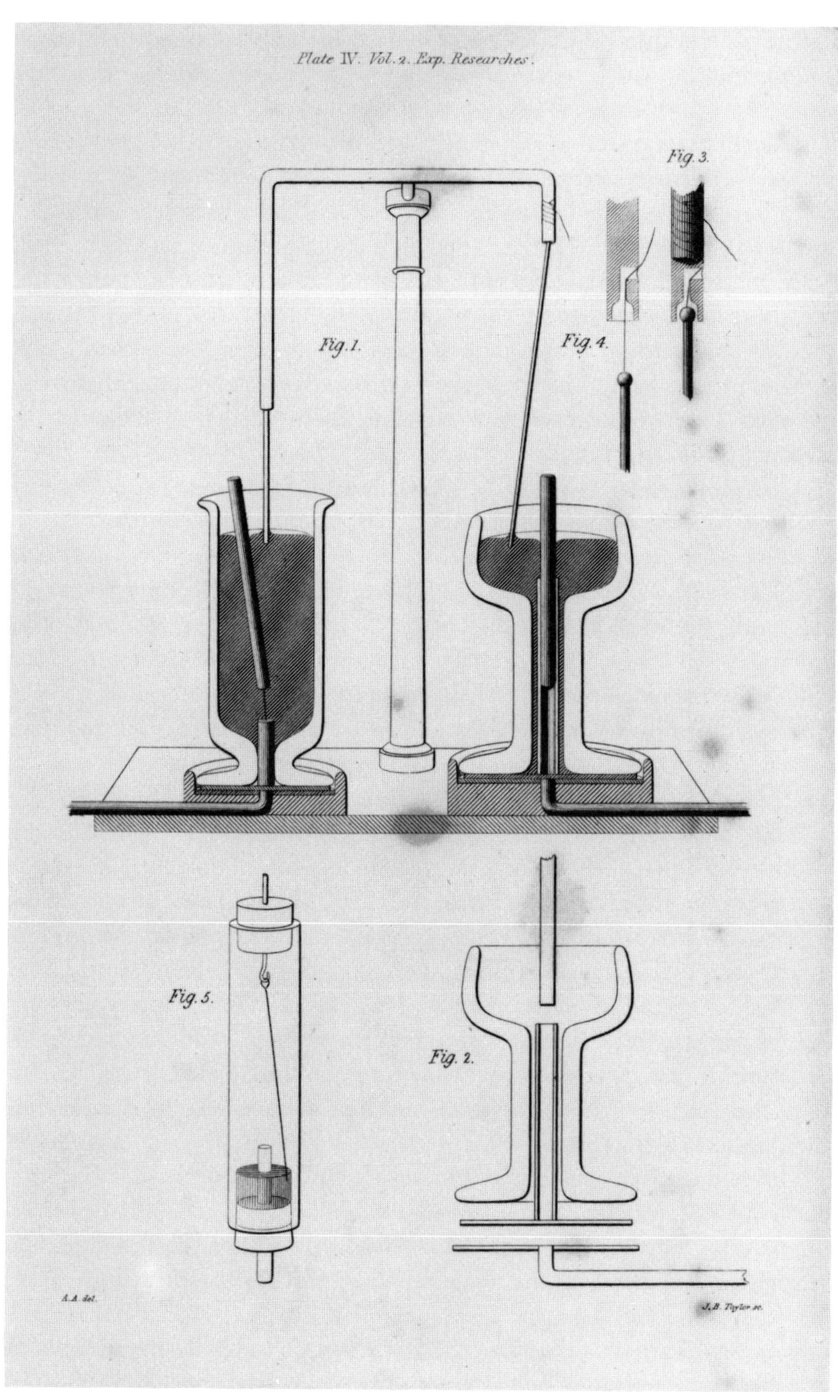

After Oersted's discovery and demonstration in 1820 that an electric current generates magnetism and the work of Ampère in the same year, which established the physical laws governing Oersted's discovery, Michael Faraday labored for ten years to produce the opposite effect—to convert magnetism into electricity. He achieved this in 1831, producing for the first time an induction current using a magnet. Faraday has been called the greatest experimental physicist of all time and in the apparatus shown here he demonstrates the conversion of electrical energy into mechanical rotation—the basis of what was to become the dynamo. In figure 1, a bar magnet in a beaker of mercury rotates around a current-carrying wire that also contacts the mercury. In figure 2, the magnet is fixed and it is the wire that rotates around the magnetic pole. In 1832, the American Joseph Henry independently discovered electromagnetic induction. *Experimental Researches in Electricity*, 1839-55. Michael Faraday.

volumes in first edition, each of which appeared in a different year—1839, 1844, and 1855.

When the Royal Society published James Clerk Maxwell's paper on "A Dynamical Theory of the Electromagnetic Field," almost two hundred years had passed since Isaac Newton had conducted his famous prism experiments on light. Upon its publication in 1865, Maxwell's seminal paper revealed an essential and heretofore unknown link between the

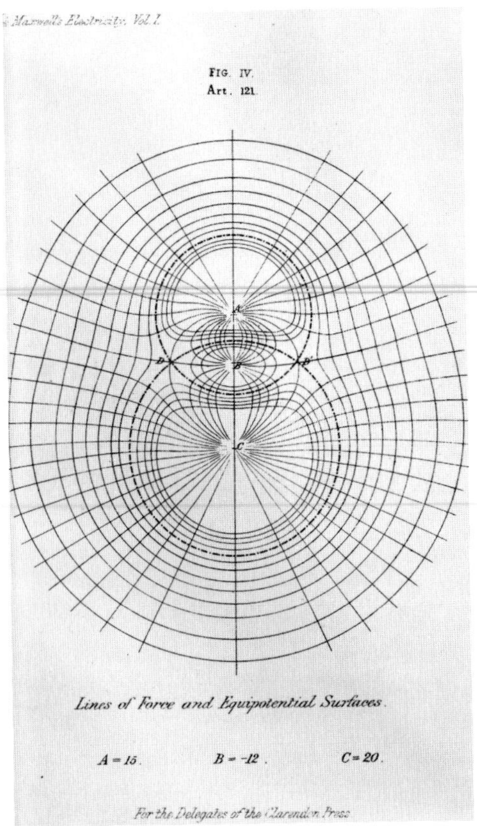

Where Faraday's intuitive genius produced what might be called concrete pictures, James Clerk Maxwell's superior gift was theoretical and mathematical. While investigating mathematically Faraday's ideas of a physical field of force, Maxwell noticed that his mathematical calculations for the transmission speed of both electromagnetic and electrostatic waves were the same as the known speed of light. He then offered the hypothesis, later proven experimentally by Heinrich Hertz, that identified light as an electromagnetic phenomenon and that brought together the three main fields of physics—electricity, magnetism, and light. Here, in a manner somewhat uncharacteristic of the mathematically inclined Maxwell, he shows the lines of force of three bodies of unequal size (*A, B,* and C) with each forming a distinct system in the ether of outer space. *A Treatise on Electricity and Magnetism*, 1873. James Clerk Maxwell.

nature of light (the subject of Newton's researches) and that of electricity and magnetism—subjects which seemingly bore no relation to each other. In the paper, which the Library has in its volumes of *Philosophical Transactions,* Maxwell achieved a classic synthesis of three main fields of physics—electricity, magnetism, and light—by showing that light was simply a form of electromagnetic radiation. Unlike the intuitive genius of his predecessor Faraday, Maxwell's genius was mathematical and theoretical. Only because he reduced Faraday's ideas and experiments to mathematical form was he able to achieve the breakthrough in knowledge that showed the way to this grandly comprehensive theory. In his mathematical construction of Faraday's theory of the field, Maxwell abandoned all mechanical models and analogies and expressed field theory instead in twenty elegant yet simple mathematical equations.

After accomplishing this in 1865, Maxwell then retreated to his estate in Scotland to devote himself to the implications and elaboration of his equations. The result was his famous two-volume opus entitled *A Treatise on Electricity and Magnetism.* The Library has the 1873 first edition of this work published in Oxford. There Maxwell pulled together and explained a vast range of diverse physical phenomena by his electromagnetic theory. In his mathematical translation and elaboration of Faraday's work, Maxwell quantified what Faraday had not, and he therefore noticed that his calculations for the speed of transmission of both electrostatic and electromagnetic waves were the same as the known speed of light. He quickly surmised that this was more than coincidence and concluded in his 1873 treatise that "light is an electromagnetic disturbance propagated according to electromagnetic laws." In 1873, Maxwell's identification of light as an electromagnetic phenomenon was still regarded as a hypothesis. Fifteen years later, it was proved experimentally by Heinrich Hertz. The Library has the 1892 Leipzig edition of Hertz's *Untersuchungen über die Ausbreitung der elektrischen Kraft,* the first published collection of his experiments. Hertz's radio apparatus not only was able to detect the predicted electromagnetic waves of Maxwell but was able to demonstrate their optical properties as well.

Maxwell's genius was broad and his contributions to physics were many. Indeed, almost the last thing he did before he died was to edit the unpublished electrical experiments of Henry Cavendish. The eccentric and reclusive Cavendish had anticipated in the early 1770s many of the major discoveries in electricity of the next half century. The Library has a first edition of Maxwell's book *The Electrical Researches of Henry Cavendish,* published in Cambridge in 1879, the year of Maxwell's death.

Like Faraday (to whom he always acknowledged a debt), Maxwell was one of science's good men—"a man of unusual sweetness and light," said C. P. Snow. He died of cancer at forty-eight. His work ranks with that of the truly great of physics, having made the physical universe more comprehensible. It also stands in transition between the old orderly Newtonian view of the universe and the new relative view of Einstein.

The electromagnetic discoveries made by Faraday and Maxwell were more than simply symbols of this transition, for although they represented one of the pinnacles of Newtonian or classical physics, they also contained the seeds of its eclipse. Ironically, it was Maxwell's theory of electromagnetism that served as the source of two new systems of physical thought—relativity and quantum physics—that were to revolutionize all of physics. Indeed, Einstein's 1905 paper on relativity was called, in translation, "On the Electrodynamics of Moving Bodies," a title that shows that Maxwell's electromagnetism was very much on his mind.

Einstein's revolutionary idea was offered seemingly in ignorance of an experiment that implied a fundamental incompatibility between Maxwell's theory and Newtonian mechanics. This discovery was revealed by a most famous failure. The traditional notion of the ether (or aether) dictated that an invisible medium was necessary for the transmission of electromagnetic waves, much as air allowed the transmission of sound waves. The inability of the Michelson-Morley experiment to detect any "ether-drift" proved conclusively that the speed of light relative to the earth was constant in all directions. The speed of the earth itself obviously had no effect on the velocity of light. Albert Michelson and Edward Morley, both Americans, published their results in the December 1887 issue of the *Philosophical Magazine,* a science journal published in London (and the very same in which Niels Bohr would publish his classic series of papers in 1913). The Library of Congress collections include a complete set of this journal. Despite its rococo title, the Michelson-Morley paper "On the Relative Motion of the Earth and the Luminiferous Aether" was destined to be a jarring harbinger of the coming modern era. Though it has become folklore that their experiments played a crucial role in the evolution of Einstein's thinking on relativity, Einstein himself denied this supposedly substantive, sequential connection.

So it was, nevertheless, that Einstein began his work not only by discarding the notion that any medium such as the traditional "ether" existed but by postulating that the velocity of light was a universal constant independent of the motion of either the observer or the source. These were daringly original hypotheses, decidedly incompatible with both the old ideas of space as a thing in itself (the ether) and the classic assumptions of Newtonian mechanics. Einstein's special theory of relativity, in dismissing the ether as a reference point through which light moved, postulated that there was nothing in the universe that could be at absolute rest or in absolute motion. Since space itself was no longer a thing and had no fixed point in it, all values became relative to the state of motion of the observer.

It was in his classic paper of 1905, "Zur Elektrodynamik bewegter Körper," published in the *Annalen der Physik* in Leipzig, that Einstein first shook the certitude of the old physics not only with his revolutionary conclusion that it was impossible to determine the absolute velocity of an isolated object (all we could know was its relative speed), but also with his realization that energy and mass are equivalent, as expressed in his famous equation $e = mc^2$. In this regard, since the only constant against which we can measure things is the velocity of light, the traditional concept of mass (which had been thought of as constant for any given body) as a fixed,

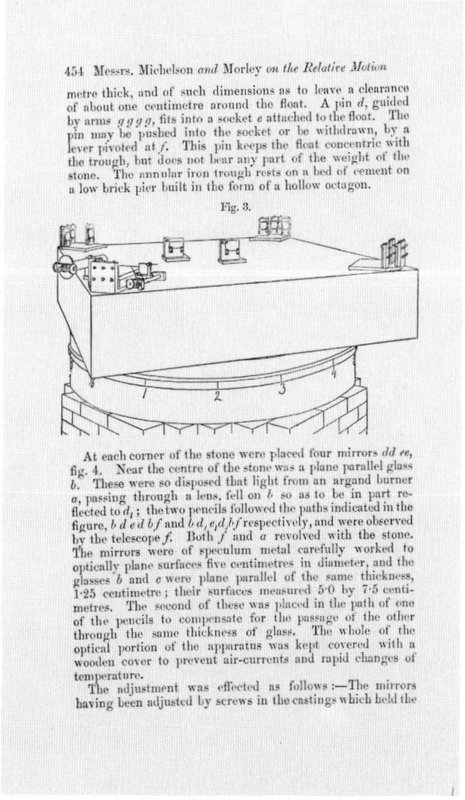

454 Messrs. Michelson *and* Morley *on the Relative Motion*

metre thick, and of such dimensions as to leave a clearance of about one centimetre around the float. A pin *d*, guided by arms *g g g g*, fits into a socket *e* attached to the float. The pin may be pushed into the socket or be withdrawn, by a lever pivoted at *f*. This pin keeps the float concentric with the trough, but does not bear any part of the weight of the stone. The annular iron trough rests on a bed of cement on a low brick pier built in the form of a hollow octagon.

Fig. 3.

At each corner of the stone were placed four mirrors *dd ee*, fig. 4. Near the centre of the stone was a plane parallel glass *b*. These were so disposed that light from an argand burner *a*, passing through a lens, fell on *b* so as to be in part reflected to d_1; the two pencils followed the paths indicated in the figure, *b d e d b f* and $b\,d_1 e_1 d_1 b f$ respectively, and were observed by the telescope *f*. Both *f* and *a* revolved with the stone. The mirrors were of speculum metal carefully worked to optically plane surfaces five centimetres in diameter, and the glasses *b* and *c* were plane parallel of the same thickness, 1·25 centimetre; their surfaces measured 5·0 by 7·5 centimetres. The second of these was placed in the path of one of the pencils to compensate for the passage of the other through the same thickness of glass. The whole of the optical portion of the apparatus was kept covered with a wooden cover to prevent air-currents and rapid changes of temperature.

The adjustment was effected as follows:—The mirrors having been adjusted by screws in the castings which held the

Like Maxwell, the scientists Michelson and Morley shared the traditional notion that as air was essential to the transmission of sound waves so the luminiferous ether was necessary for the transmission of electromagnetic waves. It was with this sensitive and highly precise interferometer and four mirrors mounted on a large stone floating on mercury (fig. 3) that Michelson and Morely planned to demonstrate the existence of this invisible medium. The experimenters assumed that the earth moved through a stable ether medium and that a light beam sent out in the same direction as the earth's motion should move faster than one sent out at right angles to it. The shocking results showed, however, not the slightest variation in the velocity of light. Accounting for this constancy of the speed of light led to both the final dismissal of the notion of the ether and the beginning of modern physics. "On the Relative Motion of the Earth and the Luminiferous Aether," *Philosophical Magazine,* 1887. Albert A. Michelson and Edward W. Morley.

The following pages are a copy of my first paper concerning the theorie of relativity. I made this copy in November 1943. The original manuscript not longer exists having been discarded by me after its publication. The publication bears the title „Zur Elektrodynamik bewegter Körper (Annalen der Physik; vierte Folge, Vol 17. 1905.)

A. Einstein. 21. XI. 1943.

inert thing now became the concept of a thing as potential energy, into which mass could be transformed under the proper circumstances. The velocity of light provided those circumstances and became the modern magic wand by which matter could disappear and change into energy or through which matter or the mass of a body would increase. If its speed, and therefore its energy, were to increase, mass would be transformed. Mass and energy were thus seen as different aspects of the same phenomenon, the transformation occurring only when the speed of an object approached that of the velocity of light. This new, dynamic concept of matter was the basis for the release of atomic energy.

In his 1905 paper, Einstein did not consider gravitation. It was only after many more years of intense research that he offered an equation that linked relativity and gravitation. So, in another landmark paper, "Die Grundlage der allgemeinen Relativitätstheorie," published in 1916 in the *Annalen der Physik,* Einstein presented what came to be known as his general theory of relativity. In the general theory, Einstein treats the universe as a whole and thus includes every type of motion. It was through the general theory, which did away with the concept of gravitation as a force, that classical physics was finally eclipsed. Whereas Newton had regarded gravity as a force, Einstein showed that the space around a planet or body is a gravitational field similar to the magnetic field around a magnet. Thus it is the matter distributed throughout space that determines the mechanical behavior of the bodies moving within it. And gravitation is not a force but a curved field in the space-time continuum, created by the presence of this mass. A great mass like our sun is surrounded by an enormous gravitational field, so large that it can even bend light. Einstein's theory was proved on March 29, 1919, when photographs taken of a total eclipse of the sun showed that starlight that passed close by the sun traveled in curves rather than in straight lines.

By 1919, the profound genius of Einstein was recognized throughout the world—his radical reformulations of physics having given the world a new

(16)

II. Elektrodynamischer Teil.

§ 6. Transformation der Maxwell-Hertz'schen Gleichungen für den leeren Raum. Über die Natur der bei Bewegung in einem Magnetfeld auftretenden elektromotorischen Kräfte.

Die Maxwell-Hertz-schen Gleichungen mögen gültig sein für das ruhende System K, so dass gelten möge,

$$\frac{1}{V}\frac{\partial X}{\partial t} = \frac{\partial N}{\partial y} - \frac{\partial M}{\partial z} \qquad \frac{1}{V}\frac{\partial L}{\partial t} = \frac{\partial Y}{\partial z} - \frac{\partial Z}{\partial y}$$
$$\frac{1}{V}\frac{\partial Y}{\partial t} = \frac{\partial L}{\partial z} - \frac{\partial N}{\partial x} \qquad \frac{1}{V}\frac{\partial M}{\partial t} = \frac{\partial Z}{\partial x} - \frac{\partial X}{\partial z}$$
$$\frac{1}{V}\frac{\partial Z}{\partial t} = \frac{\partial M}{\partial x} - \frac{\partial L}{\partial y} \qquad \frac{1}{V}\frac{\partial N}{\partial t} = \frac{\partial X}{\partial y} - \frac{\partial Y}{\partial x},$$

wobei (X, Y, Z) den Vektor der elektrischen, (L, M, N) den der magnetischen Kraft bedeutet.

Wenden wir auf diese Gleichungen die in § 3 entwickelte Transformation an, indem wir die elektromagnetischen Vorgänge auf das dort eingeführte, mit der Geschwindigkeit v bewegte Koordinatensystem beziehen, so erhalten wir die Gleichungen:

$$\frac{1}{V}\frac{\partial X}{\partial \tau} = \frac{\partial}{\partial \eta}\beta\left(N - \frac{v}{V}Y\right) - \frac{\partial}{\partial \zeta}\left\{\beta\left(M + \frac{v}{V}Z\right)\right\}$$
$$\frac{1}{V}\frac{\partial \beta\left(Y - \frac{v}{V}N\right)}{\partial \tau} = \frac{\partial L}{\partial \zeta} - \frac{\partial}{\partial \xi}\beta\left(N - \frac{v}{V}Y\right)$$
$$\frac{1}{V}\frac{\partial \beta\left(Z + \frac{v}{V}M\right)}{\partial \tau} = \frac{\partial}{\partial \xi}\beta\left(M + \frac{v}{V}Z\right) - \frac{\partial L}{\partial \eta}$$

$$\frac{1}{V}\frac{\partial L}{\partial \tau} = \frac{\partial}{\partial \zeta}\left\{\beta\left(Y - \frac{v}{V}N\right)\right\} - \frac{\partial}{\partial \eta}\beta\left(Z + \frac{v}{V}M\right)$$
$$\frac{1}{V}\frac{\partial \beta\left(M + \frac{v}{V}Z\right)}{\partial \tau} = \frac{\partial}{\partial \xi}\beta\left(Z + \frac{v}{V}M\right) - \frac{\partial X}{\partial \zeta}$$
$$\frac{1}{V}\frac{\partial \beta\left(N - \frac{v}{V}Y\right)}{\partial \tau} = \frac{\partial X}{\partial \eta} - \frac{\partial}{\partial \xi}\beta\left(Y - \frac{v}{V}N\right),$$

wobei

$$\beta = \frac{1}{\sqrt{1 - \frac{v^2}{V^2}}}.$$

In a startling mental turnabout, Albert Einstein did not seek to explain why the velocity of light remained unchanged but rather postulated it as a universal constant, independent of the state of motion of the observer or that of the light source. The revolutionary new world-picture of his special relativity theory posited no ether at all and thus no fixed point in the universe—the only absolute, invariant entity being the constancy of the velocity of light. Further, it transformed the traditional concept of mass from a fixed, inert thing into something that could be transformed into energy ($E = MC^2$)—since mass and energy were but different aspects of the same phenomenon. Einstein's theories profoundly modified the Newtonian world-picture of classical physics and established a new system of physical thought. This page is from Einstein's 1905 paper on relativity rewritten in his own hand in 1943. "Zur Elektrodynamik bewegter Körper," 1905 (1943). Albert Einstein.

and quite startling way of perceiving itself. Concepts of absolute space and time were abolished, with only the velocity of light remaining as an absolute magnitude. Intervals of time became relative quantities that depended on the relative motion of the observer and the observed. Even the concept of synchronous events became relative. These implications of Einstein's special theory were of immense importance to physics, whereas his general theory produced effects that were more philosophical or cosmological. Indeed, the general theory has yielded only three predicted effects: the deflection of light by the sun's gravitational field, the red shift of light from distant stars, and the shift of the perihelion of Mercury.

Einstein revolutionized physics to its core, altering its most primary conceptions. As a scientist, he was the most respected man of his time—the Newton of the modern era. As a man, he had a warm, almost transcendent personality, described by C. P. Snow as "rather like an Old Testament prophet, or else like a benign deity being patient with human stupidity . . . ." From an inauspicious beginning as an indifferent student and a junior official in a Swiss patent office, Einstein soared to unprecedented intellectual heights and took the rest of the world with him.

Although volume 17 (1905) and volume 19 (1916) of the *Annalen der Physik* are missing from the Library's collection of that journal, both papers are in the Library's collections reproduced in a 1922 work, *Das Relativitätsprinzip,* edited by H. A. Lorentz. The Library does, however, have a unique copy of Einstein's 1905 paper. Its collections contain a manuscript version written by Einstein himself in 1943. Einstein had discarded the original manuscript he composed in 1905, but he rewrote the paper in 1943 to assist the United States in its wartime fund-raising efforts.

Besides Einstein's relativity theory, the second great system of physical thought responsible for the major upheavals in twentieth-century physics is the quantum theory. Together these two theories describe the laws of an altogether unseen and heretofore unknown world—the world of the very fast and the very small. The laws of this micro-universe are as extraordinary as those of classical physics are commonsensible. Its domain is not that of a scaled-down version of the known physical world but rather that of a unique, abstract kingdom whose language is mathematics and whose truths are found in contradiction. The journey to this modern wonderland began in the orderly researches and discoveries of the nineteenth century—specifically those of Faraday and Maxwell—and involve one of the most important coordinating concepts in all of physics, that of the wave concept.

In 1887, when Heinrich Hertz proved by experiment Maxwell's prediction that electricity is propagated in wave formations, he began an assault on the electromagnetic spectrum that led eventually to the classification of waves (from the slowly oscillating waves of radio, to the increasingly rapid waves of television, radar, infrared, ultraviolet, X rays, gamma rays, and the stupendously high frequency oscillations of cosmic rays). Hertz's discovery led directly to radio communication (Marconi's "wireless telegraphy") but also indirectly to Röntgen's "new kind of rays," later called X

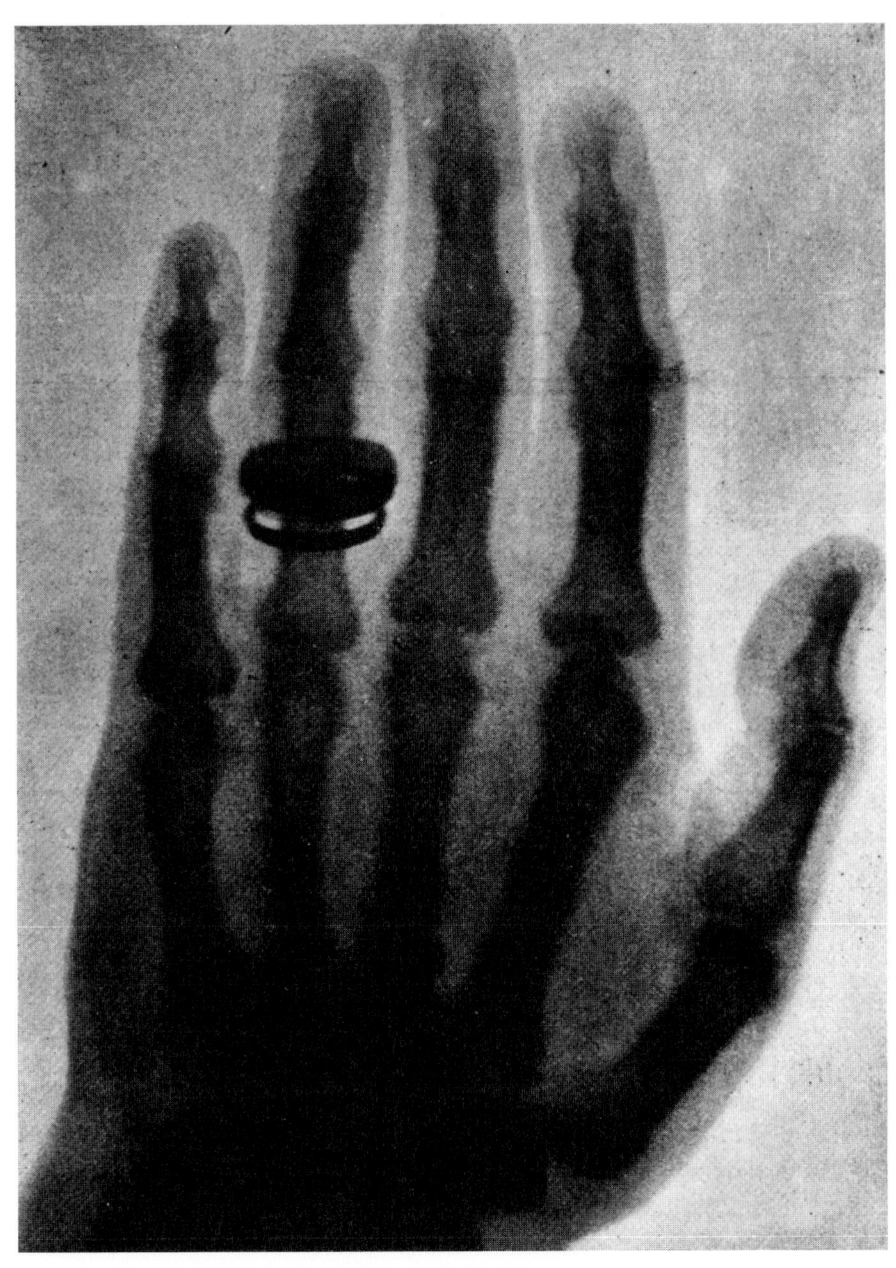

In his studies on the penetrating power of cathode rays, Wilhelm Röntgen noticed (as several had before him) that a nearby phosphorescent screen glowed and photographic plates fogged when his cathode ray tubes were operating. Upon investigating these strange and unknown rays (hence the name, X ray), he found that they traveled in straight lines for about six feet and could not be refracted or reflected or deviated by a magnet. Even more surprising were their penetrating properties, which he discovered most dramatically when the rays penetrated to the bones in his own fingers, as he held the objects he photographed with X rays. In his first public lecture after publishing his discovery, he asked for a volunteer to be X rayed. This is the historic photograph of the hand of his colleague Professor Albert von Kölliker, an octogenarian. Such a magical sight caught the imagination of a fin-de-siècle world and caused a sensation even in the scientific community. "Ueber eine neue Art von Strahlen," *Sitzungsberichte der Physikalisch-medicinische Gesellschaft,* 1895. Wilhelm Conrad Röntgen.

rays. In 1895, Wilhelm Röntgen was carrying out some experiments with cathode rays when he noticed that photographic plates fogged when placed near electrical discharges. Upon investigating, he noticed that certain objects cast a shadow from this new form of radiation, whereas others permitted it to penetrate them. Further experiments showed that these strange rays given off by the cathode ray tube passed through flesh but not through bone—making it possible to photograph bones. In December 1895 he delivered a paper entitled "Ueber eine neue Art von Strahlen" and the world first learned of X rays. The Library has Röntgen's famous paper as it first appeared in two parts in the *Sitzungsberichte der Physikalisch-medicinische Gesellschaft zu Würzburg* of 1895–96. The significance to medicine and industry of the applications of his discovery are well known, but to physics they had an even greater importance, since they would lead directly to Henri Becquerel's discovery of radioactivity and J. J. Thomson's enunciation of the electron theory.

Becquerel came from a family that had contributed much to science, and he was a professor at the École Polytechnique in Paris when he made one of the most famous and remarkable "accidental" discoveries in the history of science. Trying to determine the nature of Röntgen's new rays, Becquerel experimented to see whether the rays were at all related to the phosphorescent properties of certain substances. His planned experiment in 1896 was to place a silver coin between uranium salts and a photographic plate wrapped in thick black paper, exposing the entire package to the sun. To his astonishment, he discovered the coin's shadow on the plate despite the fact that it had not been exposed to any light. Becquerel communicated his findings in the *Comptes rendues* of 1896, which the Library has in its collections. In that paper, entitled "Sur les radiations invisibles émises par les sels d'uranium," Becquerel drew the correct conclusions that this new radiation was a characteristic atomic property of uranium and was quite independent of its chemical composition. He published four other papers in *Comptes rendus* that year. In 1903, Becquerel published a definitive summary of his work, entitled *Recherches sur une propriété nouvelle de la matière.* The Library has this work as it appeared in an extended mémoire under the same title in *Mémoires de l'Academie des Sciences* (1903).

The name Becquerel is always followed by that of Curie. As friends and colleagues of Becquerel, Pierre and Marie Curie pursued the implications of his discoveries. Marie Curie (born Marie Sklodowska) was the daughter of a physics professor in Warsaw. She married Pierre in 1895. After Becquerel's discovery, they pursued together the idea that the radiation was not contained in the uranium ore but in an unknown material or element concealed within the ore. After two years of strenuous physical labor spent refining a ton of pitchblend (the ore from which uranium is extracted), they isolated a few milligrams of two new radioactive substances, which they named polonium and radium. In 1903, Marie Curie wrote and submitted her doctoral dissertation, *Recherches sur les substances radioactives,* which contained the results of her work. It certainly must rank among the most significant contributions to science ever made toward the fulfillment of requirements for a degree. A book of the same title was published in Paris in 1903, and the Library has the 1904 edition. Madame Curie was surely one of the saints of science. She was a sincerely dedicated, selfless, and brilliant scientist who made no attempt to patent any of the practical applications of her discoveries. She died of leukemia, presumably caused by her years of exposure to radiation. The Library also has in first edition a classic account of her work in her collected papers, *Traité de radioactivité,* published in Paris in 1910.

The pioneering work of Marie Curie and her husband on the nature and origins of radioactivity generated a great deal of enthusiasm for the whole subject, and the next challenge to physics presented itself clearly—to explain the nature of radioactivity. To Ernest Rutherford, such a challenge did not seem insurmountable. A man of intuitive experimental genius, Rutherford came to study at Cambridge University from New Zealand, the son of a Scottish farmer who had emigrated there. While teaching at McGill University in Montreal early in his career, he collaborated with a pupil, Frederick Soddy, in demonstrating that radiation was the by-product of an entirely spontaneous and continuous disintegration of the atomic nuclei of certain heavy elements, like uranium. Put another way, radioactivity was the result of nature spontaneously transforming one element into another. Rutherford and Soddy's experiments were first published in the *Philosophical Magazine* of September and November 1902, in a two-part article called "The Cause and Nature of Radioactivity." During the next year, Rutherford published four articles with Soddy and three on his own, all in the *Philosophical Magazine.* The Library's set of this journal is complete.

In 1904, Rutherford's landmark work, *Radio-activity,* was published in Cambridge. The Library has this volume in first edition. Rutherford continued his work at Manchester University and later at Cambridge, and he became one of the fathers of atomic physics. It was through his famous gold foil experiment that he came to postulate a theory of the nuclear nature of the atom—namely, the theory that the greatest mass of the atom was concentrated in a minute, positively charged central nucleus, around which electrons revolved. Rutherford's model of the atom was a brilliant conception, echoing the model of the sun and its orbiting planets. In 1919, Rutherford again made history by successfully inducing the first artificial transformation of one element into another—that is, he split the atom. The release of energy observed demonstrated the equivalence of mass and energy postulated by Einstein fourteen years earlier. Rutherford published his findings, "Collision of α Particles with Light Atoms," in the *Philosophical Magazine* of 1919.

The discoveries of Einstein, Becquerel, and the Curies and the early ones of Rutherford all occurred during the first decade after Max Planck had enunciated the quantum theory. When first made public in 1900, this revolutionary idea was necessarily regarded as Planck's quantum hypothesis, and it was not considered especially radical or untraditional. Only after physicists were able to begin to probe and to understand the invisible, microworld of the atom did the validity, usefulness, and revolutionary nature of Planck's quantum theory become apparent. And once Niels Bohr incorporated the quantum theory into his model of the atom, the year 1900 was recognized as the watershed year dividing classical and modern physics. To the physicist, it is Planck's quantum theory, far more than Einstein's relativity theories, that so revolutionized physics.

As a professor of physics at the University of Berlin, Max Planck was working on one of the many riddles of physics—attempting to explain the spectrum of energy emitted by black bodies—when his mathematical conclusions repeatedly led him to contradict everything he thought to be certain. Planck's calculations could not be accounted for if he accepted the assumptions of traditional physics, and he was led to conclude that the explanation might lie outside of conventional theory. He pursued this essentially revolutionary mode of thought and published his classic paper, "Zur Theorie des Gesetzes der Energieverteilung im Normalspectrum," in *Verhandlungen der Deutsche Physikalische Gesellschaft,* in Leipzig in 1900. This journal is in the Library of Congress collections. What Planck learned that so upset both him and classical physics is now known as the discontinuity of matter. Planck discovered that light or energy is not found in nature as a continuous wave or flow but is emitted and absorbed discontinuously in little packets or quanta. Furthermore, Planck claimed, each quantum, or packet of energy, was indivisible. Different forms of electromagnetic radiation also had different size packets in proportion to their frequencies. Planck's notion of the quantum not only contradicted the mechanics of Newton and the electromagnetics of Maxwell, it seemed to contradict the continuity of nature itself. Within a short time, however, Planck's theory came to be recognized as the new rules of a new game—a new game of quantum mechanics that could be played simultaneously with but apart from the old game of classical physics.

The new game of physics, the physics of the very fast and the very small, would increasingly come to deal with and determine the effects of phenomena that are beyond the scope of everyday experience. This hidden world of electrons, protons, and other elementary particles obeys rules that are unimaginable in the concrete, visible world of classical physics. Quantum mechanics would come to reveal not only the discontinuity of matter (or energy) but its dual nature (light is both a particle and a wave). The nature of its subject matter being what it is, quantum theory is essentially abstract and capable of correct expression only through the use of mathematical symbols. Because of this nonpictorial nature, traditional methods of visualization serve no purpose—making quantum physics seem even more abstruse. Despite this, it is the language of a newly revealed world-within-a-world, which nature impels us to learn. The old rules still apply in the world we experience with our senses. In this macroworld our bridges are still built according to the laws of classical physics. But in the microworld of fast, elementary particles, these laws are useless impediments.

Max Planck survived the Nazi era in Germany, despite his outspokenness, and lived to be ninety. He is said to have almost regretted having been the instrument of the demise of classical physics, and he devoted much of his life attempting to reconcile the new and the old. Yet one does not invalidate the other as long as no crossover is attempted. In a sense, they are separate but equally valid systems.

In 1900 Planck's ideas were considered a hypothesis. Although Einstein applied the quantum to a physical phenomenon in 1905 and explained it (as classical physics could not), it was not until 1913, with the work of Niels Bohr, that the quantum became a respected theory. Bohr's achievement is often described as reconciling Rutherford's theory of the nucleus with the quantum theory of energy, and it is with Rutherford's model that Bohr began his work. The brusque and confident Rutherford had dismissed what he considered to be minor inconsistencies in his theory—he was an experimentalist after all, and no theoretician. But Bohr, the artistic and patient Dane, chose to exploit the apparent inconsistency in his mentor's model and in doing so created what has become the classic model of the structure of the atom. Bohr perceived that Rutherford's model could not be explained in full by classical physics. Since the electrons that circled the central nucleus orbited it in much the same way as planets did the sun, the known laws of physics posited that there must be some measurable energy loss by the electrons, which eventually would fall into the nucleus (as our earth and the planets will fall into the sun one day). But this was not happening, and it was to the emerging laws of the new physics—Planck's quantum theory—that Bohr turned to explain things. His results not only gave the quantum theory its first real success but stunned the world. In his classic papers, "On the Constitution of Atoms and Molecules," which appeared in the July, September, and November 1913 issues of the *Philosophical Magazine,* Bohr produced a theory of the energy status of the atom (in particular, the hydrogen atom) that showed its structure to be as mathematical as the laws of Newton's universe.

Bohr solved Rutherford's dilemma by assuming that the electrons could revolve about the nucleus in definite orbital paths, and he then established a rigorous mathematical theory to account for all those possible states. While the electrons remained in their predicted paths they emitted no energy, but when one would jump spontaneously from an outer to an inner orbit, it would emit energy as a quantum of light. In his brilliant and original synthesis, Bohr thus explained the structure of the atom and verified the quantum theory.

Aside from his work on the structure of the atom, Bohr is perhaps best known for a later theory, called the principle of complementarity. The Library has Bohr's first paper on complementarity, "The Quantum Postulate and the Recent Development of Atomic Theory," as it appeared in the British scientific journal *Nature* in 1928. It also appeared in Bohr's book, *Atomic Theory and the Description of Nature,* published in Cambridge in 1934. The Library has this volume in first edition.

This principle, which basically says that a phenomenon can be regarded in two mutually exclusive ways with both remaining valid, was one of the more profound conclusions to emerge from the research on light being done in the 1920s. By then, light phenomena had come to be regarded as both corpuscular and wavelike in nature. This coexistence of seemingly divergent characteristics would have been impossible in classical physics but was in fact demonstrable under quantum physics. It was shown that light would behave in two quite different ways depending on the type of optical experiment being performed. As with light, it was then discovered that all matter—even its elementary particles—had a dual nature. From such considerations, Bohr postulated his mind-bending complementarity principle, which in turn found its best expression in Heisenberg's fundamentally shocking principle of indeterminacy.

Werner Heisenberg was a young man of genius, coming straight from the elite of German academia. Having obtained his Ph.D. in 1923 at the age of twenty-two, Heisenberg joined in the intellectual excitement of his times, being greatly stimulated by the work of Erwin Schrödinger, Louis de Broglie, and Paul Dirac. Working as an assistant to Max Born at Göttingen and with Niels Bohr in Copenhagen, Heisenberg got to know the masters of theoretical physics. From this mix of fertile influences, Heisenberg's principles of uncertainty or indeterminacy emerged to become one of the most profound ideas in the history of science. Heisenberg published his idea in 1927 in a *Zeitschrift für Physik* article entitled "Über den anschaulichen Inhalt der quantentheoretischen Kinematik und Mechanik." The Library has a complete set of this important scientific journal. Heisenberg's principle states in general terms that on the level of subatomic events, any accurate measurement of an observable quantity necessarily produces uncertainties in one's knowledge of the other observable quantities. Related specifically to his electron studies, the principle explains that the exact position and precise speed of an electron cannot both be determined at the same time because, at this microlevel of matter, the very act of observing involves interacting. In a word, the scientist becomes as much an actor as a spectator and consequently cannot observe these microevents without interfacing with them.

Heisenberg not only stated his general principle in his article but, even more significantly, provided the mathematical equation that specified what the theoretical limits of precision would be. With this, Heisenberg's indeterminacy principle became a fundamental part of quantum mechanics, for he showed that the degree of uncertainty would always be at least on the order of Planck's quantum of action.

In 1930 Heisenberg prepared a larger work, *Die Physikalischen Prinzipien der Quantentheorie,* which contained the many themes of his early papers amplified in a separate treatise. The Library has the English translation of the same year, *The Physical Principles of the Quantum Theory,* published in Chicago. Here and in other works Heisenberg dealt with some of the philosophical implications of his work, one of which is that the traditional notion of causality breaks down since it is no longer possible to precisely and absolutely predict the pattern of quantum or microevents. At this level of scientific inquiry one must acknowledge the limits on one's accuracy and strive for the best possible statistical probability. Besides this breakdown of cause and effect, there exists the very fundamental conclusion that we can never know everything absolutely about our world. All knowledge is limited.

Such pronouncements are the stuff of a real conceptual revolution, for although the indeterminacy principle is usually confined to the microphysical sphere, it does affect all phenomena. Its application to the macroworld of classical physics has the leavening effect of calling into question assumptions based on supposed strict determinism or absolute objectivity. This lack of absolute precision or scientific certitude in no way attacks the foundation of science—its ability to predict the *probability* of future events is still very high—but rather sheds new light on the possible relations between physics and other less precise fields of human knowledge.

Although no mention has been made here of any developments beyond Heisenberg's and Bohr's work of the late 1920s, it is that period—especially wartime and the postwar era—that is best represented in the manuscript collection of the Library. Two major collections concern America's crash project to develop an atomic bomb. The papers of Vannevar Bush reflect his involvement with the Office of Scientific Research and Development (OSRD). It was Bush's report that resulted in the wartime establishment in 1942 of "The Manhattan Engineer District," which became known as the Manhattan Project. The papers of J. Robert Oppenheimer, the director of the Manhattan Project, make up a collection equal in size to Bush's. This significant collection documents not only Oppenheimer's involvement with the development of the bomb but also his later concern with the political aspects of the control of atomic energy. Also related to this era is a small collection of the papers of Enrico Fermi (on microfilm). Fermi's discoveries were essential to the success of the Manhattan Project. A related and very large collection is that of the papers of Merle A. Tuve. Tuve not only was part of the OSRD but was the creator and director of the Johns Hopkins University Applied Physics Laboratory.

Not related to wartime research but certainly very much in the forefront of contemporary physics are the Library's collections (greatly varying in size and content) of the papers of Lloyd V. Berkner, Charles S. Draper, George Gamow, George Von Bekesy, and the famed mathematician John Von Neumann.

As this century speeds through its last quarter, the fifty plus years since Heisenberg's principle of indeterminacy was enunciated have not witnessed anything to replace or to repudiate his theory. Indeed, despite the seeming ubiquitousness of physics in every aspect of modern life and science, recent decades have been more a period of consolidation than of theory-building. And despite the horrifying and perverse applications of some of its discoveries (that is, the atomic bomb and its more deadly successors), there exists for physics an apparently unending series of intellectual challenges. The quantum theory itself might someday be eclipsed, hanging as it does on the slender thread of the uncertainty principle. But until then, it will remain an inextricable part of contemporary physics.

The traditional concerns of physics have always centered around two elementary questions about the physical world: what is the basic "stuff" and structure of matter and what is responsible for the "go" of things? Thus concerned with the most fundamental aspects of its world—matter, motion, forces, and energy—the tradition of physics is that of the human intellect seeking to know nature in a most intimate and elemental manner. This elemental tradition holds true both for the large-scale, sensible, macroscopic domain of classical physics and for the high-speed, invisible, microscopic world of modern physics. This challenging tradition of asking the most penetrating, basic, yet comprehensive questions about nature exemplifies mankind at its most intellectually ambitious. Little wonder then that its theoretical correlations and explanations have come to influence not only other scientific disciplines but the way we perceive and explain the physical world as well. Finally, the power of this elemental tradition is perhaps most vivid and obvious in its shaping of the dominant concepts of our general culture.

Epilogue

Only occasionally has much been said here about twentieth-century science, primarily because of conventional historical reticence to discuss one's own times but also because this is primarily a look backward and only rarely sideways—and certainly never a look ahead. A sideways glance can be revealing, however, when compared with what has already gone before us.

Contemporary attitudes and our understanding of science are characterized by several popular conceits held by scientists and nonscientists alike. Most prevalent and certainly most obvious is today's identification of science with technology. Indeed, many do not even seem to acknowledge that a difference exists. This identity of two very different realities explains in part the existence of several other modern popular conceits: that today we actually know most of the things worth knowing; that only a few more breakthroughs are needed and we will understand the entire picture; that these breakthroughs are inevitable; that all our problems have a technological solution. These and many other popular ideas about science are based on the modern notion of progress, or secular belief in progress.

That mankind has not always believed in progress appears strange at first, and yet so it has been for the greater part of our history. Contrasting our contemporary scientific attitudes with those of nearly any part of our past shows that we not only live quite differently from any time past, but, even more importantly, we perceive the world in a different manner. Today's givens—things and ideas that any schoolchild takes for granted—are yesterday's marvels or impossibilities. The scientific and technological innocence of only a century ago appears charmingly preposterous to us today. Receding only slightly further into the past, when life was less predictable or safe and when a great deal was often lost for the want of a little knowledge, places our modern condition in even greater relief. Only one thing separates us from a wild past when nature seemed capricious, cruel, and conspiratorial. Only knowledge makes us different—makes us modern and not still medieval.

It is this knowledge of the natural world that is represented by the great works of science that are found in the collections of the Library of Congress. From this point of view, they are treasures in a most real and vivid sense, since each offers a unique intellectual richness. These books are more than any artifact or relic of the past could ever be, since each holds a physical truth and conveys it to posterity. Each work of science discussed here made a necessary contribution to the construction of the edifice of modern science. Beginning with the chance survivors of the ancient world, each is linked to the others. The irony surrounding the cumulative nature of science is that once a theory or fact is established as true, it becomes part of our information bank and joins the many scientific givens that we simply accept and no longer need to prove. In a sense, success in science means being taken for granted.

However dull or commonplace an idea may become once it is transformed into accepted wisdom, its genesis was usually quite the opposite. It is sometimes difficult to imagine a now-staid and secure theory as a fledgling idea—timorous and uncertain. But certainly many were. We would do well to study these major works of science as they were first written in hope of recapturing some sense of their excitement and curiosity about the natural world. Like the confusion and awe of primitive man or the wonder and excitement of a child, we should at least get a sense of adventure and discovery from these works. Such a feeling or impression may be even more valuable than the information they contain. Virtually every work discussed here was conceived and undertaken in that spirit—contributing in part to the enduring greatness of each. As I review these works of science, the words of Charles Lyell, who undertook a historical summary of his own field, come to mind:

> Meanwhile the charm of first discovery is our own, and as we explore this magnificent field of inquiry, the sentiment of a great historian of our times may continually be present to our minds, that "he who calls what has vanished back again into being, enjoys a bliss like that of creating."

Bibliography

The bibliography furnishes citations to the works discussed in the text, arranged alphabetically in sections corresponding to the chapters. Library of Congress call numbers are given wherever possible. Abbreviations indicate the custodial divisions holding works that are not in the Library's general collections. A key to these abbreviations follows.

Batchelder Coll.	Rare Book and Special Collections Division
Freud Coll.	Rare Book and Special Collections Division
G&M	Geography and Map Division
Jefferson Coll.	Rare Book and Special Collections Division
Law	Law Library
Micro	Microform Reading Room
MSS	Manuscript Division
Rare Bk. Coll.	Rare Book and Special Collections Division
Rosenwald Coll.	Rare Book and Special Collections Division
Thacher Coll.	Rare Book and Special Collections Division
Toner Coll.	Rare Book and Special Collections Division
Vollbehr Coll.	Rare Book and Special Collections Division

A number in parentheses following the designation "Rosenwald Coll." refers to the entry number for that work in the published catalog entitled *The Lessing J. Rosenwald Collection* (Washington: Library of Congress, 1977).

Introduction

Accademia del cimento. Saggi di naturali esperienze fatte nell'Accademia del cimento sotto la protezione del serenissimo principe Leopoldo di Toscana. [Essays on the scientific experiments conducted at the Accademia del cimento under the patronage of the Most Serene Prince Leopold of Tuscany] Firenze, G. Cocchini, 1666. 269 p.
QC17.A3 1666 Rare Bk. Coll.

Bacon, Francis, Viscount St. Albans (1561–1626). Instauratio magna. [The great restoration] Londini, Apud J. Billium, 1620. 360, 36 p.
B1165 1620 Rare Bk. Coll.

——Nevv Atlantis. A vvorke vnfinished. [London, 1628?] 46 p.
QH41.B2 1628 Rare Bk. Coll.

Beda Venerabilis (673–735). Opera. [Works] Coloniae Agrippinae, Apud I. W. Friessem, Jr., 1688. 8 v. in 4.
BR75.B38 1688 Rare Bk. Coll.

Boethius (d. 524). De consolatione philosophiae. [On the consolation of philosophy] Ghent, A. de Keysere, 1485. [360] leaves.
Incun. 1485.B64 Rosenwald Coll. (535)

——De institutione arithmetica. [On the teaching of arithmetic] Augsburg, E. Ratdolt, 1488. [48] leaves.
Incun. 1488.B66 Rosenwald Coll. (123)

Budé, Guillaume (1468–1540). Epistolae. [letters] [Parisiis] Venudantur in officina I. Badii [1520] 131 leaves.
PA8485.B6Z52 Rare Bk. Coll.

Cuningham, William (b. 1531). The cosmographical glasse, conteinyng the pleasant principles of cosmographie, geographie, hydrographie, or nauigation. Excussum Londini, In officina I. Day, 1559. 202 p.
GA6.C97 1559 Rare Bk. Coll.

Encyclopédie; ou, Dictionnaire raisonné des sciences, des arts et des métiers. [Encyclopedia; or, Classified dictionary of sciences, arts, and trades] Paris, Briasson, 1751–65. 17 v.
AE25.E53

———Supplément. Amsterdam, M. M. Rey, 1776–77. 4 v.
AE25.E53 Suppl.

———Recueil de planches. [Collection of plates] Paris, Briasson, 1762–72. 11 v.
AE25.E53 Plates

———Suite du Recueil des planches. Paris, Panckoucke, 1777. 22 p.
AE25.E53 Plates Suppl.

———Table analytique et raisonnée des matieres contenues dans les XXXIII volumes in-folio du Dictionnaire des sciences. [Analytical and classified index to the material contained in the 33 folio volumes of the Dictionary of the sciences] Paris, Panckoucke, 1780. 2 v.
AE25.E53 Table

Isidorus, Saint, Bishop of Seville (d. 636). Etymologiae. [Etymologies] [Augustae Vindelcorum] G. Zainer, 1472. 264 leaves.
Incun. 1472.I81 Rare Bk. Coll.

Journal des sçavans. Amsterdam, P. le Grand, 1665–1756, 1764–69. 228 v. in 233.
AP25.J7

Reisch, Gregor (d. 1525). Margarita philosophica. [The philosophical pearl] [Friburgi, I. Schott, 1503] [604] p.
AE3.R34 Rosenwald Coll. (595)

Royal Society of London. Philosophical transactions. London. 1665 +
Q41.L8

Vols. 1–90 in Rare Bk. Coll.

1. *Astronomy: Observing the Cosmos*

Adams, John Couch (1819–1892). An explanation of the observed irregularities in the motion of Uranus. *In* Royal Astronomical Society. Memoirs. v. 16. London, 1847. p. 427–459.

QB1.R5, v. 16

Apianus, Petrus (1495–1552). Astronomicum caesareum. [Caesar's astronomy] [Ingolstadii, 1540] [117] p.

QB41.A64 1540 Rosenwald Coll. (678)

——Cosmographia. [Cosmography] Vaeneunt Antuerpiae sub scuto Basiliensi, G. Bonito, 1545. 66 leaves.

GA6.A5 1545 Rare Bk. Coll.

Aristarchus, of Samos (ca. 310–230 B.C.). De magnitvdinibvs, et distantiis solis, et lvnae. [On the size and distance of the sun and the moon] In latunum conuersus. Pisavri, Apud C. Francischinum, 1572. 38 leaves.

PA3874.A48A2 1572 Rare Bk. Coll.

Brahe, Tyge (1546–1601). Opera omnia; sive, Astronomiae instauratae progymnasmata. [Complete works; or, Introduction to the new astronomy] Francofvrti, Impensis I. G. Schönvveteri, 1648. 470, 217 p.

QB3.B828 Rare Bk. Coll.

——Astronomiae instauratae mechanica. [Instruments of the new astronomy] Noribergae, Apud L. Hvlsivm, 1602. [107] p.

QB85.B8 1602 Rare Bk. Coll.

——De nova et nvllivs aevi memoria privs visa stella. [On the new star, never seen before] Hafniae, Impressit L. Benedictj, 1573. Bruxelles, Culture et Civilisation, 1969. [107] p.

QB41.B74 1969 Facsimile reprint.

Cassini, Giovanni Domenico (1625–1712). Abregé des observations & des reflexions svr la comete qui a paru au mois de decembre 1680. [Summary of observations and reflections on the comet which appeared in the month of December 1680] Paris, E. Michallet, 1681. xxxix, 90 p.

QB724.C34 Rosenwald Coll. (1417)

——La meridiana del tempio di S. Petronio tirata, e preparata per le osseruazioni astronomiche l'anno 1655. [The meridian of the church of San Petronio, drawn and prepared for astronomical observations in the year 1655] Bologna, Per l'erede di V. Benacci, 1695. 75 p.

QB224.C3 Rare Bk. Coll.

Copernicus, Nicolaus (1473–1543). De revolvtionibvs orbium coelestium. [On the revolutions of the heavenly spheres] Norimbergae, Apud I. Petreium, 1543. 196 leaves.

QB41.C76 1543 Rare Bk. Coll.

Flamsteed, John (1646–1719). Atlas céleste. [Celestial atlas] 2. éd. Par M. J. Fortin. Paris, Chez F. G. Deschamps [et chez] l'auteur, 1776. 40 p. 30 double maps.

QB65.F5 1776 Rare Bk. Coll.

——Historiae coelestis britannicae. [Of the British history of the heavens.] Londini, Typis H. Meere, 1725. 3 v.

QB4.F6 Rare Bk. Coll.

Fracastoro, Girolamo (1478–1553). Homocentrica. [Homocentricity] [Venetiis] 1538. [78] leaves.

QB41.F8 1538 Rare Bk. Coll.

Galilei, Galileo (1564–1642). Dialogo . . . sopra i due massimi sistemi del mondo tolemaico, e copernicano. [Dialogue on the two chief systems of the world, the Ptolemaic and the Copernican] Fiorenza, Per G. B. Landini, 1632. 458, [32] p.
QB41.G14 Rosenwald Coll. (1350)

——Siderevs nvncivs. [The starry messenger] Venetiis, Apud T. Baglionum, 1610. Bruxelles, Culture et Civilisation, 1967. 28 leaves.
QB41.G17 1610a Facsimile reprint.

——Syderevs nvncivs. *In his* Opere. v. 2. Bononiae, Ex Typ. HH. de Ducijs, 1655. p. [1]–41 (1st group)
QB3.G14 1655, v. 2 Rare Bk. Coll.

Gt. Brit. *Laws, statutes, etc., 1702–1714 (Anne).* An act for providing a publick reward for such person or persons as shall discover the longitude at sea. *In its* Anno regni Annae reginae Magnae Britanniae, Franciae, & Hiberniae, duodecimo. London, Printed by J. Baskett, 1714. p. 355–357.
Law Gt. Brit. 1

Halley, Edmond (1656–1742). Astronomiae cometicae synopsis. [A synopsis of the astronomy of comets] *In* Royal Society of London. Philosophical transactions, v. 24, Mar. 1705: 1882–1899.
Q41.L8, v. 24 Rare Bk. Coll.

——Astronomical tables with precepts . . . for computing places of the sun, moon, planets, and comets. London, Printed for W. Innys, 1752. [83] p.
QB11.H2 Rare Bk. Coll.

Harrison, John (1693–1776). The principles of Mr. Harrison's time-keeper. London, Printed by W. Richardson and S. Clark and sold by J. Nourse, 1767. 31 p.
QB107.H3 Rare Bk. Coll.

Herschel, Caroline Lucretia (1750–1848). Catalogue of stars. London, Sold by P. Elmsly, Printer to the Royal Society, 1798. 136 p.
QB6.H56 Rare Bk. Coll.

Herschel, Sir William (1738–1822). Account of a comet. *In* Royal Society of London. Philosophical transactions. v. 71, pt. 2; 1781. London, Sold by L. Davis, and P. Elmsly, Printers to the Royal Society, 1782. p. 492–501.
Q41.L8, v. 71 Rare Bk. Coll.

Hevelius, Johannes (1611–1687). Cometographia. [Cometography] Gedani, Auctoris Typis, & Sumptibus, Imprimebat S. Reiniger, 1668. 913 p.
QB721.H5 Rare Bk. Coll.

——Machinae coelestis. [Celestial machines] Gedani, Auctoris typis, & sumptibus, imprimebat S. Reiniger, 1673–79. 2 v.
QB41.H423 1673 Rosenwald Coll.

——Selenographia; sive, Lunae descriptio. [Selenography; or, Description of the moon] Gedani, Autoris Sumtibus, Typis Hünefeldianis, 1647. 563 p.
QB29.H44 Rosenwald Coll. (1321)

Horrocks, Jeremiah (1617?–1641). Opera posthuma. [Posthumous works] Londini, Typis G. Godbid, Impensis J. Martyn, 1673. 496 p.
QB41.H79 Rare Bk. Coll.

Huygens, Christiaan (1629–1695). Œuvres complètes. [Complete works] La Haye, M. Nijhoff, 1888–1950. 22 v.
Q113.H9

Index librorum prohibitorum. Index librorum prohibitorum SSm̃i D. N. Benedicti XIV. Pontificis Maximi jvssv recognitus, atque editus. [List of forbidden

books, authenticated and issued by order of Pope Benedict XIV] Romae, Ex typographia Reverendae Camerae Apostolicae, 1758. xxxix, 268 p.
Z1020.I758a Rare Bk. Coll.

Kepler, Johann (1571–1630). Astronomia nova. [The new astronomy] [Pragae] 1609. Bruxelles, Culture et Civilisation, 1968. 337 p.
QB41.K32 1968 Facsimile reprint.

——Chilias logarithmorum. [A thousand logarithms] Marpurgi, Excusa Typis C. Chemlini, 1624. 55, [53] p.
QA55.K55 Rare Bk. Coll.

——Ephemerides novae motuum coelestium. [New ephemerides of the motions of the heavenly bodies] Lincij Austriae, Sumptibus Authoris, Excudebat J. Plancvs [1617]–30. 3 pts. in 1 v.
QB7.K4 Rare Bk. Coll.

——Harmonices mvndi. [Harmony of the world] Lincii Austriae, Sumptibus G. Tampachii, Excudebat I. Plancvs, 1619. 66, 255 p.
QB41.K38 Rare Bk. Coll.

——Somnivm; seu, Opvs posthvmvm de astronomia lvnari. [The dream; or, Posthumous work on lunar astronomy] Impressum Partim Sagani Silesiorum, Absolutum Francofurti, Sumptibus Haeredum Authoris, 1634. 182 p.
QB41.K42 Rare Bk. Coll.

——Tabulae Rudolphinae. [Rudolphine tables] Vlmae, Typis J. Saurii, 1627–[30] [18], 125, 115 p.
QB41.K43 Rosenwald Coll. (1316)

Laplace, Pierre Simon, marquis de (1749–1827). Exposition du système du monde. [Explanation of the world system] 3. éd., rev. et augm. Paris, Courcier, 1808. 405 p.
QB42.L28 1808 Rare Bk. Coll.

——Traité de mécanique céleste. [Treatise on celestial mechanics] Paris, Chez J. B. M. Duprat [1798–1825] 5 v. in 3.
QB351.L29 Rare Bk. Coll.

Le Verrier, Urbain Jean Joseph (1811–1877). Recherches sur les mouvements d'Uranus. [Investigation of the motion of Uranus] *In* Académie des sciences, Paris. Comptes rendus hebdomadaires des séances, t. 22, 1 juin 1846: 907–918.
Q46.A14, v. 22

Maestlin, Michael (1550–1631). Epitome astronomiae. [Survey of astronomy] Tubingae, Excudebat T. Werlin, Impensis I. Berneri, 1624. 544 p.
QB41.M2 Rare Bk. Coll.

Newton, Sir Isaac (1642–1727). Philosophiae naturalis principia mathematica. [The mathematical principles of natural philosophy] Londini, Jussu Societatis Regiae ac Typis J. Streater, 1687. 383, 400–510 p.
QA803.A2 1687 Rare Bk. Coll.

Nicolaus Cusanus, Cardinal (1401–1464). Opuscula theologica et mathematica. [Short theological and mathematical works] [Strassburg, M. Flach, ca. 1500] 2 pts. in 1 v.
H-5893 Vollbehr Coll.

Ptolemaeus, Claudius (ca. 100–ca. 170). Almagestū. [The greatest treatise] Venetijs, P. Liechtenstein, 1515. 152 leaves.
PA4405.A4A5 1515 Rare Bk. Coll.

——Epytoma in Almagestum. [Summary of the Almagest] Venice, J. Hamman, 1496. [110] leaves.
Incun. 1496.P8 Rare Bk. Coll.

By Johannes Mueller, called Regiomontanus (1436–1476); begun by Georg von Peurbach (1423–1461).

——Omnia, qvae extant, opera, Geographia excepta. [Complete extant works, except the Geography] Basileae, Apvd H. Petrvm, 1541. 511 p.
QB41.P9 1541 G&M

Reinhold, Erasmus (1511–1553). Prvtenicae tabvlae coelestivm motvvm. [Prutenic tables of the motions of the heavenly bodies] Tvbingae, Per V. Morhardvm, 1551. 3 pts. in 1 v.
QB11.R4 Rare Bk. Coll.

Rhäticus, Georg Joachim (1514–1576). De libris revolutionum Nicolai Copernici narratio prima. [First account of Nicolaus Copernicus' books on the revolutions] Gedani, 1540. Osnabrück, Zeller, 1965. 38 leaves. (Milliaria, 6)
QB41.C815R48 1965 Facsimile reprint.

——De libris revolvtionvm Nicolai Copernici narratio prima. [First account of Nicolaus Copernicus' books on the revolutions] *In* Copernicus, Nicolaus. De revolvtionibus orbium coelestium. Basileae, Ex Officina Henricpetrina, 1566. leaves 196^{v}–213^{r}.
QB41.C76 1566 Rare Bk. Coll.

——Opvs palatinvm de triangvlis. [Palatine work on triangles] [Neostadii in Palatinatv, Excudebat M. Harnisius] 1596. 1 v. (various pagings)
QA33.R5 Rare Bk. Coll.

Riccioli, Giovanni Battista (1598–1671). Almagestvm novvm astronomiam veterem novamqve complectens. [The new almagest, containing the old and the new astronomy] Bononiae, Ex Typographia Haeredis V. Benatij, 1651. 1 v. in 2.
QB41.R4 Rare Bk. Coll.

Sacro Bosco, Joannes de (fl. 1230). Sphaera mundi. [The sphere of the world] [Venice] E. Ratdolt, 1485. [58] leaves.
Incun. 1485.S3 Rosenwald Coll.

Shapley, Harlow (1885–1972). Starlight. New York, G. H. Doran Co. [1926] 143 p. (The Humanizing of knowledge series)
QB44.S55

Zacuto, Abraham ben Samuel (b. ca. 1450). Almanach perpetuum. [Perpetual almanac] [Leiria, S. Dortas, 1496] [172] leaves.
Incun. 1496.Z3 Thacher Coll.

2. *Botany: From Herbalism to Science*

Apuleius Barbarus (ca. 5th cent.). Herbarium. [Herbal] [Rome, J. P. de Lignamine, ca. 1483–84] [108] leaves.
Incun. X.A7 Rosenwald Coll. (237)

Bartholomaeus Anglicus (13th cent.). De proprietatibus rerum. [On the properties of things] French. Lyons, J. Siber [after Jan. 26, 1486] [252] leaves.
Incun. X.B28 Rosenwald Coll. (394)

Captain Cook's florilegium. London, Lion and Unicorn Press, 1973. 1 v. (unpaged)
QK5.C36 1973b Rosenwald Coll.
Engravings based on drawings and sketches made during the voyage by Sydney Parkinson.

Darwin, Erasmus (1731–1802). Phytologia; or, The philosophy of agriculture and gardening. Dublin, P. Byrne, 1800. 556, [11] p.
QK45.D23 Rare Bk. Coll.

——Zoonomia; or, The laws of organic life. Dublin, P. Byrne, and W. Jones, 1794–96. 2 v.
QP29.D22 1794 Toner Coll.

Dioscorides, Pedanius, of Anazarbos (fl. ca. 60). De materia medica. [On medicinal substances] Colle, J. de Medemblick, 1478. [104] leaves.
Incun. 1478.D5516 Thacher Coll.

Fuchs, Leonhart (1501–1566). De historia stirpivm commentarii. [Commentaries on the history of plants] Basileae, In Officina Isingriniana, 1542. 896 p.
QK41.F7 1542 Rosenwald Coll. (905)

——New Kreüterbůch. [New herbal] Basell, Durch M. Isingrin, 1543. [888] p.
QK41.F8 Rare Bk. Coll.

Gesner, Konrad (1516–1565). Opera botanica. [Botanical works] Norimbergae, Impensis I. M. Seligmanni, Typis I. I. Fleischmanni, 1751–71. 2 v.
QK41.G39 Rare Bk. Coll.

Goethe, Johann Wolfgang von (1749–1832). Versuch die Metamorphose der Pflanzen zu erklären. [Attempt to explain the metamorphosis of plants] Gotha, C. W. Ettinger, 1790. 86 p.
QK641.G6 1790 Rare Bk. Coll.

Grew, Nehemiah (1641–1712). The anatomy of plants. [London] Printed by W. Rawlins, for the author, 1682. 24, [300] p.
QK41.G82 Rare Bk. Coll.

Hales, Stephen (1677–1761). Vegetable staticks; or, An account of some statical experiments on the sap in vegetables. London, W. and J. Innys, 1727. 376 p.
QK710.H3 Rare Bk. Coll.

Haller, Albrecht von (1708–1777). Historia stirpium indigenarum Helvetiae inchoata. [History of Swiss plants] Bernae, Sumptibus Societatis Typographicae, 1768. 3 v. in 2.
QK331.H18 Rare Bk. Coll.

Herbarius. Herbarius Latinus. [The Latin herbal] Venice, S. Bevilaqua, 1499. cl, [18] leaves.
Incun. 1499.H4 Rosenwald Coll. (339)

Hooke, Robert (1635–1703). Micrographia; or, Some physiological descriptions of minute bodies made by magnifying glasses. London, Printed by J. Martyn and J. Allestry, 1665. 246, [10] p.
QH271.H79 Rosenwald Coll. (1511)

Hortus sanitatis [maior] Hortus sanitatis. [The garden of health] Mainz, J. Meydenbach, 1491. [454] leaves.
Incun. 1491.H75 Rosenwald Coll. (152)

Humboldt, Alexander, Freiherr von (1768–1859). De distributione geographica plantarum. [On the geographical distribution of plants] Lutetiae Parisorum, In Libraria Graeco-Latino-Germanica, 1817. 249 p.
QK101.H92 Rare Bk. Coll.

——Essai sur la géographie des plantes. [Essay on the geography of plants] Par Alexandre de Humboldt et Aimé Bonpland. Éd. facsimilaire. México, Institut panaméricain de géographie et d'histoire, 1955. 155, [17] p. (Instituto Panamericano de Geografía e Historia. Publication no. 200)

F1401.P153, no. 200

Reproduces the edition published in Paris by Levrault, Schoell in 1805.

——Ideen zu einer Geographie der Pflanzen. [Thoughts on the geography of plants] Von Al. von Humboldt und A. Bonpland. Tübingen, F. G. Cotta, 1807. 182 p.

QK101.H933 Rare Bk. Coll.

Ingenhousz, Jan (1730–1799). Experiments upon vegetables. London, Printed for P. Elmsly and H. Payne, 1779. lxviii, 302, [17] p.

QK867.I38 Rare Bk. Coll.

Jussieu, Antoine Laurent de (1748–1836). Genera plantarum, secundum ordines naturales disposita, juxta methodum in Horto regio parisiensi exaratam. [Genera of plants, arranged by their natural orders, corresponding to the system cultivated at the Jardin du Roi in Paris] Parisiis, Apud Viduam Herissant, 1789. 24, lxxii, 498 p.

QK93.J9 Rare Bk. Coll.

Konrad von Megenberg (14th cent.). Buch der Natur. [Book of nature] Augsburg, J. Bämler, 1481. [194] leaves.

Incun. 1481.K6 Rosenwald Coll. (80)

Linné, Carl von (1707–1778). Systema naturae. [The system of nature] 1735. Facsimile of the 1st ed. Nieuwkoop, B. de Graaf, 1964. 30, [19] p. (Dutch classics on history of science, v. 8)

QH43.S73 1735a Rare Bk. Coll.

————Ed. 2., auctior. Stockholmiae, Apud G. Kiesewetter, 1740. 80 p.

QH43.S12 Rare Bk. Coll.

————Ed. 10., reformata. Holmiae, Impensis L. Salvii, 1758–59. 2 v.

QH43.S48 Rare Bk. Coll.

Malpighi, Marcello (1628–1694). Anatome plantarum. [The anatomy of plants] Londini, Impensis J. Martyn, 1675–79. 2 v. in 1.

QK671.M26 Rare Bk. Coll.

Mattioli, Pietro Andrea (1500–1577). Commentarii secvndo avcti, in libros sex Pedacii Dioscoridis Anazarbei De medica materia. [Commentaries, enlarged for the second time, on the six books of Materia medica by Pedanius Dioscorides of Anazarbos] Venetiis, Ex officina Erasmiana, V. Valgrisij, 1558. 776, 50 p.

R126.D7M42 Rare Bk. Coll.

Mendel, Gregor (1822–1884). Versuche über Pflanzen-Hybriden. [Investigations of plant hybrids] *In* Naturforschender Verein in Brünn. Verhandlungen. 4. Bd.; 1865. Brünn, 1866. Abhandlungen, p. [3]–47.

Q44.B89, v. 4

Parkinson, Sydney (1745?–1771). A journal of a voyage to the South seas, in His Majesty's ship, the Endeavour. London, 1773. xxiii, 212, 22 p.

G420.C65P3 Rare Bk. Coll.

Plinius Secundus, C. (ca. A.D. 23–79). Historia naturalis. [Natural history] Venice, J. de Spira, 1469. [356] leaves.

Incun. 1469.P55 Rosenwald Coll. (212)

—— Natvrae historiarum libri XXXVII e castigationibus Hermolai Barbari. [The 37 books of natural history, from the corrections of Ermolao Barbaro] [Vene-

tiis, Sumptibus L. A. de Giunta; impressum in aedibus G. de Rusconibus, 1519] 286, [79] leaves.
PA6611.A2 1519 Rare Bk. Coll.

Ray, John (1627–1705). Historia plantarum. [The history of plants] London, Typis M. Clark, prostant apud H. Faithorne, 1686–1704. 3 v.
QK41.R2 Rare Bk. Coll.

Redouté, Pierre Joseph (1759–1840). Les roses. [Roses] Paris, Impr. de F. Didot, 1817–24. 3 v.
QK495.R78R248 Rosenwald Coll. (1892)

Robin, Jean (1550–1629). Histoire des plantes, novvellement trouuées en l'isle Virgine, & autres lieux. [History of the plants recently found in the island of Virginia and other places] Paris, G. Macé, 1620. 16 p.
QK41.R7 Rosenwald Coll. (1378)

Rousseau, Jean Jacques (1712–1778). La botanique de J. J. Rousseau. [J. J. Rousseau's botany] Paris, Delachaussée, 1805. 122 p.
QK98.R6 Rosenwald Coll. (1881)

Sachs, Julius von (1832–1897). Lehrbuch der Botanik. [Textbook of botany] Leipzig, W. Engelmann, 1873. xvi, 848 p.
QK45.S12

Schleiden, Matthias Jacob (1804–1881). Principles of scientific botany. Translated by Edwin Lankester. London, Longman, Brown, Green, and Longmans, 1849. 616 p.
QK45.S3413 1849

Sprengel, Christian Konrad (1750–1816). Das entdeckte Geheimniss der Natur im Bau und in der Befruchtung der Blumen. [The discovered secret of nature in the structure and fertilization of flowers] (1793.) Leipzig, W. Engelmann, 1894. 4 v. (Ostwald's Klassiker der exakten Wissenschaften, Nr. 48–51)
QK692.S77

Theophrastus (ca. 371-ca. 287 B.C.). De historia et causis plantarum. [On the history and origins of plants] Treviso, B. Confalonerius, 1483. [156] leaves.
Incun. 1483.T39 Rare Bk. Coll.

Thornton, Robert John (1768–1837), ed. New illustration of the sexual system of Carolus von Linnaeus . . . and The temple of Flora, or Garden of nature. London, 1807 [i.e. 1799–1810] 3 pts. in 2 v.
QK92.T59 Rosenwald Coll. (1778)

Tournefort, Joseph Pitton de (1656–1708). Elemens de botanique. [Elements of botany] Paris, De l'Impr. royale, 1694. 3 v.
QK41.T7 Rare Bk. Coll.

——The compleat herbal; or, The botanical institutions of Mr. Tournefort. Carefully translated from the original Latin. With large additions from Ray, Gerarde, Parkinson, and others. London, Printed for R. Bonwicke, 1719–30. 2 v.
QK41.T74 Rare Bk. Coll.

Vries, Hugo de (1848–1935). Die Mutationstheorie. Versuche und Beobachtungen über die Entstehung von Arten im Planzenreich. [The mutation theory. Experiments and observations on the origin of species in the plant kingdom] Leipzig, Veit, 1901–3. 2 v.
QH406.V95

3. *Zoology: Our Shared Nature*

Albertus Magnus, Saint, Bp. of Ratisbon (1193?–1280). De animalibus. [On animals] Venice, J. and G. de Gregoriis, de Forlivio, 1495. 254 leaves.
Incun. 1495.A48 Rare Bk. Coll.

Aldrovandi, Ulisse (1522–1605?). De piscibus. [On fishes] Bononiae, Apud Bellagambam, 1613. 732, [26] p.
QL41.A36 Rare Bk. Coll.

——De quadropedibus solidipedibus. [On hoofed quadrupeds] Bononiae, Apud N. Thebaldinum, 1639. 495, [30] p.
SF283.A383 Rare Bk. Coll.

——Ornithologiae, hoc est, de avibvs historiae. [Of ornithology, that is, on the history of birds] Bononiae, Apud F. de Franciscis Senensem, 1599–1603. 3 v.
QL673.A36 Rare Bk. Coll.

——Serpentvm, et draconv̄ historiae. [Of the history of serpents and snakes] Bononiae, Apud C. Ferronium, 1640. 427, [30] p.
QL41.A37 Rare Bk. Coll.

Aristoteles (ca. 384–322 B.C.). De animalibus. [On animals] Latin. Venice, J. de Colonia and J. Manthen, 1476. [252] leaves.
Incun. 1476.A7 Rare Bk. Coll.

Audubon, John James (1785–1851). The birds of America. London, 1827–38. 4 v.
QL674.A9 1827 Rare Bk. Coll.

——Ornithological biography . . . accompanied by descriptions of the objects represented in the work entitled The birds of America. Edinburgh, A. Black, 1831–[39] 5 v.
QL674.A9 1831 Rare Bk. Coll.

Baer, Karl Ernst von (1792–1876). De ovi mammalium et hominis genesi. [On the mammalian egg and the genesis of man] Lipsiae, Sumptibus L. Vossii, 1827. [Bruxelles, Culture et Civilisation, 1966] 40 p.
QL965.B24 1827a Facsimile reprint.

——Über Entwickelungsgeschichte der Thiere. [On the embryology of animals] Königsberg, Bei den Gebr. Bornträger, 1828–37. 2 v.
QL959.B3 Rare Bk. Coll.
LC holds v. 1 only. A facsimile reprint of both volumes, published in Bruxelles by Culture et Civilisation in 1967, is available under call no. QL959.B3 1828a.

Bartram, William (1739–1823). Travels through North & South Carolina, Georgia, East & West Florida . . . and the country of the Chactaws. Philadelphia, Printed by James & Johnson, 1791. xxxiv, 522 p.
F213.B28 Rare Bk. Coll.

Belon, Pierre (1517?–1564). L'histoire de la natvre des oyseavx. [The history of the nature of birds] Paris, G. Cauellat, 1555. 381 p.
QL673.B45 Rare Bk. Coll.

Bonaparte, Charles Lucien Jules Laurent, prince de Canino (1803–1857). American ornithology; or, The natural history of birds inhabiting the United States, not given by Wilson. Philadelphia, Carey, Lea & Carey, 1825–33. 4 v.
QL674.W76 Rare Bk. Coll.

Borelli, Giovanni Alfonso (1608–1679). De motv animalivm. [On the motion of animals] Romae, Ex Typographia A. Bernabò, 1680–81. 2 v.
QP301.B6 1680 Rare Bk. Coll.

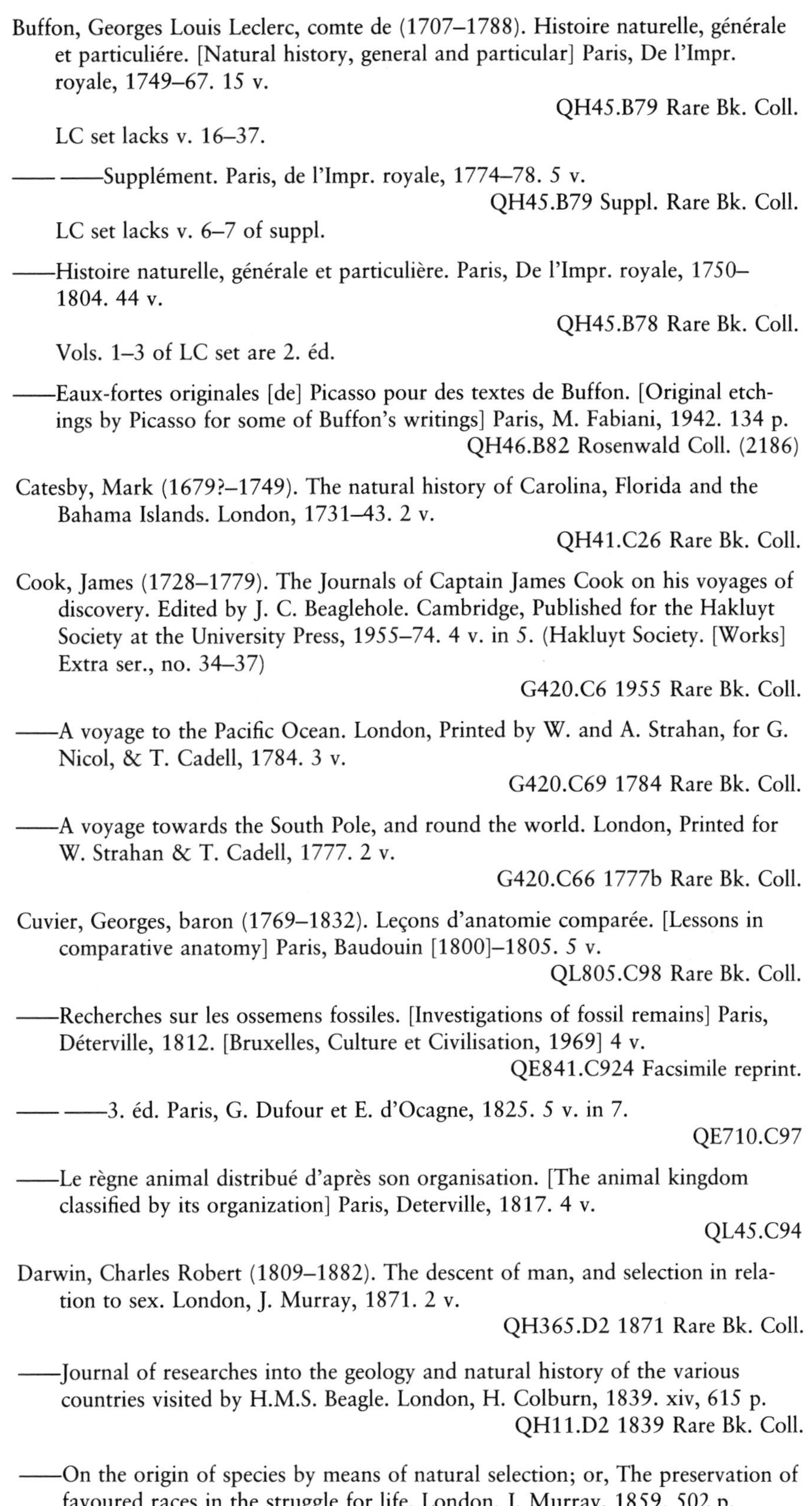

Buffon, Georges Louis Leclerc, comte de (1707–1788). Histoire naturelle, générale et particuliére. [Natural history, general and particular] Paris, De l'Impr. royale, 1749–67. 15 v.
QH45.B79 Rare Bk. Coll.

LC set lacks v. 16–37.

——Supplément. Paris, de l'Impr. royale, 1774–78. 5 v.
QH45.B79 Suppl. Rare Bk. Coll.

LC set lacks v. 6–7 of suppl.

—Histoire naturelle, générale et particulière. Paris, De l'Impr. royale, 1750–1804. 44 v.
QH45.B78 Rare Bk. Coll.

Vols. 1–3 of LC set are 2. éd.

—Eaux-fortes originales [de] Picasso pour des textes de Buffon. [Original etchings by Picasso for some of Buffon's writings] Paris, M. Fabiani, 1942. 134 p.
QH46.B82 Rosenwald Coll. (2186)

Catesby, Mark (1679?–1749). The natural history of Carolina, Florida and the Bahama Islands. London, 1731–43. 2 v.
QH41.C26 Rare Bk. Coll.

Cook, James (1728–1779). The Journals of Captain James Cook on his voyages of discovery. Edited by J. C. Beaglehole. Cambridge, Published for the Hakluyt Society at the University Press, 1955–74. 4 v. in 5. (Hakluyt Society. [Works] Extra ser., no. 34–37)
G420.C6 1955 Rare Bk. Coll.

—A voyage to the Pacific Ocean. London, Printed by W. and A. Strahan, for G. Nicol, & T. Cadell, 1784. 3 v.
G420.C69 1784 Rare Bk. Coll.

—A voyage towards the South Pole, and round the world. London, Printed for W. Strahan & T. Cadell, 1777. 2 v.
G420.C66 1777b Rare Bk. Coll.

Cuvier, Georges, baron (1769–1832). Leçons d'anatomie comparée. [Lessons in comparative anatomy] Paris, Baudouin [1800]–1805. 5 v.
QL805.C98 Rare Bk. Coll.

—Recherches sur les ossemens fossiles. [Investigations of fossil remains] Paris, Déterville, 1812. [Bruxelles, Culture et Civilisation, 1969] 4 v.
QE841.C924 Facsimile reprint.

——3. éd. Paris, G. Dufour et E. d'Ocagne, 1825. 5 v. in 7.
QE710.C97

—Le règne animal distribué d'après son organisation. [The animal kingdom classified by its organization] Paris, Deterville, 1817. 4 v.
QL45.C94

Darwin, Charles Robert (1809–1882). The descent of man, and selection in relation to sex. London, J. Murray, 1871. 2 v.
QH365.D2 1871 Rare Bk. Coll.

—Journal of researches into the geology and natural history of the various countries visited by H.M.S. Beagle. London, H. Colburn, 1839. xiv, 615 p.
QH11.D2 1839 Rare Bk. Coll.

—On the origin of species by means of natural selection; or, The preservation of favoured races in the struggle for life. London, J. Murray, 1859. 502 p.
QH365.O2 1859 Rare Bk. Coll.

——On the tendency of species to form varieties; and on the perpetuation of varieties and species by natural means of selection. By Charles Darwin and Alfred Wallace. *In* Linnean Society of London. Journal of the proceedings. Zoology. v. 3. London, Longman, Brown, Green, Longmans & Roberts and Williams and Norgate, 1859. p. 45–62.

QH1.L54, v. 3

——The zoology of the voyage of H.M.S. Beagle. London, Smith, Elder, 1839–43. 5 v.

QL5.B3 Rare Bk. Coll.

Friedrich II, Emperor of Germany (1194–1250). De arte venandi cum auibus. [On the art of hunting with birds] Avgvstae Vindelicorvm, Apud I. Praetorium, 1596. 414 p.

SK321.F86 Rare Bk. Coll.

Gesner, Konrad (1516–1565). Historiae animalium. [Of the history of animals] Tigvri, Apvd C. Froschovervm, 1551–58. 4 v. in 3.

QL41.G37 Rare Bk. Coll.

The complete work includes a fifth volume, lacking in this set.

Gt. Brit. Challenger Office. Report on the scientific results of the voyage of H.M.S. Challenger during the years 1873–76. [Edinburgh] Printed for H.M. Stationery Off. [by Neill] 1880–95. 40 v. in 44.

Q115.C4

Hawkesworth, John (1715?–1773), comp. An account of the voyages undertaken by the order of His present Majesty for making discoveries in the southern hemisphere. London, Printed for W. Strahan & T. Cadell, 1773. 3 v.

G420.C65H3 1773 Rare Bk. Coll.

Hooke, Robert (1635–1703). Micrographia; or, Some physiological descriptions of minute bodies made by magnifying glasses. London, Printed by J. Martyn and J. Allestry, 1665. 246, [10] p.

QH271.H79 Rosenwald Coll. (1511)

Konrad von Megenberg (14th cent.). Buch der Natur. [Book of nature] Augsburg, J. Bämler, 1481. [194] leaves.

Incun. 1481.K6 Rosenwald Coll. (80)

Lamarck, Jean Baptiste Pierre Antoine de Monet de (1744–1829). Histoire naturelle des animaux sans vertèbres. [Natural history of invertebrates] Paris, Verdière, 1815–22. 7 v. in 8.

QL362.L225

——Philosophie zoologique. [Zoological philosophy] Nouv. éd. Paris, G. Baillière, 1830. 2 v.

QL45.L21

——Système des animaux sans vertèbres. [System of invertebrates] Paris, Chez Deterville, 1801. 432 p.

QL362.L22 Rare Bk. Coll.

Leeuwenhoek, Anthony van (1632–1723). Arcana naturae detecta. [Secrets of nature discovered] Delphis Batavorum, Apud H. a Krooneveld, 1695. 568, [14] p.

QH271.L48 Rare Bk. Coll.

——Ontledingen en ontdekkingen . . . Vervat in verscheide brieven. [Analysis and discoveries . . . Contained in sundry letters] Leiden, C. Boutestein, 1696–1718. 5 v.

QH9.L4 Rare Bk. Coll.

LC copy of v. 1 is 2. druk. Title of v. 2–4 is *Vervolg der brieven* [Continuation of the letters]; of v. 5, *Send-brieven* [Circular letters]

Linné, Carl von (1707–1778). Systema naturae. [The system of nature] 1735. Facsimile of the 1st ed. Nieuwkoop, B. de Graaf, 1964. 30, [19] p. (Dutch classics on history of science, v. 8)
QH43.S73 1735a Rare Bk. Coll.

———Ed. 2., auctior. Stockholmiae, Apud G. Kiesewetter, 1740. 80 p.
QH43.S12 Rare Bk. Coll.

———Ed. 10., reformata. Holmiae, Impensis L. Salvii, 1758–59. 2 v.
QH43.S48 Rare Bk. Coll.

Persius Flaccus, Aulus (A.D. 34–62). Persio, tradotto in verso sciolto e dichiarato da Francesco Stellvti. [Persius, translated into blank verse and made known by Francesco Stelluti] Roma, G. Mascardi, 1630. 218 p.
PA6555.A2 1630 Rosenwald Coll. (1348)

Physiologus. Peri ton physiologon. [On natural philosophy] Romae, Apud Zannettum et Ruffinellum, 1587. 122 p.
PA4273.P8 1587 Rosenwald Coll. (878)

Plinius Secundus, C. (ca. A.D. 23–79). Historia naturalis. [Natural history] Venice, J. de Spira, 1469. [356] leaves.
Incun. 1469.P55 Rosenwald Coll. (212)

Ray, John (1627–1705). Synopsis methodica animalium quadrupedum et serpentini generis. [Classified outline of the genera of quadruped and serpentine animals] Londini, Impensis S. Smith & B. Walford Societatis Regiae Typographorum, 1693. 336 p.
QL41.R26 Rare Bk. Coll.

Redi, Francesco (1626–1698). Esperienze intorno alla generazione degl'insetti. [Experiments dealing with the generation of insects] Firenze, All'insegna della Stella, 1668. 228 p.
QL496.R35 1668 Rare Bk. Coll.

Rondelet, Guillaume (1507–1566). Libri de piscibus marinis. [Books on marine fishes] Lugduni, Apud M. Bonhomme, 1554–55. 2 v. in 1.
QL41.R68 Rare Bk. Coll.
Title of v. 2 is *Uniuersae aquatilium historiae pars altera* [The second part of the history of the aquatic universe]

Ruini, Carlo (fl. 1598). Dell'anotomia et dell'infirmità del cavallo. [On the anatomy and illness of the horse] Bologna, Presso gli heredi di G. Rossi, 1598. 295, 386 p.
SF765.R8 1598 Rosenwald Coll.

Schwann, Theodor (1810–1882). Microscopical researches into the accordance in the structure and growth of animals and plants. Translated from the German. London, Sydenham Society, 1847. xx, 268 p.
QH581.S39

Swammerdam, Jan (1637–1680). Bybel der natuure. [Bible of nature] Leyden, I. Severinus, 1737–38. 2 v.
QL362.S96 Rare Bk. Coll.

Topsell, Edward (1572–1625?). The historie of fovre-footed beastes. London, Printed by W. Iaggard, 1607. 757, [10] p.
QL41.T66 Rare Bk. Coll.

——The history of four-footed beasts and serpents. London, Printed by E. Cotes, for G. Sawbridge, 1658. 2 v. in 1.

QL41.T68 Rare Bk. Coll.

——The historie of serpents. London, Printed by W. Jaggard, 1608. 315 p.

QL41.T66 copy 2 Rare Bk. Coll.

Vincent de Beauvais (d. 1264). Speculum naturale. [Mirror of nature] [Strassburg, Printer of the Legenda aurea, 1485?] 2 v.

Incun. X.V774 Rare Bk. Coll.

Wilkes, Charles (1798–1877). Narrative of the United States Exploring Expedition. During the years 1838, 1839, 1840, 1841, 1842. Philadelphia, Printed by C. Sherman, 1844. 5 v. and atlas.

Q115.W6, v. 1–5 Rare Bk. Coll.
Q115.W6, Atlas G&M

Wilson, Alexander (1766–1813). American ornithology; or, The natural history of the birds of the United States. Philadelphia, Bradford and Inskeep, 1808–14. 9 v.

QL674.W73 Rare Bk. Coll.

4. *Medicine: The Healing Science*

Albertus Magnus, Saint, Bp. of Ratisbon (1193?–1280). Spurious and doubtful works. Liber aggregationis; seu, Liber secretorum de virtutibus herbarum, lapidum et animalium quorundam. [The book of the collection; or, Book of secrets of the virtues of herbs, stones, and certain animals] [Rome, G. Herolt, ca. 1481] [24] leaves.

Incun. X.A4335 Rare Bk. Coll.

————Bologna, Petrus de Heidelberga, 1482. [30] leaves.

Incun. 1482.A5 Thacher Coll.

Avenzoar (d. ca. 1162). Liber Teisir; sive, Rectificatio medicationis et regiminis. [Practical manual; or, Rectification of medication and regimen] Latin. Venice, J. and G. de Gregoriis, de Forlivio, 1491. 40, 63 leaves.

Incun. 1491.G23 copy 2 Rare Bk. Coll.

Averroës (1126–1198). Colliget. [Book of universals] *In* Avenzoar. Liber Teisir. Latin. Venice, J. and G. de Gregoriis, de Forlivio, 1491. leaves 2–63 (2d group)

Incun. 1491.G23 copy 2 Rare Bk. Coll.

Avicenna (980–1037). Canon medicinae. [Canon of medicine] Latin. Padua [J. Herbort, de Seligenstadt] 1479. [434] leaves.

Incun. 1479.A96 Thacher Coll.

——Cantica de medicina. [Canticle of medicine] Latin. Venice, A. de Soziis, Parmensis, 1484. [34] leaves.

Incun. 1484.A94 Rare Bk. Coll.

Bartisch, George (1535–ca. 1607). Ophthalmodouleia; das ist, Augendienst. [Eye-service] [Dressden, Gedruckt durch M. Stöckel] 1583. 274 leaves, [15] p.

RE41.B3 Rosenwald Coll. (718)

Beaumont, William (1785–1853). Experiments and observations on the gastric juice, and the physiology of digestion. Plattsburgh, Printed by F. P. Allen, 1833. 280 p.
QP151.B36 Rare Bk. Coll.

Berengario, Jacopo (d. 1550). Commentaria . . . super Anatomia Mūdini. [Commentaries on the Anatomy of Mondino] [Bononiae, Impressum per H. de Benedictis, 1521] cccccxxviii leaves.
QM21.B43 1521 Rosenwald Coll. (802)

Bernard, Claude (1813–1878). Leçons de physiologie expérimentale appliquée à la médecine. [Lessons of experimental physiology applied to medicine] Paris, J. B. Baillière, 1855–56. 2 v.
QP31.B46

Bigelow, Henry Jacob (1816–1890). Insensibility during surgical operations produced by inhalation. Boston medical and surgical journal, v. 35, Nov. 18, Dec. 9, 1846: 309–317, 379–382.
R11.B7, v. 35

Boccaccio, Giovanni (1313–1375). Decamerone. Venice, J. and G. de Gregoriis, 1492. 137 leaves.
Incun. 1492.B65 Rosenwald Coll. (269)

Boerhaave, Herman (1668–1738). Boerhaave's aphorisms; concerning the knowledge and cure of diseases. Translated from the . . . Latin. London, Printed for W. and J. Innys, 1724. 444 p.
R128.7.B65 Rare Bk. Coll.

——Institutiones medicae. [Institutions of medicine] Lugduni Batavorum, Ex Officina Boutesteniana, 1727. 526, [34] p.
R128.7.B67 1727 Toner Coll.

Borelli, Giovanni Alfonso (1608–1679). De motv animalivm. [On the motion of animals] Romae, Ex Typographia A. Bernabò, 1680–81. 2 v.
QP301.B6 1680 Rare Bk. Coll.

Breuer, Josef (1842–1925). Studien über Hysterie. Von Jos. Breuer und Sigm. Freud. Leipzig, F. Deuticke, 1895. 269 p.
RC532.B7 Freud Coll.

Brunschwig, Hieronymus (ca. 1450–ca. 1512). Pestbuch. [Book of the plague] [Strassburg] J. Grüninger, 1500. xxxvi leaves.
Incun. 1500. B8 Rosenwald Coll. (204)

Celsus, Aulus Cornelius (fl. A.D. 25). De medicina. [On medicine] Venice, P. Pincius, for B. Fontana, 1497. xci leaves.
Incun. 1497.C4 Rare Bk. Coll.

Claudon, Émile. Fabrication du vinaigre fondée sur les études de M. Pasteur. [The making of vinegar, founded on the studies of M. Pasteur] Paris, F. Savy, 1875. 60 p.
TP445.C55 Rare Bk. Coll.

Defoe, Daniel (1661?–1731). The dreadful visitation in a short account of the progress and effects of the plague. Germantown [Pa.] Printed by C. Sower, 1763. 16 p.
RC178.G7L Rare Bk. Coll.

Abridged from his *Journal of the Plague Year.*

[Encyclopedic manuscript containing allegorical and medical drawings. South Germany, ca. 1410] [8] leaves.
Rosenwald Coll., ms. no. 3

Falloppius, Gabriel (1523–1562). Lectiones de partibvs similaribvs hvmani corporis. [Lectures on the similar parts of the human body] Noribergae, In Officina T. Gerlachii, 1575. [76] p.
QM21.F28 1575 Rosenwald Coll. (710)

Fracastoro, Girolamo (1478–1553). Opera omnia. [Complete works] Venetiis, Apvd Ivntas, 1555. [281], 32 leaves.
PA8520.F7 1555 Rare Bk. Coll.

Freud, Sigmund (1856–1939). Die Traumdeutung. [The interpretation of dreams] Leipzig, F. Deuticke [1899] 371 p.
BF1078.F7 1899 Freud Coll.

Galenus (ca. 130-ca. 200). Extra ordinem classium libri. [Books in categories outside the regular arrangement] Venetiis, Apud Haeredes L. Iuntae, 1541. [81] leaves.
R126.G315 1541 Rosenwald Coll. (838)

Harvey, William (1578–1657). Exercitatio anatomica de motv cordis et sangvinis in animalibvs. [Anatomical exercise on the motion of the heart and the blood in animals] Francofvrti, Sumptibus G. Fitzeri, 1628. [Florence, R. Lier, 1928] 72 p. (Monumenta medica, v. 5)
QP101.H35 1628b Facsimile edition.

——De motu cordis et sanguinis in animalibus. [On the motion of the heart and the blood in animals] Patavii, Apud S. Sardum, Sumptibus D. Ricciardi, 1643. 227 p.
QP101.H35 1643 Rare Bk. Coll.

——The anatomical exercises of Dr. William Harvey . . . concerning the motion of the heart and blood. London, Printed by F. Leach, for R. Lowndes, 1653. 111, 123, 86 p.
QP101.H36 1653 Rare Bk. Coll.

Helmont, Jean Baptiste van (1577–1644). Ortus medicinae. [The beginning of medicine] Amsterodami, Apud L. Elzevirium, 1652. [884, 48] p.
R128.7.H475 1652 Rare Bk. Coll.

Hippocrates (460–375 B.C.). Opera quae ad nos extant omnia. [Complete surviving works] Basileae [Per H. Frobenium et N. Episcopivm] 1546. 695 p.
R126.H51 1546 Rare Bk. Coll.

——Opera omnia qvae extant. Francofvrti, In Officina D. ac D. Aubriorum & C. Schleichij, 1624. 1344, [45] p.
R126.H5 1624 Rare Bk. Coll.

——Œuvres complètes. [Complete works] Paris, J. B. Baillière, 1839–61. 10 v.
R126.H55 1839

Translation, introduction, commentaries, and notes by Émile Littré. LC set incomplete; v. 7–10 lacking.

——Aphorismi. [Aphorisms] Parisiis, Apud S. Colinaeum, 1530. 187 leaves.
R126.H6A65 1530 Rare Bk. Coll.

Holmes, Oliver Wendell (1809–1894). On the contagiousness of puerperal fever. New England quarterly journal of medicine and surgery, v. 1, Apr. 1843: 503–530.
Micro 01104, reel 816

——Puerperal fever, as a private pestilence. Boston, Ticknor and Fields, 1855. 60 p.
RG811.H75 Toner Coll.

Jenner, Edward (1749–1823). An inquiry into the causes and effects of the Variolae vaccinae, 1798. Facsimile reprint. London, Dawsons, 1966. 75 p.
RM276.J4 1798b

———London, Printed for the author, by S. Low, and sold by Law, 1800. 182 p.
RM786.J53 1800 Rare Bk. Coll.

Kalendrier des bergers. Calendrier des bergers. [Shepherds' calendar] Paris, G. Marchant for J. Petit, 1497. [91] leaves.
Incun. 1496.K3 Rosenwald Coll. (436)

——Der scaepherders kalengier. [Shepherds' calendar] [Antwerpen, Gheprent bij W. Vorsterman, 1516] [108] p.
AY831.K317 1516 Rosenwald Coll. (1137)

Ketham, Joannes de (15th cent.). Fasciculus medicinae. [Medical miscellany] Venice, J. and G. de Gregoriis, de Forlivio, 1495. [40] leaves.
Incun. 1491.G23 Rosenwald Coll. (303)

——Fasciculꝰ medicine. [Tantwerpen, Geprint bi C. die Graue, 1529] [lxxxv] leaves.
R128.6.K393 1529 Rosenwald Coll. (1150)

Koch, Robert (1843–1910). Die Aetiologie der Tuberculose. [The etiology of tuberculosis] *In* Germany. Reichsgesundheitsamt. Mittheilungen aus dem Kaiserlichen Gesundheitsamte. 2. Bd. Berlin, A. Hirschwald, 1884. p. 1–88.
R111.G4, v. 2.

——Investigations into the etiology of traumatic infective diseases. Translated by W. Watson Cheyne. London, New Sydenham Society, 1880. 74 p. (The New Sydenham Society. [Publications] v. 88)
RC112.K77

Le Boë, Frans de (1614–1672). Opera medica. [The medical works] Amstelodami, Apud D. Elsevirium et A. Wolfgang, 1679. 934, [26] p.
R128.7.L5

Leonardo da Vinci (1452–1519). I manoscritti di Leonardo da Vinci della Reale biblioteca di Windsor. [The manuscripts of Leonardo da Vinci in the Royal Library at Windsor] Parigi, E. Rouveyre, 1898. 202 p.
QM21.L5

Lister, Joseph Lister, 1st Baron (1827–1912). On a new method of treating compound fracture, abscess, &c. with observations on the conditions of suppuration. Lancet, Mar. 16–30, Apr. 27, 1867: 326–329, 357–359, 387–389, 507–509; July 27: 95–96.
R31.L3, 1867

Mesmer, Franz Anton (1734–1815). Mémoire sur la découverte du magnétisme animal. [Report on the discovery of animal magnetism] Geneve, P. F. Didot le jeune, 1779. 85 p.
BF1132.M37 Rare Bk. Coll.

Morgagni, Giovanni Battista (1682–1771). The seats and causes of diseases investigated by anatomy. Translated from the Latin. London, Printed for A. Millar, and T. Cadell, 1769. 3 v.
RB24.M7 Toner Coll.

Paracelsus (1493–1541). Opera, Bücher vnd Schrifften. [Works, books and pamphlets] Strassburg, L. Zetzners seligen Erben, 1616.
R128.6.P2 Rare Bk. Coll.

——Opera omnia: medico-chemico-chirvrgica. [Complete works: medical-chemical-surgical] Genevae, Sumptibus I. Antonij, & S. De Tournes, 1658. 3 v. in 2.
R128.6.P3 1658 Rare Bk. Coll.

——Een excellent tracktaet leerende hoemen alle ghebreken der pocken sal moghen ghenesen. [An excellent treatise teaching how one may cure all the infirmities of the pox] Thantwerpen, Gheprint by J. Roelants, 1553. 32 leaves.
R128.6.P315 Rosenwald Coll. (1177)

Paré, Ambroise (1510?–1590). Oevvres. [Works] Paris, G. Buon, 1579. [xicvi, 108], lxxxix p.
R128.6.P35 1579 Rosenwald Coll. (1080)

——The workes . . . translated out of Latine and compared with the French. By Tho: Johnson. London, Printed by R. Cotes, and W. Dugard, 1649. 787 (i.e. 795), 50 p.
R128.6.P37 1649 Rare Bk. Coll.

Pasteur, Louis (1822–1895). Études sur le vin. [Studies on wine] 2. éd., rev. et augm. Paris, F. Savy, 1873. 344 p.
QR151.P32

——Études sur le vinaigre. [Studies on vinegar] Paris, Gauthier-Villars, 1868. 119 p.
TP445.P3

——Sur les maladies virulentes. [On virulent diseases] *In* Académie des sciences, Paris. Comptes rendus hebdomadaires des séances, t. 90, 9 fév. 1880: 239–248.
Q46.A14, v. 90

Pavlov, Ivan Petrovich (1849–1936). Lektsīi o rabotie glavnykh pishchevaritel'nykh zhelez. [Lectures on the function of the main food-digesting glands] S.-Peterburg, 1897. 223 p.
QP190.P3 1897 Rare Bk. Coll.

Pinel, Philippe (1745–1826). Traité médico-philosophique sur l'aliénation mentale. [Medical-philosophical treatise on mental derangement] 2. éd., entièrement refondue et très-augm. Paris, J. A. Brosson, 1809. xxxij, 496 p.
RC601.P65

Ramazzini, Bernardino (1633–1714). A treatise on the diseases of tradesmen, shewing the various influence of particular trades upon the state of health. Written in Latin . . . and now done in English. London, Printed for A. Bell, 1705. 274 p.
RA787.R16 Rare Bk. Coll.

Raynalde, Thomas (fl. 1540–1551). The byrth of mankind, otherwyse named the womans booke. [Imprynted at London by T. Ray] 1545. [162] leaves.
RG91.R27 1545 Rare Bk. Coll.

al-Rāzī, Abū Bakr Muhammad ibn Zakarīyā (865?–925?). Clarificatorium Iohānis de Tornamira super nono Almansoris. [Commentary of Joannes de Tornamira on the ninth book of the Almansor] [Lugd', Imp̄ssum per J. Bachalariū, 1501] clix leaves.
R143.R386 Rosenwald Coll. (915)

Regimen sanitatis Salernitanum. Regimen sanitatis Salernitanum. [The Salernitan regimen of health] Louvain, J. de Paderborn [ca. 1483–96] [136] leaves.
Incun. X.R33 Rosenwald Coll. (532)

Roeslin, Eucharius (d. 1526). Libro nel qual si tratta del parto de lhuomo. [Book in which human birth is treated] [In Vinegia, Per G. A. Vauassore detto

Guadagnino, 1538] [126] p.
RG91.R715 Rare Bk. Coll.

——De partv hominis. [On human birth] Franc[ofurti] Apud C. Egenolphum [1551] 62 leaves.
RG91.R72 1551 Toner Coll.

——Eucharius Rösslin's "Rosengarten," gedruckt im Jahre 1513. München, C. Kuhn, 1910. 110, xvii p. (Alte Meister der Medizin und Naturkunde in Facsimile-Ausgaben und Neudrucken nach Werken des 15.–18. Jahrhunderts, 2)
RG91.R7 1513ba Rare Bk. Coll.
Facsimile of the third of the three editions of 1513, published by H. Gran in Hagenau, entitled *Der Swangern Frawen vnd Heb Ammē Roszgartē* [The rose garden of pregnant women and midwives]

Rush, Benjamin (1745–1813). Medical inquiries and observations, upon the diseases of the mind. Philadelphia, Kimber & Richardson, 1812. 367 p.
RC601.R95

Santorio, Santorio (1561–1636). De statica medicina aphorismorvm. [Aphorisms on medical measurement] Lugduni Batavorum, Apud C. Boutesteyn, 1703. 231 p.
R128.7.S2 Rare Bk. Coll.

——Medicina statica; or, Rules of health, in eight sections of aphorisms. English'd by J. D. London, J. Starkey, 1676. 180 p.
RA775.S21

Semmelweis, Ignác Fülöp (1818–1865). Die Aetiologie, der Begriff und die Prophylaxis des Kindbettfiebers. [The etiology, the concept, and the prophylaxis of puerperal fever] Reprinted from the 1861 ed. New York, Johnson Reprint Corp., 1966. xxxii, 543 p. (The Sources of science, no. 19)
RG811.S43 1966

Sydenham, Thomas (1624–1689). Opera medica. [The medical works] Genevae, Apud fratres De Tournes, 1757. 2 v.
R114.S94 1757 Rare Bk. Coll.

——The whole works of that excellent practical physician, Dr. Thomas Sydenham. London, M. Wellington, 1717. xv, 447 p.
R114.S96 1717 Toner Coll.

Vesalius, Andreas (1514–1564). De humani corporis fabrica. [On the structure of the human body] Basileae [Ex Officina I. Oporini, 1543] [663] leaves.
QM21.V418 Rosenwald Coll. (907)

——De humani corporis fabrica librorum epitome. [Summary of the books on the structure of the human body] Basileae [Ex Officina I. Oporini, 1543] [27] p.
QM21.V425 Rosenwald Coll. (906)

Virchow, Rudolf Ludwig Karl (1821–1902). Die Cellularpathologie in ihrer Begründung auf physiologische und pathologische Gewebelehre. [Cellular pathology as based upon physiological and pathological histology] Berlin, A. Hirschwald, 1858. xvi, 440 p.
RB25.V81 Rare Bk. Coll.

5. *Chemistry: Fertile Alchemy*

Agricola, Georg (1494–1555). De re metallica. [On metals] Basileae [Apud H. Frobenivm et N. Episcopivm] 1556. [502, 74] p.
TN617.A25 1556 Rosenwald Coll. (910)

Arrhenius, Svante August (1859–1927). Försök att beräkna dissociationen (aktivitetskoefficienten) hos i vatten lösta kroppar. [Attempt to calculate the dissociations (activity coefficients) of bodies dissolved in water] *In* Svenska vetenskapsakademien, Stockholm. Öfversigt af förhandlingar, 44. årg., no. 6, 1887: 405-414.
Q64.S87, v. 44

Artis auriferae. [On the art of producing gold] Basileae, Excudebat C. Vvaldkirch, Expensis C. de Marne, & I. Avbry, 1593. 2 v.
QD25.A2A8 1593 Rare Bk. Coll.

——Accessit nouiter volumen tertium. Basileae, Typis C. Waldkirchi, 1610. 3 v.
Micro QD-17

Ashmole, Elias (1617–1692), comp. Theatrum chemicum britannicum. Containing severall poeticall pieces of our famous English philosophers, who have written the hermetique mysteries in their owne ancient language. The first part. London, Printed by F. Grismond for N. Brooke, 1652. 486 p.
QD25.A78 Rare Bk. Coll.
No more published.

Avogadro, Amedeo (1776–1856). Essai d'une manière de déterminer les masses relatives des molécules élémentaires des corps. [An essay on a method of determining the relative masses of the elementary molecules of substances] Journal de physique, de chimie, d'histoire naturelle et des arts, t. 73, juil. 1811: 58–76.
Q2.J79, v. 73

Bacon, Roger (1214?–1294). Le miroir d'alquimie. [The mirror of alchemy] Lyon, Macé Bonhomme, 1557. 4 pts. in 1 v.
QD25.B12 Rare Bk. Coll.

Berzelius, Jöns Jakob, Friherre (1779–1848). Lehrbuch der Chemie. [Textbook of chemistry] Übersetzt von F. Wöhler. Dresden, Arnold, 1825–31. 4 v. in 8.
QD28.B567

————4. verb. Original-Aufl. Dresden, Arnold, 1835–41. 10 v.
QD28.B57

Biringucci, Vannuccio (1480–1539?). De la pirotechnia. [On pyrotechnics] [In Venetia, Per V. Roffinello] 1540. 168 leaves.
TN144.B45 1540 Rare Bk. Coll.

Black, Joseph (1728–1799). Experiments upon magnesia alba, quicklime, and some other alcaline substances. *In* Philosophical Society of Edinburgh. Essays and observations, physical and literary. v. 2. Edinburgh, Printed by G. Hamilton and J. Balfour, 1756. p. 157–225.
AS122.E5, v. 2

——Lectures on the elements of chemistry. Edinburg, Printed by Mundell for Longman and Rees, London, and W. Creech, Edinburgh, 1803. 2 v.
QD28.B7 1803 Rare Bk. Coll.

Boyle, Robert (1627–1691). Chymista scepticus. [The skeptical chemist] Ed. 2. priori emendiator. Roterodami, A. Leers, 1668. [28], 392 p.
QD27.B7816 1668 Rare Bk. Coll.

——The sceptical chymist. Oxford, Printed by H. Hall for R. Davis and B. Took, 1680. 440, [27], 268 p.
QD27.B75 Rare Bk. Coll.

Brunschwig, Hieronymus (ca. 1450-ca. 1512). Kleines Distillierbuch. [Little distillation book] Strassburg, J. Grüninger, 1500. [18], ccix leaves.
Incun. 1500.B78 Rosenwald Coll. (202)

Cavendish, Henry (1731–1810). The electrical researches of Henry Cavendish, written between 1771 and 1781. Edited . . . by J. Clerk Maxwell. Cambridge [Eng.] University Press, 1879. lxvi, 454 p.
QC517.C35

——Experiments on air. *In* Royal Society of London. Philosophical transactions. v. 74–75; 1784–85. London, Sold by L. Davis, and P. Elmsly, Printers to the Royal Society. p. 119–153; p. 372–384.
Q41.L8, v. 74–75 Rare Bk. Coll.

Dalton, John (1766–1844). A new system of chemical philosophy. Manchester, Printed by S. Russell for R. Bickerstaff, London, 1808–27. 2 v. in 3.
QD28.D15 Rare Bk. Coll.

Davy, Sir Humphry, Bart. (1778–1829). Elements of chemical philosophy. pt. 1, v. 1. Philadelphia, Bradford and Inskeep, 1812. 296 p.
QD28.D27 Toner Coll.
No more published.

De alchemia. [On alchemy] Norimbergae, Apud I. Petreium, 1541. 373 p.
Micro QD-6

Gay-Lussac, Joseph Louis (1778–1850). Recherches physico-chimiques. [Physico-chemical researches] Paris, Deterville, 1811. 2 v.
QD3.G28 Rare Bk. Coll.

Geber (13th cent.). The works of Geber, the most famous Arabian prince and philosopher. Faithfully Englished. London, Printed for N. E. by T. James, 1678. 302 p.
QD25.G3613 1678 Rare Bk. Coll.

——Chimia. [Chemistry] [Lugduni Batavorum, Apud A. Doude, 1668] [279] p.
QD25.G367 1668 Rare Bk. Coll.

Gibbs, Josiah Willard (1839–1903). On the equilibrium of heterogeneous substances. *In* Connecticut Academy of Arts and Sciences, New Haven. Transactions. v. 3. New Haven, 1876–78. p. 108–248, 343-524.
Q11.C9, v. 3

Guyton de Morveau, Louis Bernard, baron (1737–1816). Méthode de nomenclature chimique, proposée par MM. de Morveau, Lavoisier, Bertholet, & de Fourcroy. [Method of chemical nomenclature, proposed by MM. de Morveau, Lavoisier, Bertholet, & de Fourcroy] Paris, Cuchet, 1787. 314 p.
QD7.G85 Batchelder Coll.

Lavoisier, Antoine Laurent (1743–1794). Traité élémentaire de chimie. [Elementary treatise on chemistry] Paris, Cuchet, 1789. 2 v.
QD28.L4 1789 Rare Bk. Coll.

Lémery, Nicolas (1645–1715). Cours de chymie. [A course of chemistry] Paris, J.-B. Delespine, 1730. 938, [58] p.
QD27.L4 1730 Rare Bk. Coll.

——A course of chymistry. London, Printed for W. Kettilby, 1680. 323, [35], 140 p.
QD27.L55 1680 Rare Bk. Coll.

Libavius, Andreas (d. 1616). Alchymia. [Alchemy] Francofvrti, Excudebat J. Saurius, Impensis P. Kopffii, 1606. 2 v. in 1.

QD25.L5 Rare Bk. Coll.

Liebig, Justus, Freiherr von (1803–1873). Anleitung zur Analyse organischer Körper. [Introduction to the analysis of organic substances] Braunschweig, F. Vieweg, 1837. 72 p.

QD3.N31 Rare Bk. Coll.

——Organic chemistry in its applications to agriculture and physiology. Cambridge, J. Owen; Boston, J. Munroe, 1841. xx, 435 p.

S585.L72 1841

——Die organische Chemie in ihrer Anwendung auf Physiologie und Pathologie. [Organic chemistry in its application to physiology and pathology] Braunschweig, F. Vieweg, 1842. xvi, 342 p.

QP514.L718

Lull, Ramón (d. 1315), supposed author. De secretis naturae. [On the secret of nature] [Argentorati, B. Beck] 1541. 56 leaves.

Micro QD-15

—— ——Venetijs, Apud P. Schoeffer, 1542. 324 p.

QD25.L82 1542 Rare Bk. Coll.

Maier, Michael (1568?–1622). Atalanta fugiens, hoc est, Emblemata nova de secretis naturae chymica. [The fleeing Atalanta, that is, new emblems of the chemical secrets of nature] Oppenheimii, Ex typographia H. Galleri, sumptibus J. T. de Bry, 1618. 211 p.

QD25.M2A8 1618 Rare Bk. Coll.

Manget, Jean Jacques (1652–1742). Bibliotheca chemica curiosa. [A thoroughgoing chemical library] Genevae, Sumpt. Chouet, G. De Tournes, Cramer, Perachon, Ritter, & S. De Tournes, 1702. 2 v.

QD25.M27 Rare Bk. Coll.

Mendeleev, Dmitriĭ Ivanovich (1834–1907). O soedineniī spirta s vodoiu. [On the compounds of alcohol with water] Sanktpeterburg, Obshchestvennaia pol'za, 1865. 119 p.

QD305.A4M45 Rare Bk. Coll.

——Osnovy khimiī. [The principles of chemistry] 6. izd. S.-Peterburg, Tip. V. Demakova, 1895. 778 p.

QD31.M528 1895

——Sootnosheniē svoĭstv s atomnym viesom elementov. [Relation of the properties to the atomic weights of the elements] Zhurnal Russkago khimicheskago obshchestva, t. 1, vyp. 2/3, 1869: 60–77.

Micro 04854

Musaeum hermeticum. [The hermetic museum] Francofurti, Sumptibus L. Jennisii, 1625. 445, 35 p.

QD25.M75 1625 Rare Bk. Coll.

Musaeum hermeticum reformatum et amplificatum. [The hermetic museum, improved and enlarged] Francofurti, Apud Hermannum à Sande, 1678. 863 p.

QD25.M75 1678 Rare Bk. Coll.

Priestley, Joseph (1733–1804). The doctrine of phlogiston established, and that of the composition of water refuted. Northumberland [Pa.] Printed for the author by A. Kennedy, 1800. xv, 90 p.

QD27.P67 Toner Coll.

——Experiments and observations on different kinds of air. London, J. Johnson, 1774–77. 3 v.

QD27.P7 1774 Rare Bk. Coll.

Rayleigh, John William Strutt, Baron (1842–1919). Argon, a new constituent of the atmosphere. By Lord Rayleigh and Professor William Ramsay. *In* Royal Society of London. Philosophical transactions. v. 186A, pt. 1; 1895. London, Printed by Harrison. p. 187–241.

Q41.L8, v. 186A, pt. 1

—— ——City of Washington, Smithsonian Institution, 1896. 43 p. (Smithsonian contributions to knowledge, [v. 29, art. 4]) (Smithsonian Institution publication 1033)

Q11.S68, v. 29

——Density of nitrogen. Nature, v. 46, Sept. 29, 1892: 512–513.

Q1.N2, v. 46

Roth-Scholtz, Friedrich (1687–1736). Bibliotheca chemica; oder, Catalogus von Chymischen-Büchern. [The chemical library; or, Catalog of chemistry books] [Nürnberg, Bey J. D. Taubers seel. Erben, 1727–29] Hildesheim, New York, G. Olms, 1971. 5 v. in 1.

Z5526.R65 1971 Facsimile reprint.

——Deutsches Theatrum chemicum. [The German chemical theater] Nürnberg, Bey A. J. Felsseckern, 1728–32. 3 v.

QD25.R6 Rare Bk. Coll.

Scheele, Karl Wilhelm (1742–1786). Chemische Abhandlung von Luft und Feuer. [A chemical essay on air and fire] Leipzig, Bey S. L. Crusius, 1782. 32, 286 p.

QD27.S38 1782 Rare Bk. Coll.

Stahl, Georg Ernst (1660–1734). Zufällige Gedancken und nützliche Bedencken über den Streit, von dem so genannten Svlphvre. [Random thoughts and useful reflections on the dispute about the so-called sulfur] Halle, In Verlegung des Wäysenhauses, 1718. 373 p.

QD181.S1S7 Rare Bk. Coll.

Theatrum chemicum. [Chemical theater] Argentorati, Sumptibus Heredum E. Zetzneri, 1659–61. 6 v.

QD25.T4 Rare Bk. Coll.

Trismosin, Salomon (pseud.?). Avrevm vellvs. [The golden fleece] Erstlich gedruckt zu Rorschach am Bodensee, 1599. 214, 165, [702] p.

QD25.A2T7 1599 Rare Bk. Coll.

Verae alchemiae . . . doctrina. [Science of the true alchemy] Basileae [Per H. Petri & P. Pernam] 1561. 244, 299 p.

QD25.V4 1561 Rare Bk. Coll.

Edited by Guglielmo Grataroli.

Wöhler, Friedrich (1800–1882). Ueber künstliche Bildung des Harnstoffs. [On the synthetic preparation of urea] Annalen der Physik und Chemie, 12. Bd., 2. Stück, 1828: 253–256.

QC1.A6, v. 12

6. *Geology: The Secret in the Stone*

Agassiz, Louis (1807–1873). Études sur les glaciers. [Studies on the glaciers] Neuchâtel, Jent et Gassmann, 1840. 346 p. and atlas.
QE576.A26 Rare Bk. Coll.

Agricola, Georg (1494–1555). De ortu & causis subterraneorum . . . De natura fossilium. [On the origin and causes of things underground. On the nature of minerals] Basileae [Per H. Frobenivm et. N. Episcopivm] 1546. 487, [52] p.
QE25.A34 Rare Bk. Coll.

——De re metallica. [On metals] Basileae [Apud H. Frobenivm et N. Episcopivm] 1556. [502, 74] p.
TN617.A25 1556 Rosenwald Coll. (910)

—— ——Translated . . . by Herbert Clark Hoover and Lou Henry Hoover. London, Mining Magazine, 1912. xxxi, 640 p.
TN617.A4 Rare Bk. Coll.

Albertus Magnus, Saint, Bp. of Ratisbon (1193?–1280). De mineralibus. [On minerals] Pavia, C. de Canibus, 1491. [28] leaves.
Incun. 1491.V47 Rare Bk. Coll.

—— ——Coloniae, Apud I. Birckmannum & T. Baumium, 1569. 391, [11] p.
QE362.A6 Rare Bk. Coll.

Beringer, Johann Bartholomäus Adam (d. 1740). The lying stones of Dr. Johann Bartholomew Adam Beringer, being his Lithographiae Wirceburgensis. Translated and annotated by Melvin F. Jahn and Daniel J. Woolf. Berkeley, University of California Press, 1963. xiv, 221 p.
QE714.3.B413

Burnet, Thomas (1635?–1715). Telluris theoria sacra. [The sacred theory of the earth] Amstelaedami, Apud J. Wolters, 1699. 558 p.
BL224.B8 Rare Bk. Coll.

——The sacred theory of the earth. 6th ed. London, Printed for J. Hooke, 1726. 2 v.
BL224.B82 1726 Rare Bk. Coll.

Cuvier, Georges, baron (1769–1832). Discours sur les révolutions de la surface du globe. [A discourse on the revolutions of the surface of the globe] Paris, G. Dufour et E. d'Ocagne, 1826. 196 p.
QE501.C99

——Essai sur la géographie minéralogique des environs de Paris. [Essay on the mineralogical geography of the environs of Paris] Par G. Cuvier et Alexandre Brongniart. Paris, Baudouin, 1811. 278 p.
QE268.C97 G&M

——Recherches sur les ossemens fossiles. [Investigations of fossil remains] Paris, Déterville, 1812. [Bruxelles, Culture et Civilisation, 1969] 4 v.
QE841.C924 Facsimile reprint.

—— ——3. éd. Paris, G. Dufour et E. d'Ocagne, 1825. 5 v. in 7.
QE710.C97

Dana, James Dwight (1813–1895). Manual of geology. Philadelphia, T. Bliss, 1863. xvi, 798 p.
QE26.D2 1863

——A system of mineralogy. New Haven, Durrie & Peck, and Herrick & Noyes, 1837. xiv, 452, 119 p.
QE372.D23 1837

Desmarest, Nicolas (1725–1815). Mémoire sur la détermination de trois époques de la nature par les produits des volcans. [Report of the determination of three epochs of nature by the products of volcanos] *In* Académie des sciences, Paris. Mémoires de l'Institut des sciences, lettres et arts. Sciences mathématiques et physiques. t. 6. Paris, Baudouin, 1806. p. 219–289.
Q46.A13, v. 6

Eratosthenes (ca. 276-ca. 196 B.C.). Eratosthenica. [Fragments of the writings of Eratosthenes] Composuit Godefredus Bernhardy. Berolini, Impensis G. Reimeri, 1822. xvi, 272 p.
PA3970.E4 1822

Ercker, Lazarus (d. 1593). Beschreibung aller fürnemisten mineralischen Ertzt vnnd Bergkwercks Arten. [Description of all the principal mineral ores and mining methods] Franckfurt am Mayn, J. Feyerabendt, 1598. 134 leaves.
TN664.E7 1598 Rare Bk. Coll.

Faujas de Saint-Fond, Barthélémy (1741–1819). Recherches sur les volcans éteints du Vivarais et du Velay. [Researches on the extinct volcanos of Vivarais and Velay] Grenoble, J. Cuchet, 1778. xviii, 460 p.
QE527.F3 Rare Bk. Coll.

Guettard, Jean Étienne (1715–1786). Mémoire et carte minéralogique sur la nature & la situation des terreins qui traversent la France & l'Angleterre. [Mineralogical memoir and map on the nature and location of the terrains that span France and England] *In* Académie des sciences, Paris. Mémoires de mathématique et de physique, tirés des registres de l'Académie royale des sciences, de l'année 1746. [A Paris, De l'Impr. royale, 1751] p. 363–392.
Q46.A13, 1746 Rare Bk. Coll.

Haüy, René Just (1743–1822). Essai d'une théorie sur la structure des crystaux. [Attempt at a theory on the structure of crystals] Paris, Chez Gogué & Mée de la Rochelle, 1784. 236 p.
QC516.H33 1787 Rare Bk. Coll.

——Traité de minéralogie. [Treatise on mineralogy] Paris, Chez Louis, 1801. 4 v. and atlas.
QE363.H2 Jefferson Coll.

Hutton, James (1726–1797). Theory of the earth; or, An investigation of the laws observable in the composition, dissolution, and restoration of land upon the globe. *In* Royal Society of Edinburgh. Transactions. v. 1, pt. 2. Edinburgh, Printed for J. Dickson, Bookseller to the Royal Society, 1788. p. 209–304.
Q41.E2, v. 1

Lyell, Sir Charles, Bart. (1797–1875). Principles of geology, being an attempt to explain the former changes of the earth's surface, by reference to causes now in operation. London, J. Murray, 1830–33. 3 v.
QE26.L956 Rare Bk. Coll.

Murchison, Sir Roderick Impey, Bart. (1792–1871). The Silurian system. London, J. Murray, 1839. 2 v. in 1.
QE661.M94

Palissy, Bernard (1510?–1590). Œuvres. [Works] Paris, Ruault, 1777. [lxxvi], 734 p.
AC21.P24 Rare Bk. Coll.

Playfair, John (1748–1819). Illustrations of the Huttonian theory of the earth. Edinburgh, Printed for W. Creech, 1802. xx, 528 p.
QE26.P64 1802 Rare Bk. Coll.

Reisch, Gregor (d. 1525). Margarita philosophica noua. [The new philosophical pearl] [Ex Argentoraco veteri, J. Grüningerus, 1515] 324 leaves.
AE3.R384 Rare Bk. Coll.

Saussure, Horace Bénédict de (1740–1799). Voyages dans les Alpes. [Travels in the Alps] Neuchatel, S. Fauche, 1779–90. 4 v.
DQ823.S245

Smith, William (1769–1839). A delineation of the strata of England and Wales with part of Scotland. [London, W. Cary] 1815. 1 map on 15 sheets, index map.
G1811.C5S6 1815 G&M Vault

Steno, Nicolaus (1638–1686). The prodromus of Nicolaus Steno's dissertation concerning a solid body enclosed by process of nature within a solid. An English version. New York, Macmillan Co., 1916. 169–283 p. (University of Michigan studies. Humanistic series, v. 11. Contributions to the history of science, pt. 2)
QE709.S85

Strabo (ca. 63 B.C.-A.D. 20). De situ orbis. [On the geography of the world] [Venetiis, Impressum per B. de Zanis de Portesio, 1502] 150 leaves.
G87.S88 1502 Rare Bk. Coll.

——[Venetiis, a P. Pincio Mantuano Impressum, 1510] cl leaves.
G87.S88 1510 Rare Bk. Coll.

——Geographica, latine. [Geography, in Latin] [Venetiis] V. de Spira, 1472. 219 leaves.
Incun. 1472.S89 Rare Bk. Coll.

Theophrastus (ca. 374-ca. 286 B.C.). Peri lithōn. [On stones] *In* Aristoteles. Opera. v. 2. Venice, Aldus Manutius, Romanus, 1497. leaves 254–260.
Incun. 1495.A7, v. 2 Rare Bk. Coll.

——Traité des pierres. [Treatise on stones] Traduit du grec. A Paris, Chez J.-T. Herissant, 1754. xxiv, 287 p.
QE362.T52 Rare Bk. Coll.

Ussher, James, Abp. of Armagh (1581–1656). Annales Veteris Testamenti. [Annals of the Old Testament] Londini, Ex Officina J. Flesher, & prostant apud J. Crook & J. Baker, 1650. 554, [10] p.
D57.U8 Rare Bk. Coll.

——The annals of the world. London, Printed by E. Tyler for J. Crook and G. Bedell, 1658. 907, [49] p.
D57.U87 Rare Bk. Coll.

Wegener, Alfred Lothar (1880–1930). Die Entstehung der Kontinente und Ozeane. [The origin of the continents and oceans] 3., gänzlich umgearb. Aufl. Braunschweig, F. Vieweg, 1922. 144 p. (Die Wissenschaft, Bd. 66)
GB60.W4 1922

Werner, Abraham Gottlob (1749–1817). A treatise on the external characters of fossils. Translated from the German. Dublin, M. N. Mahon, 1805. xx, 312 p.
QE367.W545

Whiston, William (1667-1752). A new theory of the earth. London, Printed by R. Roberts, for B. Tooke, 1696. 95, 388 p.
BL224.W5 1696 Rare Bk. Coll.

Woodward, John (1665–1728). An essay toward a natural history of the earth and terrestrial bodies. London, Printed for R. Wilkin, 1695. 277 p.
QE25.W58 Rare Bk. Coll.

Alberti, Leone Battista (1404–1472). Della pittura e della statua. [On painting and on statuary] *In* Leonardo da Vinci. Trattato della pittvra. In Parigi, Appresso G. Langlois, 1651. p. 1–62 (2d group)
ND1130.L5 1651 Rare Bk. Coll.
Translated from the Latin by Cosimo Bartoli.

——Milano, Società tip. de' Classici italiani, 1804. xxvi, 136 p.
N7420.A4

Apollonius Pergaeus (ca. 212–190 B.C.). Conicorum. [Of conics] Oxoniae, E Theatro Sheldoniano, 1710. 3 pts. in 1 v.
QA31.A4 1710 Rare Bk. Coll.
Edmond Halley's edition.

Archimedes (ca. 287–212 B.C.). Opera, quae quidem extant, omnia. [Complete surviving works] Basileae, I. Heruagius, 1544. 139, 65 p.
QA31.A681 Rare Bk. Coll.

Boethius (d. 524). De institutione arithmetica. [On the teaching of arithmetic] Augsburg, E. Ratdolt, 1488. [48] leaves.
Incun. 1488.B66 Rosenwald Coll. (123)

Bólyai, Farkas (1775–1856). Tentamen iuventutem studiosam in elementa matheseos purae elementaris ac sublimioris methodo intuitiva evidentiaque huic propria introducendi. [An attempt to introduce studious youth to the elements of pure mathematics, by an intuitive method and appropriate evidence] Ed. 2. Budapestini, Sumptibus Academiae Scientiarum Hungaricae, 1897–1904. 2 v. and atlas.
QA36.B6

Bólyai, János (1802–1860). Appendix scientiam spatii absolute veram exhibens. [Appendix explaining the absolutely true science of space] Ed. nova. Lipsiae, In Ædibus B. G. Teubneri, 1903. 40 p.
QA685.B68

Boole, George (1815–1864). An investigation of the laws of thought, on which are founded the mathematical theories of logic and probabilities. London, Walton and Maberly, 1854. 424 p.
BC135.B7

Calandri, Filippo (15th cent.). Aritmetica. [Arithmetic] Florence, L. Morgiani and J. Petri, 1492. [104] leaves.
Incun. 1492.C3 Rosenwald Coll. (266)

Cardano, Girolamo (1501–1576). Artis magnae. [The great skill] *In his* Opvs novvm de proportionibvs nvmerorvm. [New work on the proportions of numbers] Basileae [Ex Officina Henricpetrina, 1570] p. 1–163 (2d group)
QA33.C27 Rare Bk. Coll.

Descartes, René (1596–1650). Discours de la methode pour bien conduire sa raison, & chercher la verité dans les sciences. [Discourse on the method of rightly conducting the reason and seeking truth in the sciences] A Leyde, De l'Impr. de I. Maire, 1637. 78, 413, [34] p.
Q155.D43 Rare Bk. Coll.

——Geometria. [Geometry] Lvgdvni Batavorum, Ex Officinâ I. Maire, 1649. 336 p.
QA33.D43 1649

Dodgson, Charles Lutwidge (1832–1898). Symbolic logic. pt. 1. Elementary. By Lewis Carroll. London, New York, Macmillan, 1896. xxxi, 188 p.
BC135.D67 Rare Bk. Coll.
No more published.

Dürer, Albrecht (1471–1528). Vnderweysung der Messung. [Instruction in measurement] [Nüremberg] 1525. [178] p.
NC765.D8 Rosenwald Coll. (655)

Euclides (fl. 300 B.C.). Elementa geometria. [Elements of geometry] Venice, E. Ratdolt, 1482. [138] leaves.
Incun. 1482.E8616 Rare Bk. Coll.

Euler, Leonhard (1707–1783). Methodus inveniendi lineas curvas maximi minimive proprietate gaudentes. [The art of finding curved lines which enjoy some maximum or minimum property] Lausannae, Apud M.-M. Bousquet, 1744. 322 p.
QA315.E88 Rare Bk. Coll.

Fermat, Pierre de (1601–1665). Varia opera mathematica. [Diverse mathematical works] Tolosae, Apud J. Pech, 1679. 210 p.
QA33.F33 Rare Bk. Coll.

Fibonacci, Leonardo (fl. 1220). Scritti di Leonardo Pisano. [Writings of Leonardo of Pisa] Roma, Tipografia delle scienze mathematiche e fisiche, 1857–62. 2 v.
QA32.F5

Gauss, Karl Friedrich (1777–1855). Disqvisitiones arithmeticae. [Arithmetical investigations] Lipsiae, In Commission apvd G. Fleischer, 1801. xviii, 668, [10] p.
QA241.G26 Rare Bk. Coll.

Gödel, Kurt (1906–1978). Über formal unentscheidbare Sätze der Principia mathematica und verwandter Systeme I. [On formally undecidable propositions of Principia mathematica and related systems] Monatshefte für Mathematik und Physik, 38. Bd., 1. Heft, 1931: 173–198.
QA1.M877, v. 38

Isidorus, Saint, Bp. of Seville (d. 636). De vocabvlo arithmetice disciplinae. [On the name of the science of arithmetic] *In his* Etymologiae. [Etymologies] [Augustae Vindelicorum] G. Zainer, 1472. leaves 43^{v}–46^{v}.
Incun. 1472.I81 Rare Bk. Coll.

Lagrange, Joseph Louis, comte (1736–1813). Mécanique analytique. [Analytical mechanics] Nouv. éd. rev. et augm. Paris, Ve Courcier, 1811–15. 2 v.
QA804.L17 Rare Bk. Coll.

Leibniz, Gottfried Wilhelm (1646–1716). Nova methodus pro maximis et minimis. [A new method for maxima and minima] Acta eruditorum, Oct. 1684: 467–473.
Z1007.A18, 1684 Rare Bk. Coll.

Lobachevskiĭ, Nikolaĭ Ivanovich (1792–1856). Études géométriques sur la théorie des paralleles. [Geometric studies on the theory of parallels] Traduit de l'allemand. Paris, Gauthier-Villars, 1866. 42 p.
QA685.L785

——Ob izchezaniī trigonometricheskikh strok. [On the disappearance of trigonometric lines] Kazan, V Univ. tip., 1834. 62 p.
QA404.L76 Rare Bk. Coll.

——Pangéométrie; ou, Précis de géométrie fondée sur une théorie générale et rigoreuse des paralleles. [Pangeometry; or, A summary of geometry founded

on a general and rigorous theory of parallels] *In* Kazan. Universitet. Sbornik uchenykh stateĭ. t. 1. Kazan, 1856. p. 277–340.
Q60.K3, v. 1

——Polnoe sobranie sochineniĭ. [Complete works] Moskva, Gos. izd-vo tekhniko-teoreticheskoi lit-ry, 1946–51. 5 v.
QA3.L6

LC set lacks v. 2.

Mueller, Johannes, Regiomontanus (1436–1476). De triangvlis omnímodis. [On all classes of triangles] Norimbergae, In Aedibus I. Pe[tr]ei, 1533. 2 pts. in 1 v.
QA33.M88 Rare Bk. Coll.

Napier, John (1550–1617). Mirifici logarithmorum canonis descriptio. [Description of the wonderful canon of logarithms] Edinbvrgi, Ex Officinâ A. Hart, 1614. 57, [91] p.
QA33.N44 Rare Bk. Coll.

——The construction of the wonderful canon of logarithms. Translated from Latin into English. Edinburgh, Blackwood, 1889. xix, 169 p.
QA33.N46 Rare Bk. Coll.

Newton, Sir Isaac (1642–1727). Analysis per quantitatum series, fluxiones, ac differentias. [Analysis by quantitative series, fluxions, and also differences] Londini, Ex Officina Pearsoniana, 1711. [14], 101 p.
QA35.N5 Rare Bk. Coll.

——The method of fluxions and infinite series. London, Printed by H. Woodfall, and sold by J. Nourse, 1736. xxiv, 339 p.
QA35.N565 Rare Bk. Coll.

——Opticks; or, A treatise of the reflexions, refractions, inflexions and colours of light. London, Printed for S. Smith, and B. Walford, 1704. 144, [213] p.
QC353.N556 Rare Bk. Coll.

——Philosophiae naturalis principia mathematica. [The mathematical principles of natural philosophy] Londini, Jussu Societatis Regiae ac Typis J. Streater, 1687. 383, 400–510 p.
QA803.A2 1687 Rare Bk. Coll.

Paccioli, Luca (d. ca. 1514). Somma di aritmetica, geometria, proporzione e proporzionalità. [The whole of arithmetic, geometry, proportion and proportionality] Venice, Paganinus de Paganinis, 1494. 224, 76 leaves.
Incun. 1494.P3 Rosenwald Coll. (294)

Recorde, Robert (1510?–1558). The grounde of artes: teaching the worke and practise of arithmetike. [London, Imprinted by R. Wolfe, 1561] [400] p.
QA33.R3 1561 Rare Bk. Coll.

——The pathway to knowledg, containing the first principles of geometrie. [London, Imprinted by R. Wolfe, 1551] [182] p.
QA33.R31 Rare Bk. Coll.

——The whetstone of witte, whiche is the seconde part of arithmetike. [London] J. Kyngstone [1557] [328] p.
QA33.R32 Rare Bk. Coll.

Reisch, Gregor (d. 1525). Margarita philosophica. [The philosophical pearl] [Friburgi, I. Schott, 1503] [604] p.
AE3.R34 Rosenwald Coll. (595)

Russell, Bertrand Russell, 3d Earl (1872–1970). The principles of mathematics. v. 1. Cambridge [Eng.] University Press, 1903. xxix, 534 p.
QA9.R88

Tartaglia, Niccolò (d. 1557). La noua scientia. [The new science] [Stampata in Venetia per N. de Bascarini, 1550] 32 leaves.
QC123.T3 1550 Rare Bk. Coll.

——Quesiti et inventioni diverse. [Various problems and inventions] [In Venetia, per N. de Bascarini] 1554. 128 leaves.
U101.T3 Rare Bk. Coll.

Tunstall, Cuthbert, Bp. of Durham (1474–1559). De arte svppvtandi. [On the art of computation] [Londini, Impress. in Ædibvs R. Pynsoni, 1522] [407] p.
QA35.T9 Rosenwald Coll. (1218)

Viète, François (1540–1603). Opera mathematica. [Mathematical works] Lvgdvni Batavorvm, Ex Officinâ B. & A. Elzeviriorum, 1646. 554 p.
QA33.V5 Rare Bk. Coll.

Whitehead, Alfred North (1861–1947). Principia mathematica. By Alfred North Whitehead and Bertrand Russell. 2d ed. Cambridge [Eng.] University Press, 1925–27. 3 v.
QA9.W5 1925 Rare Bk. Coll.

8. *Physics: The Elemental Why*

Alhazen (965–1039). Opticae thesavrvs. [Treasury of optics] Basileae, Per Episcopios, 1572. 288 p.
QC353.A3316 Rare Bk. Coll.

Ampère, André Marie (1775–1836). De l'action mutuelle de deux courans électriques. [On the mutual action of two electric currents] Annales de chimie et de physique, [2. sér.], t. 15, sept.-oct. 1820: 59–76, 170–218.
QD1.A7, s. 2, v. 15

——Théorie des phénomènes électro-dynamiques. [Theory of electrodynamic phenomena] Paris, Méquignon-Marvis, 1826. 226 p.
QC517.A63 1826 Rare Bk. Coll.

Archimedes (ca. 287–212 B.C.). Opera, quae quidem extant, omnia. [Complete surviving works] Basileae, I. Heruagius, 1544. 139, 65 p.
QA31.A681 Rare Bk. Coll.

Aristoteles (ca. 384–322 B.C.). Opera. [Works] Venice, Aldus Manutius, Romanus, 1495–98. 5 v. in 6.
Incun. 1495.A7 Rare Bk. Coll.

——De caelo. [On the heavens] Latin. Venice, B. Locatellus for O. Scotus, 1495. 76 leaves.
Incun. 1495.A73 Rare Bk. Coll.

——De generatione et corruptione. [On generation and corruption] *In* Paulus Venetus (d. 1429). Expositio in Aristotelem De generatione et corruptione. Venice, B. Locatellus, for O. Scotus, 1498. 118 leaves.
Incun. 1498.P35 Rosenwald Coll. (333)

——Physica. [Physics] Latin. [Louvain, C. Braem, ca. 1475] [68] leaves.
Incun. 1475.A7 Rosenwald Coll. (514)

Becquerel, Antonie Henri (1852–1908). Recherches sur une propriété nouvelle de la matière; activité radiante spontanée, ou radioactivité de la matière. [Investigation of a new property of matter; spontaneous radiant activity, or radioactivity of matter] [Paris, Firmin-Didot, 1903] 360 p. (Mémoires de l'Académie des sciences de l'Institut de France, [2. sér.], t. 46)
Q46.A13, s. 2, v. 46

——Sur les radiations invisibles émises par les sels d'uranium. [On the invisible radiation emitted by uranium salts] *In* Académie des sciences, Paris. Comptes rendus hebdomadaires des séances, t . 122, 23 mars 1896: 689–694.
Q46.A14, v. 122

Bohr, Niels Henrik David (1885–1962). Atomic theory and the description of nature. v. 1. New York, Macmillan Co., 1934. 119 p.
QC173.B535

No more published.

——On the constitution of atoms and molecules. Philosophical magazine, 6th ser., v. 26, July, Sept., Nov. 1913: 1–25, 476–502, 857–875.
Q1.P5, s. 6, v. 26

——The quantum postulate and the recent development of atomic theory. Nature, v. 121, Apr. 14, 1928: 580–590.
Q1.N2, v. 121

Boyle, Robert (1627–1691). The works. Epitomiz'd . . . by Richard Boulton. London, Printed for J. Phillips, and J. Taylor, 1699–1700. 4 v.
QC3.B8 1699 Rare Bk. Coll.

——The works. A new ed. London, Printed for J. and F. Rivington, 1772. 6 v.
QC3.B9 Rare Bk. Coll.

——Nova experimenta physico-mechanica de vi aëris elastica et eivsdem effectibvs. [New physico-mechanical experiments on the elastic force of the air and its effects] Genevae, Apud S. de Tovrnes, 1680. 154 p.
QC161.B792 Rare Bk. Coll.
A translation of *New Experiments Physico-Mechanicall* (1660).

——A continuation of New experiments, physico-mechanical, touching the spring and weight of the air, and their effects. Oxford, Printed by H. Hall, for R. Davis, 1669–82. 2 v. in 1.
QC161.B793 Rare Bk. Coll.

——Opera varia. [Diverse works] Genevae, Apud S. de Tovrnes, 1680. 19 v. in 1.
QC3.B7 Rare Bk. Coll.

Cardano, Girolamo (1501–1576). De svbtilitate. [On the subtlety of things] Basileae, Per L. Lvcivm, 1554. 561 p.
Q155.C26 Rosenwald Coll. (689)

Cavendish, Henry (1731–1810). The electrical researches of Henry Cavendish, written between 1771 and 1781. Edited . . . by J. Clerk Maxwell. Cambridge [Eng.] University Press, 1879. lxvi, 454 p.
QC517.C35

Coulomb, Charles Augustin de (1736–1806). Construction et usage d'une balance électrique. [Construction and use of an electric balance] *In* Société française de physique. Collection de mémoires relatifs a la physique. t. 1. Mémoires de Coulomb. Paris, Gauthier-Villars, 1884. p. [107]–115.
QC3.S67, v. 1

——Où l'on détermine suivant quelles lois le fluide magnétique ainsi que le fluide électrique agissent soit par répulsion, soit par attraction. [How to determine the laws governing the action of the magnetic fluid as well as the electrical fluid, either by replusion or by attraction] *In* Société française de physique. Collection de mémoires relatifs a la physique. t. 1. Mémoires de Coulomb. Paris, Gauthier-Villars, 1884. p. 116–146.

QC3.S67, v. 1

Curie, Marie Sklodowska (1867–1934). Recherches sur les substances radioactives. 2. éd., rev. et corr. [Researches on radioactive substances] Paris, Gauthier-Villars, 1904. 155 p.

QC721.C95 Rare Bk. Coll.

——Traité de radioactivité. [Treatise on radioactivity] Paris, Gauthier-Villars, 1910. 2 v.

QC721.C98 Rare Bk. Coll.

Descartes, René (1596–1650). Discours de la methode pour bien conduire sa raison, & chercher la verité dans les sciences. [Discourse on the method of rightly conducting the reason and seeking truth in the sciences] A Leyde, De l'Impr. de I. Maire, 1637. 78, 413, [34] p.

Q155.D43 Rare Bk. Coll.

——Principia philosophiae. [Principles of philosophy] Amstelodami, Apud L. Elzevirium, 1644. 310 p.

B1860 1644 Rosenwald Coll. (1431)

Einstein, Albert (1879–1955). Die Grundlage der allgemeinen Relativitätstheorie. [The foundation of the general theory of relativity] *In* Lorentz, Hendrik Antoon. Das Relativitätsprinzip. 4., verm. Aufl. Leipzig, B. G. Teubner, 1922. (Fortschritte der mathematischen Wissenschaften in Monographien, Heft 2) p. 81–124.

QC6.L6 1922

An English translation appears in Lorentz's *The Principle of Relativity* (London, Methuen [1923] QC6.L63), p. 111–164.

——Zur Elektrodynamik bewegter Körper. [On the electrodynamics of moving bodies] *In* Lorentz, Hendrik Antoon. Das Relativitätsprinzip. 4., verm. Aufl. Leipzig, B. G. Teubner, 1922. (Fortschritte der mathematischen Wissenschaften in Monographien, Heft 2) p. 26–50.

QC6.L6 1922

An English translation appears in Lorentz's *The Principle of Relativity* (London, Methuen [1923] QC6.L63), p. 37–65.

————Nov. 1943. [1], 28 leaves. MSS

Holograph signed. Gift of the Kansas City Life Insurance Company.

Faraday, Michael (1791–1867). Experimental researches in electricity. London, R. and J. E. Taylor, 1839–55. 3 v.

QC503.F21 Rare Bk. Coll.

Franklin, Benjamin (1706–1790). Experiments and observations on electricity made at Philadelphia in America. London, Printed and sold by E. Cave, 1751–53. 2 v.

QC516.F85 1751 Rare Bk. Coll.

Galilei, Galileo (1564–1642). Discorsi e dimostrazioni matematiche, intorno à due nuoue scienze. [Mathematical discourses and demonstrations, relating to two new sciences] Leida, Appresso gli Elsevirii, 1638. [314] p.

QA33.G28 Rosenwald Coll. (1430)

——Le operazioni del compasso geometrico, et militare. [The operations of the geometric and military compass] Padova, In casa dell'autore, per P. Marinelli, 1606. [32] leaves.
QA33.G3 1606 Rosenwald Coll. (1335)

Galvani, Luigi (1737–1798). Commentary on the effects of electricity on muscular motion; translated into English . . . Together with a facsim. of Galvani's De viribus electricitatis in motu musculari commentarius (1791). Norwalk, Conn., Burndy Library [1954] 176 p. (Burndy Library. Publication no. 10)
QC517.G1713 1954

——Abhandlung über die Kräfte der thierischen Elektrizität auf die Bewegung der Muskeln. [Essay on the effects of animal electricity on the motion of the muscles] Eine Uebersetzung. Prag, J. G. Calve, 1793. xxviii, 183 p.
QC517.G18

Gilbert, William (1540–1603). De magnete. [On the magnet] Londini, Excvdebat P. Short, 1600. 240 p.
QC751.G44 Rare Bk. Coll.

Guericke, Otto von (1602–1686). Experimenta nova (ut vocantur) magdeburgica de vacuo spatio. [New (so-called) Magdeburg experiments on void space] Amstelodami, Apud J. Janssonium à Waesberge, 1672. 244 p.
Q155.G93 Rare Bk. Coll.

Heisenberg, Werner (1901–1976). The physical principles of the quantum theory. Translated into English. Chicago, Ill., University of Chicago Press [1930] 186 p. (The University of Chicago science series)
QC174.1.H4

——Über den anschaulichen Inhalt der quantentheoretischen Kinematik und Mechanik. [On the intuitive content of quantum kinematics and mechanics] Zeitschrift für Physik, 43. Bd., 3./4. Heft, 1927: 172–198.
QC1.Z44, v. 43

Hero, of Alexandria (fl. A.D. 62). Gli artifitiosi et cvriosi moti spiritali di Herrone. [The artful and curious pneumatic devices of Hero] Ferrara, V. Baldini, stampator ducale, 1589. 103 p.
QC142.H527 Rare Bk. Coll.
Translated by Giovanni Battista Aleotti.

——De gli avtomati. [On the automata] In Venetia, Appresso G. Porro, 1589. 47 leaves.
TJ215.H4 1589 Rare Bk. Coll.
Translated by Bernardino Baldi.

——Spiritalivm liber. [Book of pneumatics] Urbini, 1575. [80] leaves.
QC142.H54 Rosenwald Coll. (875)
Translated by Federico Commandino.

Hertz, Heinrich Rudolph (1857–1894). Untersuchungen ueber die Ausbreitung der elektrischen Kraft. [Researches on the propagation of electrical force] Leipzig, J. A. Barth, 1892. 295 p.
QC661. H59 Rare Bk. Coll.

Huygens, Christiaan (1629–1695). Œuvres complètes. [Complete works] La Haye, M. Nijhoff, 1888–1950. 22 v.
Q113.H9

——Horologivm oscillatorivm. [The oscillating clock] Parisiis, Apud F. Muguet, 1673. [14], 161 p.
TS545.H88 1673 Rare Bk. Coll.

——Opuscula postuma. [Short posthumous works] Lugduni Batavorum, Apud C. Boutesteyn, 1703. 460 p.

QC19.H87 Rare Bk. Coll.

——Opuscula posthuma. Amstelodami, Apud Janssonio-Waesbergios, 1728. 2 v. in 1.

QC3.H8 Rare Bk. Coll.

Maxwell, James Clerk (1831–1879). A dynamical theory of the electromagnetic field. *In* Royal Society of London. Philosophical transactions. v. 155; 1865. London, Printed by Taylor and Francis. p. 459–512.

Q41.L8, v. 155

——A treatise on electricity and magnetism. Oxford, Clarendon Press, 1873. 2 v. (Clarendon Press series)

QC518.M46

Michelson, Albert Abraham (1852–1931). On the relative motion of the earth and the luminiferous aether. By Albert A. Michelson and Edward W. Morley. Philosophical magazine, 5th ser., v. 24, Dec. 1887: 449–463.

Q1.P5, s. 5, v. 24

Newton, Sir Isaac (1642–1727). A letter . . . containing his new theory about light and colors. *In* Royal Society of London. Philosophical transactions, v. 6, Feb. 1672: 3075–3087.

Q41.L8, v. 6 Rare Bk. Coll.

——Opticks; or, A treatise of the reflexions, refractions, inflexions and colours of light. London, Printed for S. Smith, and B. Walford, 1704. 144, [213] p.

QC353.N556 Rare Bk. Coll.

——Philosophiae naturalis principia mathematica. [The mathematical principles of natural philosophy] Londini, Jussu Societatis Regiae ac Typis J. Streater, 1687. 383, 400–510 p.

QA803.A2 1687 Rare Bk. Coll.

Ørsted, Hans Christian (1777–1851). Experiments on the effect of a current of electricity on the magnetic needle. Annals of philosophy, v. 16, Oct. 1820: 273–277.

Q1.A6, v. 16

Translated from the Latin.

——Expériences sur l'effet du conflict électrique sur l'aiguille aimantée. Annales de chimie et de physique, [2 sér.], t. 14, août 1820: 417–425.

QD1.A7, s. 2, v. 14

——Versuche über die Wirkung des electrischen Conflicts auf die Magnetnadel. Annalen der Physik und der physikalischen Chemie, 6. Bd., 11. Stück, 1820: 295–304.

QC1.A6, v. 66

——Esperienze intorno all'effetto del conflitto elettrico sull'ago calamitato. Giornale di fisica, chimica, storia naturale, medicina ed arti, decade 2, t. 3, sett./ott. 1820: 335–339.

Q4.G4, 1820

Pascal, Blaise (1623–1662). Traitez de l'eqvilibre des liqvevrs, et de la pesantevr de la masse de l'air. [Treatise on the equilibrium of liquids, and on the weight of the mass of the air] A Paris, Chez G. Desprez, 1663. 232 p.

QC143.P3 Rare Bk. Coll.

Planck, Max Karl Ernst Ludwig (1858–1947). Zur Theorie des Gesetzes der Energieverteilung im Normalspectrum. [Toward a theory of the law of the distribution of energy in the normal spectrum] *In* Deutsche Physikalische Gesellschaft, Berlin. Verhandlungen, 2. Jahrg., Nr. 17, 1900: 237–245.
QC1.D41, v. 2

Röntgen, Wilhelm Conrad (1845–1923). Ueber eine neue Art von Strahlen. [On a new kind of ray] *In* Physikalisch-medicinische Gesellschaft, Würzburg. Sitzungsberichte, No. 8, 1895: 132–141; No. 1, 1896: 11–19.
Q49.W62, 1895–96

Rutherford, Ernest Rutherford, Baron (1871–1937). The cause and nature of radioactivity. By E. Rutherford and F. Soddy. Philosophical magazine, 6th ser., v. 4, Sept., Nov. 1902: 370–396, 569–585.
Q1.P5, s. 6, v. 4

——Collision of α particles with light atoms. Philosophical magazine, 6th ser., v. 37, June 1910: 537–587.
Q1.P5, s. 6, v. 37

——Radio-activity. Cambridge [Eng.] University Press, 1904. 399 p. (Cambridge physical series)
QC721.R97 1904 Rare Bk. Coll.

Schott, Gaspar (1608–1666). Mechanica hydraulico-pnevmatica. [Hydraulic-pneumatic mechanics] [Francofurti ad M.] Sumptu Heredum J. G. Schönweteri, Excudebat H. Pigrin Typographus, Herbipoli, 1657. 488, [14] p.
QC143.S3 Rare Bk. Coll.

Stevin, Simon (1548–1620). Les oeuvres mathematiques. [The mathematical works] Leyde, Chez B. & A. Elsevier, imprimeurs, 1634. 222, 678 p.
QA33.S84 Rare Bk. Coll.

Tartaglia, Niccolò (d. 1557). La noua scientia. [The new science] [Stampata in Venetia per N. de Bascarini, 1550] 32 leaves.
QC123.T3 1550 Rare Bk. Coll.

Torricelli, Evangelista (1608–1647). Lezioni accademiche. [Academic lectures] Firenze, Nella stamp. di S.A.R. per J. Guiducci, e S. Franchi, 1715. xlix, 96 p.
Q155.T69 Rare Bk. Coll.

——Opera geometrica. [Geometrical works] [Florentiae, Typis A. Masse & L. de Landis, 1644] 243, [151] p.
QA33.T69 Rare Bk. Coll.

Volta, Alessandro Giuseppe Antonio Anastasio, conte (1745–1827). On the electricity excited by the mere contact of conducting substances of different kinds. *In* Royal Society of London. Philosophical transactions. v. 90, pt. 2; 1800. London, Printed by W. Bulmer and sold by Peter Elmsly, Printer to the Royal Society. p. 403–431.
Q41.L8, v. 90 Rare Bk. Coll.

In French.

Index